Target Organ Toxicology Series

Toxicant–Receptor Interactions

Modulation of Signal Transduction and Gene Expression

Target Organ Toxicology Series

Series Editors
A. Wallace Hayes, John A. Thomas, and Donald E. Gardner

Toxicant–Receptor Interactions: Modulation of Signal Transductions and Gene Expression
Michael S. Denison and William G. Helferich, editors, 243 pp., 1998

Toxicology of the Liver, Second Edition
Gabriel L. Plaa and William R. Hewitt, editors, 431 pp., 1997

Free Radical Toxicology
Kendall B. Wallace, editor, 442 pp., 1997

Endocrine Toxicology, Second Edition
John A. Thomas and Howard D. Colby, editors, 352 pp., 1997

Reproductive Toxicology, Second Edition
Raphael J. Witorsch, editor, 336 pp., 1995

Carcinogenesis
Michael P. Waalkes and Jerrold M. Ward, editors, 496 pp., 1994

Developmental Toxicology, Second Edition
Carole A. Kimmel and Judy Buelke-Sam, editors, 496 pp., 1994

Immunotoxicology and Immunopharmacology, Second Edition
Jack H. Dean, Michael I. Luster, Albert E. Munson, and Ian Kimber, editors, 784 pp., 1994

Nutritional Toxicology
Frank N. Kotsonis, Maureen A. Mackey, and Jerry J. Hjelle, editors, 336 pp., 1994

Toxicology of the Kidney, Second Edition
Jerry B. Hook and Robin J. Goldstein, editors, 576 pp., 1993

Toxicology of the Lung, Second Edition
Donald E. Gardner, James D. Crapo, and Roger O. McClellan, editors, 688 pp., 1993

Cardiovascular Toxicology, Second Edition
Daniel Acosta, Jr., editor, 560 pp., 1992

Neurotoxicology
Hugh A. Tilson and Clifford L. Mitchell, editors, 416 pp., 1992

Ophthalmic Toxicology
George C. Y. Chiou, editor, 352 pp., 1992

Toxicology of the Blood and Bone Marrow
Richard D. Irons, editor, 192 pp., 1985

Toxicology of the Eye, Ear, and Other Special Senses
A. Wallace Hayes, editor, 264 pp., 1985

Cutaneous Toxicity
Victor A. Drill and Paul Lazar, editors, 288 pp., 1984

Target Organ Toxicology Series

Toxicant–Receptor Interactions

Modulation of Signal Transduction and Gene Expression

Editors

Michael S. Denison, Ph.D.
Department of Environmental Toxicology
University of California
Davis, California

William G. Helferich, Ph.D.
Department of Food Science and Human Nutrition
University of Illinois
Urbana, Illinois

USA	Publishing Office	Taylor & Francis 325 Chestnut St., # 800 Philadelphia, PA 19106 Tel: (215) 625-8900 Fax: (215) 625-2940
	Distribution Center	Taylor & Francis 1900 Frost Road, Suite 101 Bristol, PA 19007-1598 Tel: (215) 785-5800 Fax: (215) 785-5515
UK		Taylor & Francis Ltd. 1 Gunpowder Square London EC4A 3DE Tel: 0171 583 0490 Fax: 0171 583 0581

TOXICANT-RECEPTOR INTERACTIONS: Modulation of Signal Transduction and Gene Expression

1 2 3 4 5 6 7 8 9 0 BRBR 9 0 9 8

This book was set in Times Roman. Composition and editorial services by TechBooks. Cover design by Michelle Fleitz.

A CIP catalog record for this book is available from the British Library.
∞ The paper in this publication meets the requirements of the ANSI Standard Z39.48-1984 (Permanence of Paper).

Library of Congress Cataloging-in-Publication Data

Toxicant-Receptor interactions : modulation of signal transduction and gene
expression / edited by Michael S. Denison and William G. Helferich.
p. cm. — (Target organ toxicology series)
ISBN 1-56032-633-6 (cloth : alk. paper)
1. Molecular toxicology. 2. Cellular signal transduction.
3. Genetic regulation. 4. Xenobiotics—Physiological effect.
I. Denison, Michael S. II. Helferich, William. III. Series
RA1220.3.R43 1988
571.9'5–dc21 97-49036
CIP

To Grace and Nicki
for their love, support, and unending patience

Contents

Part IV: Analysis of Xenobiotic-Inducible Responses

Contributing Authors

Alan L. Blankenship, Ph.D. *Department of Zoology, National Food Safety and Toxicology Center, and Institute for Environmental Toxicology, Michigan State University, East Lansing, Michigan 48824*

Michael S. Denison, Ph.D. *Department of Environmental Toxicology, University of California, Davis, California 95616-8588*

Cornelius J. Elferink, Ph.D. *Institute of Chemical Toxicology, Wayne State University, Detroit, Michigan 48201*

Armand J. Fulco, Ph.D. *Department of Biological Chemistry, UCLA School of Medicine, Los Angeles, California 90095-1737*

Kevin W. Gaido, Ph.D. *Chemical Industry Institute of Toxicology, Research Triangle Park, North Carolina 27709*

John P. Giesy, Ph.D. *Department of Zoology, National Food Safety and Toxicology Center, and Institute for Environmental Toxicology, Michigan State University, East Lansing, Michigan 48824*

Martin Göttlicher, M.D. *Forschungszentrum Karlsruhe GmbH, Institute of Genetics, Karlsruhe, Federal Republic of Germany*

William G. Helferich, Ph.D. *Department of Food Science and Human Nutrition, University of Illinois, Urbana, Illinois 61801*

Debie Hoivik, Ph.D. *Department of Veterinary Physiology and Pharmacology, Texas A&M University, College Station, Texas 77843*

Anil K. Jaiswal, Ph.D. *Department of Pharmacology, Baylor College of Medicine, One Baylor Plaza, Houston, Texas 77030-3498*

Shinji Koizumi, Ph.D. *Department of Experimental Toxicology, National Institute of Industrial Health, Nagao, Kawasaki 214, Japan*

Burra V. Madhukar, Ph.D. *Department of Pediatrics and Human Development, Michigan State University, East Lansing, Michigan 48824*

Fuminori Otsuka, Ph.D. *Department of Environmental Toxicology, Faculty of Pharmaceutical Sciences, Teikyo University, Sagamiko, Kanagawa 199-01, Japan*

Issac N. Pessah, Ph.D. *Department of Molecular Biosciences, School of Veterinary Medicine, University of California, Davis, California 95616*

Dorothy Phelan, Ph.D. *Department of Environmental Toxicology, University of California, Davis, California 95616-8588*

Stephen H. Safe, Ph.D. *Department of Veterinary Physiology and Pharmacology, Texas A&M University, College Station, Texas 77843*

John P. Vanden Heuvel, Ph.D. *Department of Veterinary Science, Pennsylvania State University, University Park, Pennsylvania 16802-3500*

Daniel L. Villeneuve, Ph.D. *Department of Zoology, National Food Safety and Toxicology Center, and Institute for Environmental Toxicology, Michigan State University, East Lansing, Michigan 48824*

Patty W. Wong, Ph.D. *Department of Molecular Biosciences, School of Veterinary Medicine, University of California, Davis, California 95616*

Preface

For cells to respond to external stimuli, there must be a mechanism by which the external signals are transferred from the environment surrounding the cell to the genetic material located within the nucleus of the cell. These signals are commonly transmitted via receptor-mediated responses and signal-transduction pathways and, most commonly, these pathways ultimately result in specific alterations in gene expression. Many of the toxic effects elicited by xenobiotics (foreign chemicals) can be explained at the molecular level by the interaction of the chemicals with receptors or by disruption or interference with receptor-mediated signal-transduction pathways.

The idea for this book arose out of discussions with fellow toxicologists regarding the application of cutting-edge molecular biologic techniques to the study of the mechanism by which xenobiotics adversely affect signal transduction processes and the regulation of gene expression. The past 10 years have witnessed a tremendous increase in the utilization of these techniques in toxicology, pharmacology, and related research areas and the resulting emergence of a new research area, molecular toxicology. The rapid growth in these areas is not surprising, especially given the power of these technologies, which now allow researchers to address some very basic mechanistic questions. The molecular events by which xenobiotics can modulate signal-transduction processes and gene expression are critical elements basic to our understanding of the mechanism by which a chemical or class of chemicals can exert its actions. Given that an ever-increasing number of researchers are beginning to apply these molecular techniques to the analysis of the mechanisms by which xenobiotics alter signal transduction and gene expression, the lack of an up-to-date book describing molecular approaches and reviews of current research in this area was somewhat surprising. The current volume fills that void by providing reviews of numerous research areas that are directly related to the analysis of the molecular mechanism by which xenobiotics modulate signal transduction and gene expression.

This book is divided into four discrete sections, with three parts devoted to signal-transduction systems affected by xenobiotics and one describing methodologic approaches for identification of xenobiotic-inducible genes. The first part of this book focuses on the interaction of xenobiotics with specific soluble receptor proteins and a review of the mechanism by which these interactions result in induction or repression of gene expression or toxicity. Topics covered within this section include the Ah-receptor signal-transduction system, which mediates the biochemical and toxic action of dioxin-like chemicals (chapter 1); the peroxisome proliferator–activated receptor, an orphan steroid-hormone receptor whose transcriptional activation function is positively modulated both by xenobiotics and endogenous lipids (chapter 2); and steroid-hormone receptors, whose transcriptional activity can be positively and negatively regulated by a variety of environmental contaminants, including discussion of the controversial "endocrine disrupters" (chapters 3 and 4).

The next section focuses on the effect of xenobiotics on transcription factors, and it includes reviews of the molecular mechanisms by which barbiturate-like chemicals induce gene expression in prokaryotes and eukaryotes (chapter 5), the induction of expression of xenobiotic metabolizing enzymes by antioxidants and electrophilic xenobiotics (chapter 6), and the mechanisms by which heavy metals, such as cadmium and zinc, induce gene expression (chapter 7).

The third section focuses on the ability of xenobiotics to modulate second-messenger signal-transduction systems, and it includes chapters on the mechanisms by which xenobiotics modulate protein kinases (chapter 8) and on cellular calcium signaling systems (chapter 9). The final section emphasizes methodologies, and it reviews current molecular biologic approaches for the identification and cloning of xenobiotic-inducible genes (chapter 10).

The overall goal in bringing these various xenobiotic-regulated receptor and signal-transduction systems together is to provide the reader with information as to the spectrum of mechanisms by which chemicals can modulate signal transduction and gene expression. In addition, similarities and differences in the methodologic approaches described in the chapters, as well as the final chapter, which describes specific methodologies that have application for all regulated gene-expression systems, hopefully will provide other researchers with ideas and avenues of experimentation that they can apply to their specific research questions. The information provided in this book will be of interest to toxicologists, pharmacologists, and all other researchers who are examining the mechanisms by which xenobiotics alter normal cellular functions.

The editors wish especially to thank the authors for their cooperation and contributions, and Dr. Dorothy Phelan for her invaluable editorial assistance. Finally, the contributions of our past and present students, research associates, and colleagues also are acknowledged gratefully, for it is these interactions that continue to make the science exciting and fun.

PART I

Xenobiotics and Soluble Receptors

Toxicant–Receptor Interactions
Edited by Michael S. Denison and William G. Helferich

1

The Ah Receptor Signal Transduction Pathway

Michael S. Denison and Dorothy Phelan

Department of Environmental Toxicology, University of California, Davis, California, USA

Cornelius J. Elferink

Institute of Chemical Toxicology, Wayne State University, Detroit, Michigan, USA

Halogenated aromatic hydrocarbons (HAHs), such as polychlorinated dibenzo-p-dioxins (PCDDs), biphenyls (PCBs), and dibenzofurans (PCDFs) and related compounds represent a diverse group of widespread environmental contaminants. PCDD and PCDF formation has been demonstrated to occur as a result of chlorine bleaching of wood pulp; during synthesis of various organochlorine products (such as the herbicide 2,4,5-T); during municipal, hospital, and industrial waste incineration; during metal production and fossil fuel or wood burning as well as from a variety of other sources (1,2). PCBs, on the other hand, were produced commercially for use in transformers, capacitors, heat transfer and hydraulic fluids, and other applications. It has been estimated that greater than 1.5 million metric tons of PCBs have been produced worldwide, with 20%-30% of this amount in the environment (1, 3). HAHs represent a class of environmental chemicals that, because of their ubiquitous distribution, toxicity, fat solubility, resistance to biological and chemical degradation, and potential for bioaccumulation and biomagnification, can persist for long periods of time and thus could have a significant impact on the health and well-being of humans and animals (2, 3).

2,3,7,8-Tetrachlorodibenzo-p-dioxin (TCDD), the prototypical and most potent member of this class of compounds, has gained widespread attention in recent years as one of the most toxic man-made chemicals known. The long biological half-life of TCDD (up to 10 years in humans) and other HAHs, combined with their extremely high toxicity in animals (2, 4), has made them the focus of intensive research for more than 20 years. Exposure to TCDD and related HAHs produces a wide variety of species- and tissue-specific effects, including tumor promotion, teratogenicity, immuno- and dermal toxicity, wasting, lethality, modulation of cell growth, proliferation and differentiation, alterations in endocrine homeostasis, reduction in steroid hormone–dependent responses, and induction of numerous enzymes (4–7). Although all of these effects have been documented to occur in animals, the persistent affliction chloracne is the best-characterized response to TCDD and related HAHs in humans. Recent epidemiological evidence, however, has suggested an increase in the number of cancers in exposed human populations and the occurrence of some developmental learning defects in children exposed to HAHs in utero (8, 9). Although the toxicological and biological effects of HAHs in humans are still a matter of intense debate, ongoing epidemiological studies of human populations in Japan and Taiwan (where human consumption of PCB-contaminated rice cooking oil containing low levels of PCDFs was documented) as well as those in Seveso, Italy (where an industrial accident resulted in human exposure to TCDD), should provide more definitive information as to the human health effects of HAHs (9).

One consistent response observed in all species exposed to TCDD and related chemicals is the induction of gene expression. TCDD induces expression of a wide variety of genes, the most studied of which are those involved in xenobiotic metabolism, such as cytochromes P450 1A1/2 and 1B1, glutathione S-transferase Ya, NAD(P)H:quinone reductase, aldehyde dehydrogenase 3, and UDP-glucuronosyltransferase-6. Induction of many of these genes occurs in a species- and tissue-specific manner. The induction of cytochrome P4501A1 and its associated monooxygenase activity, aryl hydrocarbon hydroxylase (AHH), however, is one response common to most species and tissues, and thus it has been used as a model system to examine the mechanism of action of HAHs (10–12). Early experiments examining the induction of AHH activity by a series of PCDDs, PCDFs, and PCBs revealed a relationship between HAH structure and ability to induce AHH (1, 5, 13). These results, in combination with those obtained using various inbred strains of mice (5, 14), suggested the presence of a specific receptor that recognized these compounds. Subsequently, a cytosolic protein that bound TCDD saturably and with high affinity and exhibited the properties of a receptor was identified (15). Although quantitative structure–activity relationship studies revealed that the ability of a compound to bind to this receptor was well correlated with its ability to induce AHH activity, the positive correlation with its toxic potency (i.e., ability to produce thymic involution, wasting, and epidermal keratinization) also suggested that this receptor also mediated the toxicity of these compounds (4, 5, 13). Overall, these observations suggested a common mechanism of action of HAHs. This "TCDD receptor" has been identified and characterized in a wide variety of species and tissues (16–19) and has been designated as the aryl hydrocarbon receptor (AhR). There are many excellent reviews on the biochemistry and molecular biology of the AhR (7, 10–12, 20); hence, this chapter does not attempt to cover all aspects of this system in detail. Given the pace of research on the AhR signal transduction system, we provide an overview of this area, with emphasis on recent developments and aspects not covered in detail elsewhere.

BIOCHEMICAL CHARACTERIZATION OF THE AhR COMPLEX

Ligand Binding

For the AhR to become activated, exogenous ligands must enter the cell, presumably by passive diffusion, where they bind to the AhR in the cytosol. All of the high-affinity AhR ligands identified to date are planar, hydrophobic molecules and include several classes of HAHs (PCDDs, PCDFs, PCBs, etc.) as well as polycyclic aromatic hydrocarbons (PAHs, such as benzo[a]pyrene, 3-methylcholanthrene, flavones, rutacarpine alkaloids and aromatic amines) and other chemicals (Table 1). Although the affinity of binding of TCDD and other HAHs is in the picomolar to nanomolar range, PAHs bind with significantly lower affinity (in the high nanomolar to micromolar range) and this difference has been positively correlated with observed differences in the toxic and biological potency of these classes of chemicals, with HAHs significantly more active (1, 5, 13). Available evidence suggests that the differences in potency between HAHs and PAHs are likely due to a combination of the higher AhR binding affinity of HAHs and the increased resistance of HAHs to metabolic degradation, which results in sustained AhR occupancy by HAHs and persistent activation of gene expression (1, 5, 7, 13). Given that many of the toxic effects of HAHs are not observed until several days following TCDD exposure (4), it has been proposed that the adverse effects of these chemicals result from the continuous and inappropriate expression of specific genes in responsive cells rather than an acute response to these chemicals. This hypothesis also suggests that if the concentrations of PAHs are maintained at relatively high levels in an organism, toxic effects similar to that produced by TCDD should be observed. In fact, continual dietary exposure of juvenile catfish to the prototypical PAHβ-naphthoflavone for 90 days resulted in a variety of morphological and toxicological outcomes similar to those effects previously observed following TCDD

TABLE 1. *Known AhR ligands and/or inducers of gene products regulated by the AhR*

Chemicals	References
AhR ligands	
Halogenated and nonhalogenated aromatic hydrocarbons	
Dibenzo-p-dioxins	1
Dibenzofurans	1
Biphenyls	1
Diphenyl ethers	132
Naphthalenes	133
Trans stilbenes	134
Substituted PAHs and nitroarenes	23, 135
β-Naphthoflavone and substituted flavones	136, 137
6-Substituted benzocoumarins	138
Hydroxylated benzo(a)pyrenes	16
Methylenedioxybenzenes	139[a]
Indoles	31, 140, 141
Tryptophan-derived products	33, 34, 26, 38
Miscellaneous	
2-(4′-Chlorophenyl) benzothiazole	142
1,3-Diaryltriazenes (SKF71739, 72298)	143
Thiazolium Compound, YH439	144
2,3-Diaminotoluene	145
Staurosporin	146
Ah Inducers	
Imidazoles and pyridines	25, 27, 147–150
Oxidized carotinoids	32, 151
Heterocyclic amines	152
Miscellaneous	
Cinnamyl anthranilate	153
2,5-Diphenyloxazole	149
Fumonisin B1	154
Picrotoxin	155
GABA analogs	156

[a]Denison, unpublished data.

exposure (21). These data are consistent with the hypothesis that AhR-dependent toxic effects require sustained AhR occupancy and activation.

The physicochemical characteristics necessary for binding of a chemical to the AhR have been examined by numerous investigators over the past 20 years. These investigations have provided some detailed information about the structure of AhR ligands and have allowed predictions of the structural and electronic characteristics of the AhR ligand-binding pocket (22, 23). Three-dimensional molecular-volume mapping studies of a large number of HAH and PAH AhR ligands (22, 23) suggest that the ligand-binding pocket of the AhR can accommodate planar ligands with maximal dimensions of 14 Å × 12 Å × 5 Å. In addition to limitations on the physical characteristics of AhR ligands, high-affinity binding of chemicals to the AhR also is critically dependent on key electronic and thermodynamic properties of the ligand (22–24). Although these modeling studies can be used to predict whether a given chemical could bind to the AhR, however, the constraints of the currently defined model are too simplistic, especially given the limited number of molecules used in these studies and the recent identification of numerous "nonclassical" AhR ligands.

The majority of studies characterizing AhR ligands and Ah inducers have focused on HAHs and a limited number of PAHs, but recent studies have described numerous chemicals that can induce CYP1A1 and whose structures are inconsistent with currently defined structural requirements for AhR ligands (Table 1). Although direct AhR binding has not been demonstrated for many of these chemicals, their ability to induce CYP1A1 or activate AhR and AhR-dependent gene expression in cultured cells is consistent with their interaction with the AhR (although the possibility of a metabolite or secondary inducer or ligand cannot be ruled out). For example, several substituted benzimidazole drugs such as omeprazole have been observed to induce CYP1A1/2 in human cells, although these chemicals could not competitively displace [^{3}H]TCDD from the AhR (25, 26). The ability of these chemicals to activate the AhR in cells in culture, combined with their inability to induce in AhR-defective cells (26, 27), suggest that these chemicals actually do bind to the AhR; however, given their extremely low binding affinity, they are unable to effectively compete with [^{3}H]TCDD, a extremely high-affinity ligand. The ability of a wide range of structurally dissimilar chemicals to induce CYP1A1 expression or bind to the AhR indicate that a greater spectrum of chemicals can interact with this receptor than previously thought; moreover, they provide us with clues with which to identify more "natural" and endogenous AhR ligands. Although several naturally occurring AhR ligands have been identified, no high-affinity physiologic endogenous ligand has been found. Recent observations demonstrating AhR-mediated responses that occur in the absence of exogenous ligand (28–30) suggest that endogenous ligands must exist.

The ability of several dietary compounds (such as indole-3-carbazole (23, 31) and the oxidized carotinoids, canthaxanthin and astaxanthin (32)) to bind to the AhR or induce gene expression in an AhR dependent manner has been reported. Conversion of dietary indoles (including tryptophan) in the mammalian digestive tract to significantly more toxic AhR ligands (33), as well as conversion of tryptophan by UV light into several products that bind to the AhR with high affinity (34–36) have also recently been described. The observation that UV irradiation of animals induces CYP1A1-dependent monooxygenase activity in the liver and skin suggests that this activation may also occur in vivo (37). More recently, we have observed that endogenous metabolites of tryptophan, namely tryptamine and indole acetic acid, not only can bind to the AhR but also exhibit full AhR agonist activity (38). The relatively weak affinity and low cellular concentrations of these latter chemicals suggest that they are unlikely "the" endogenous ligands for the AhR. These results, combined with the identification of other novel structurally distinct AhR ligands (See Table 1), provide us with unique insight into the spectrum of chemicals that can interact with the AhR. The availability of several sensitive AhR-dependent screening bioassays (35, 39, 40), should allow us to identify and characterize the natural physiological AhR ligand or ligands in the near future.

Finally, although it can be generalized that ligand-binding specificity and rank-order potency of HAH and PAH ligands are similar for AhRs between species and tissues, they are not identical and significant species differences do exist (16, 41–43). For example, 2,5,2′,5′-tetrachlorobiphenyl is a weak ligand (antagonist) for the murine AhR, is somewhat less potent for the rat AhR, and fails to bind to the human and guinea pig AhRs (43); omeprazole on the other hand, is reported to be a human-specific AhR ligand (44). Additional evidence for species-specific differences in ligand binding comes from studies using a series of PAHs (41) and singly hydroxylated benzo(a)pyrene molecules (16) in which significant differences in rank-order potency of ligand binding were observed. Although it can be generalized that AhR ligand-binding specificity is similar among species, significant differences in specificity and affinity do exist and these may contribute to some of the differential species responsiveness that is observed.

Overall, the results of available binding studies suggest that the AhR contains a "sloppy" or "promiscuous" ligand binding pocket. The broad ligand-binding specificity actually may confer a selective advantage to the organism. If exposure to AhR ligands induces a variety of distinct detoxification enzymes (each of which exhibits its own broad substrate specificity), then a regulatory protein that can recognize and be activated by a spectrum of structurally diverse chemicals would greatly increase the rate at which these chemicals could be metabolized and detoxified, thus providing the organism with a greater range of protection from toxic xenobiotics. In fact, it is reasonably well established that many of the AhR ligands identified here are known substrates for P4501A1 or other members of the Ah gene battery.

AhR Structure

Prior to ligand binding, the AhR is localized in the cytosolic fraction of cells, in which it exists at least as a tetrameric complex (250–300 kDa) consisting of one ligand-binding (AhR) subunit, two molecules of hsp90 (a heat-shock protein of 90 kDa) and one as-yet-uncharacterized 43-kDa protein identified in protein cross-linking studies (41, 45). Cloning of mouse cDNAs that encode a 330-amino-acid protein of approximately 37 to 43 kDa, which appears to be contained within the unliganded cytosolic AhR complex, recently has been reported (46, 47). The recombinant AhR interacting protein (AIP) has sequence similarity to FKBP52, an immunophilin chaperone-like protein that is thought to play a role in protein folding and targeted movement of the glucocorticoid receptor (48). Overexpression of AIP in cells in culture appears to augment the induction of CYP1A1, primarily by influencing ligand receptivity or nuclear AhR targeting (46). The identity of AIP as the 43-kDa protein present in the cytosolic AhR complex as determined by chemical cross-linking experiments remains to be confirmed. In the cytosol, AhR complexes sediment at approximately 9 to10S (250 to 300 kDa) in low ionic strength and are readily dissociated in high ionic strength conditions with release of the monomeric ligand binding 4-5S (approximately 100 kDa) AhR subunit (42). It is interesting to note that salt-induced dissociation of the heteromeric complex to monomeric AhR occurs readily in all species with the exception of mouse AhR complexes, which are extremely resistant to salt-mediated dissociation and inactivation, the basis of which is not known (25, 49). UV crosslinking using an azido-labeled AhR ligand has demonstrated that the molecular mass of the AhR ranges from approximately 95 kDa in the mouse to about 146 kDa in rainbow trout (50, 51). This large variation in size is atypical of most soluble ligand-dependent receptors that have been characterized to date.

The association of the cytosolic AhR with hsp90, a molecular chaperone protein, is thought to be important for protein folding as well as regulating nuclear shuttling of receptors (48, 52). Recent studies in yeast have highlighted the absolute requirement for hsp90 for the formation of fully functional AhR, with little functional AhR produced in yeast strains containing reduced amounts of hsp90 (53). Consistent with this observation, Pongratz and coworkers (54) reported that AhR ligand-binding activity of rat hepatic AhR requires prior association with hsp90, and dissociation of unliganded AhR from hsp90 abolishes AhR ligand-binding activity, analogous to the behavior of the glucocorticoid receptor (55, 56). Given hsp90's role as a molecular chaperone, these results suggest that hsp90 keeps the AhR folded into its correct ligand-binding conformation. In contrast, recent data from our lab (Denison, unpublished results) suggest that the presence of bound hsp90 is not absolutely required for ligand-binding activity, because high salt dissociation of hsp90 from the cytosolic AhR prior to ligand binding resulted in varying degrees of AhR inactivation (as measured by loss of ligand-binding activity); binding to the human AhR was completely eliminated and the rat AhR was partially affected (57), whereas ligand binding of the guinea pig AhR was unaffected. Although the yeast results clearly demonstrate that hsp90 is required to produce a functional AhR, the apparent discrepancy of our data from those of others can be reconciled if one envisions that hsp90 functions primarily to direct the folding of the nascent AhR and the absence of hsp90 results in production of an incorrectly folded and inactive AhR protein. After AhR synthesis and correct folding, hsp90 continues to function in AhR action by maintaining the high-affinity ligand-binding conformation of AhR in some species and to keep the unliganded AhR complex contained within the cytosol. Thus, hsp90 plays several important roles in the AhR signaling pathway.

The AhR is a ligand-activated transcription factor that binds to a specific DNA enhancer element linked to target genes, and its DNA binding activity is dependent upon its dimerization with its nuclear protein partner, the AhR nuclear translocator (ARNT). In the past few years both proteins have been cloned from several species (7, 10, 12), facilitating an extensive analyses of functionally important domains (Fig. 1). Both the AhR and ARNT are basic helix-loop-helix (bHLH) proteins distinguishable from other bHLH-containing transcription factors such as MyoD, E12, and E47 by an additional PAS domain (7, 10, 12) (*PAS* is the designation for a region of structural homology contained within ARNT and two *Drosophila* proteins, period [Per] and single-minded (Sim)). The bHLH region functions in both AhR and ARNT dimerization and DNA binding, while the PAS domain facilitates stable AhR-ARNT dimerization and, in the AhR protein, contains the ligand and hsp90 binding sites.

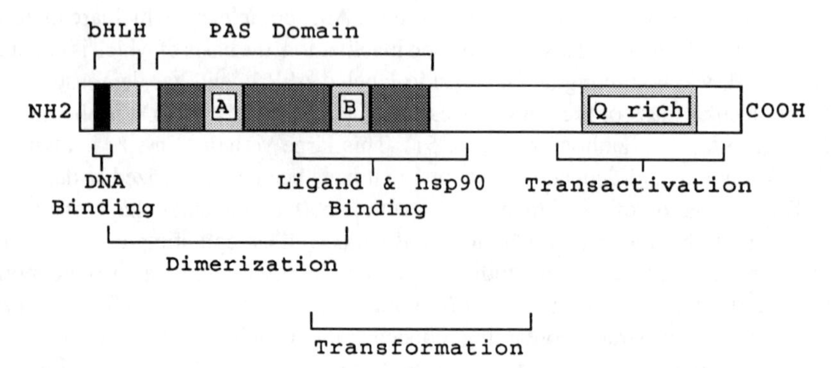

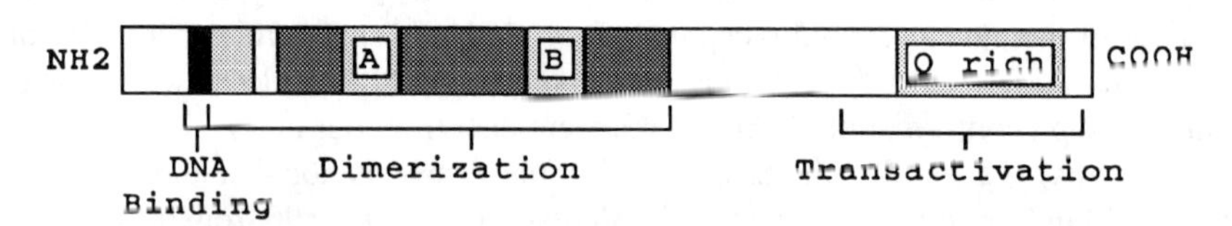

FIG. 1. Domain structures of the AhR and Arnt. The relative position of the basic helix-loop-helix (bHLH) and PAS domains as well as the regions of the proteins that appear to play a role in selected AhR and Arnt functions are indicated.

Each protein also contains a C-terminal Q-rich transactivation domain, but on DNA binding, the AhR–ARNT complex activates transcription by a mechanism that seems to require only the AhR transactivation domain (58). The reader is directed to several excellent recent reviews (7, 10, 12) for a more detailed description of the domain structure of the AhR and ARNT proteins.

LIGAND-DEPENDENT AhR TRANSFORMATION AND NUCLEAR TRANSLOCATION

TCDD enters cells by passive diffusion and binds to the cytosolic AhR complex. Ligand binding by the cytosolic AhR complex leads to release of hsp90 and other associated proteins, presumably because of a conformational change in the AhR protein. The liganded AhR complex then translocates into the nucleus, by a process that still requires elucidation but that may rely in part on the unmasking of a cryptic nuclear localization sequence on the AhR previously concealed by hsp90 (59) and the subsequent association of the liganded AhR with nuclear transport proteins. Once in the nucleus, the TCDD:AhR complex dimerizes with ARNT and becomes associated with DNA, with release from the nucleus requiring high ionic strength conditions. AhR nuclear translocation was examined using immunofluorescence microscopy (60, 61), which revealed not only that ARNT was localized solely in the nucleus in intact cells but also that nuclear translocation of the ligand:AhR complex did not require a functional ARNT protein. Subcellular fractionation experiments using an ARNT-defective mouse hepatoma variant cell line revealed that the liganded AhR was not retained in the nuclear compartment (60, 61). These data implied that the ARNT protein functions to facilitate AhR nuclear retention by dimerizing with the receptor and presumably forming stable high-affinity AhR:ARNT:DNA complexes. Thus, the designation of ARNT as the AhR nuclear translocator is actually a misnomer in that it does not actually facilitate AhR nuclear translocations but rather nuclear retention (i.e., DNA binding). Recognizing that the AhR and ARNT proteins form dimers also helps explain earlier results that indicated that the AhR extracted from nuclei had an apparent molecular mass of about 180 kDa, much larger than the 100-kDa AhR monomer (62). Whether hsp90 or the 43-kDa immunophilin-like protein play any role in the nuclear shuttling process, as suggested for glucocorticoid receptors (48, 52), remains to be determined.

AhR DNA BINDING

To characterize the mechanism by which TCDD and the nuclear AhR complex stimulate gene transcription, Whitlock and coworkers (63, 64) isolated the 5′-flanking region of the mouse CYP1A1 gene and inserted it directly upstream of the bacterial chloramphenicol acetyltransferase (CAT) gene. Transfection of the recombinant plasmid into wild-type and receptor-defective variant mouse hepatoma cells revealed that CAT expression was TCDD-inducible in wild-type cells, but absent or diminished in the AhR- or ARNT-defective cells (63, 64). Thus, the DNA insert contained a domain or domains with properties expected of a TCDD-responsive DNA element. Deletion analysis (63–65) revealed that the 5′-flanking region of the CYP1A1 gene contained a relatively strong basal promoter, an inhibitory region, and a TCDD-responsive domain (Fig. 2), the latter of which contained the binding site or sites for TCDD:AhR complex. Detailed deletion mapping and linker scanner analyses using a transient transfection-based assay (63, 66–68) revealed the presence of at least four distinct TCDD-responsive elements (also called *dioxin-responsive elements* or DREs) upstream of the murine CYP1A1 gene (see Fig. 2). Subsequent transfection studies revealed that each DRE exhibited the properties of an inducible transcriptional enhancer in that it required transformed nuclear TCDD:AhR complexes for functionality and conferred TCDD-responsiveness on an adjacent promoter and gene independent of orientation or position (64–67, 69). Similar DREs have been identified in the upstream flanking region and promoter of a variety of TCDD-responsive genes (Fig. 3), and they appear to be functionally similar to those characterized in the CYP1A1 gene (70, 71).

Utilizing a sensitive gel retardation assay (66, 69) we demonstrated that nuclear extracts from TCDD-treated Hepa1 cells contained a protein or proteins that bound to a [^{32}P]-DRE-containing oligonucleotide in a TCDD-inducible, AhR-dependent, and DNA-sequence-specific manner. The presence of the AhR in the TCDD-inducible

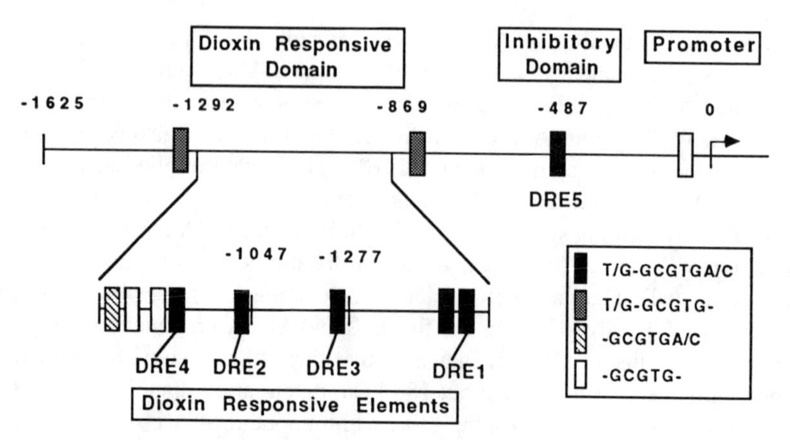

FIG. 2. Regulatory domains of the upstream region of the mouse CYP1A1 gene. The relative location of the four functional dioxin responsive elements (DREs 1–4) are indicated by black boxes; the position of the nonfunctional perfect consensus DRE 5 is contained within the inhibitory region. In addition, the position of numerous degenerate DRE-like sequences contained within this region and the degree of homology of each element to the core DRE consensus are indicated.

protein DNA complex was demonstrated using a photoaffinity TCDD agonist (72) and in supershift experiments using an AhR antibody (73, 74). In addition, double-label gel retardation assay experiments ([^{125}I] 2-azido-3-iodo-7, 8-dibromodibenzo-p-dioxin-labeled nuclear AhR and [^{32}P]-labeled DRE oligonucleotide) as well as UV cross-linking studies revealed that one transformed ligand:AhR:ARNT complex bound to a single DRE recognition site (75). Transformed TCDD:AhR complexes also bound to double-stranded DRE-containing DNA kinetically as a single species with an apparent affinity of 1.0 to 2.5 NM, about 1000-fold higher than to double-stranded nonspecific DNA (76, 77).

Sequence alignment of the currently identified DREs reveals the presence of an invariant core sequence, GCGTG, flanked by several partially conserved nucleotides (see Fig. 3). The results of methylation interference studies (77, 78) demonstrated that several of these “core” nucleotides were critical for TCDD:AhR:DRE complex formation. Extensive gel retardation analyses and DRE mutagenesis experiments (77, 79) have been carried out to determine the nucleotide sequence requirements for both formation of the TCDD:AhR:DRE complex and TCDD-inducible enhancer activity (Table 2). The putative TCDD:AhR DNA-binding consensus sequence of T/GNG/TCGTGA/CG/TA/TA/G has been derived from the mutagenesis experiments and was in reasonably good agreement with the consensus derived from alignment of currently identified DREs (see Fig. 3). The four core nucleotides, CGTG, appear to be critical for TCDD-inducible protein–DNA complex formation because their substitution decreased AhR binding affinity by 100- to 800-fold; the remaining conserved bases also were important, albeit to a lesser degree (three- to five-fold) (75, 77, 80). The contribution, if any, of the “nonconsensus” nucleotides to the high-affinity AhR:DRE interaction remains to be determined. Although the primary interaction of transformed

mDREa	C	A	A	G	C	T	C	G	C	G	T	G	A	G	A	A	G	C
mDREb	C	C	T	G	T	G	T	G	C	G	T	G	C	C	A	A	G	C
mDREc	G	A	G	G	C	T	A	G	C	G	T	G	C	G	T	A	A	C
mDREd	C	G	G	A	G	T	T	G	C	G	T	G	A	G	A	A	G	A
mDREe	C	C	A	G	C	T	A	G	C	G	T	G	A	C	A	G	C	A
mDREf	C	G	G	G	T	T	T	G	C	G	T	G	C	G	A	T	G	C
GSTYa	G	C	A	T	G	T	T	G	C	G	T	G	C	A	T	C	C	C
NQO1	T	C	C	C	C	T	T	G	C	G	T	G	C	A	A	A	G	G
rXRE2	G	A	T	C	C	T	A	G	C	G	T	G	A	C	A	G	C	A
hXRE1	C	C	G	G	C	T	C	G	C	G	T	G	A	G	A	A	G	C
hXRE2	A	G	G	C	G	T	T	G	C	G	T	G	A	G	A	A	G	G
hXRE3	C	C	C	C	C	T	C	G	C	G	T	G	A	C	T	G	C	G
rUGT	A	G	A	A	T	G	T	G	C	G	T	G	A	C	A	A	G	G
hUGT	G	T	A	G	T	T	G	G	C	G	T	G	A	C	T	G	T	G
rALDH3	C	A	C	T	A	(A)	T	G	C	G	T	G	C	C	C	C	A	T
H1A2	A	G	G	T	A	G	T	G	C	G	T	G	(T)	C	A	G	G	T

Position		1	2	3	4	5	6	7	8	9	10	11	12	13	14
	G	4	3	1	16	0	16	0	16	0	6	0	5	10	5
	A	2	1	2	0	0	0	0	0	10	2	12	8	1	4
	T	4	12	10	0	0	0	16	0	1	0	3	1	1	2
	C	6	0	3	0	16	0	0	0	5	8	1	2	4	5
1 Base	%	38	75	62	100	100	100	100	100	62	50	75	50	62	31
2 Base	%		94	81						94	88	94	81	88	62

Derived DRE Consensus Sequence

N	T	T	G	C	G	T	G	A	C	A	A	G	N
	G							C	G	T	G	C	

FIG. 3. Sequence alignment of 16 known DREs from TCDD-inducible genes. *A*, Alignment of DRE sequences from the following genes: mouse CYP1A1 (mDRE), rat glutathione S-transferase Ya (GSTYa), rat NAD(P)H:quinone oxidoreductase 1 (NQO1), rat CYP1A1 (rXRE), human CYP1A1 (hXRE), rat UDP-glucuronosyl transferase (rUGT), human UDP-glucuronosyl transferase (hUGT), rat aldehyde dehydrogenase 3 (rALDH3) and human CYP1A2. The boxed areas indicates the highly conserved core nucleotides of the DREs, with the circled bases representing bases that differ from the conserved bases. *B*, Comparison of nucleotide base utilization at each position within the 16 known DREs. The number of times a specific base appears at each position is indicated. In addition, the relative percentage of times (out of 16) that the most commonly used base is present in a given position as well as the relative percentage of times (out of 16) that the two most commonly used bases are present in a given position are calculated, and the DRE consensus derived from this analysis is shown.

TABLE 2. *Relative AhR binding and enhancer function of the DRE derived from mutagenesis experiments*

Mutant oligo	DRE nucleotide position																			Binding[c]	Function[c]
	1 C	2 G	3 G	4 A	5 G	6 T	7 T	8 G	9 C	10 G	11 T	12 G	13 A	14 G	15 A	16 A	17 G	18 A	19 G		
WT	.													.		.			.	+++	+++
m1[a]	T					.						.							.	+++	+++
m2	A					.						.							.	+++	+++
W9[b]	T					.						.							.	+++	+++
W10	.		T			.						.							.	+++	+++
m3	.				A	.						.							.	+++	+++
m4	.				T	.						.							.	+++	+++
W11	.				A	.						.							.	+++	+++
m5	.					G						.							.	+++	+++
W6	.					C						.							.	+/−?	−
W17	.					.	C					.							.	+++	+++
m6	.					.		T				.							.	++	+++
W5	.					.		A				.							.	++	−
m7	.					.			A			.							.	−	−
m8	.					.			T			.							.	−	ND
W1	.					.			T			.							.	−	ND
W13	.					.			G			.							.	−	ND
m9	.					.				T		.							.	−	−
W2	.					.				A		.							.	−	−
m10	.					.					G	.						.		−	−
W7	.					.					C	.							.	−	−
W14	.					.					A	.						.		−	ND
m11	.					.						T							.	−	−
W3	.					.						A							.	−	ND
W15	.					.						C							.	−	ND
m12	.					.						.	G						.	−	−
W16	.					.						.	G						.	+/−	−
m13	.					.						.		T					.	+++	+++
W4	.					.						.		A					.	+/−	−
m14	.					.						.			C				.	++	−
W12	.					.						.			G				.	+/−	−
m15	.					.						.			T				.	++	++
m16	.					.						.					A		.	++	++
m17	.					.						.							T	++	+++
m18	T				A	G						.							.	+++	ND
m19	.					.						.		T	C				.	++	ND
m20	.					.						.		T					T	++	ND
DRE CONSENSUS	N	N	N	N	N	T G	N	G T	C	G	T	G	A C	G T	A T	A G	N	N	N		

Note. ND, not determined.

[a]Data taken from Yao and Denison (77) and unpublished results.

[b]Data taken from Shen and Whitlock (78).

[c]Relative magnitude of binding on function is indicated by +'s, while the lack of activity is indicated by a minus (−). The dots are included to help in visual alignment.

TCDD:AhR complex occurs within the CGTG "core" sequence, flanking nucleotides also contribute to DRE activity, because the core nucleotides by themselves fail to confer enhancer activity (69). More recently, binding-site selection analysis has been carried out to more fully define the nucleotide specificity of AhR DNA binding (81, 82). Although the results of these studies have provided some additional refinement of the nucleotide consensus sequence of the DRE with regard to DNA binding, the importance of these additionally identified bases has not been examined. Finally, sequence analysis of the upstream region of the mouse CYP1A1 gene also has revealed the presence of seven additional DRE core consensus sequences (see Fig. 2). In fact, because we have determined that many of these elements bind transformed TCDD:AhR complexes with varying affinities (they lack enhancer function (Denison, unpublished observations)), it is likely that these elements increase the ability of transformed nuclear AhR complexes to find the CYP1A1 upstream region within the genome by providing additional high-affinity DNA-binding sites for the receptors. Subsequent to binding, one can envision that the complexes can migrate to the higher-affinity DRE enhancer sequences. The rapid time course of induction of CYP1A1 gene expression by TCDD (11, 68) is consistent with this possibility.

Overall, the results of the DRE mutagenesis experiments reveal that DNA binding affinity is a poor predictor of enhancer function, because certain DREs that bound AhR complexes tightly proved to be marginal transcriptional enhancers. The reason for this discrepancy remains unclear, but it may reflect minor alterations in AhR–DRE nucleotide contacts that affect AhR conformation in ways that impact on transcriptional activation. Alternatively, mutations in this region might not alter the affinity of AhR–DNA binding but may alter the ability of the AhR to bend the DNA, which appears to be an important step in signal transduction by the promoter region (12, 68). Thus, just because the ligand:AhR complex can bind to a given DRE sequence in vitro does not mean a priori that the element will confer TCDD responsiveness. Analysis of the functionality of a DRE element taken out of context (i.e., separated from sequences to which it is normally adjacent in the intact flanking region) does not guarantee that a similar function will be demonstrated when the element is tested in context. For example, a DRE element present within the inhibitory domain of the 5′-flanking region of the mouse CYP1A1 gene (designated as *DRE5* or *DRE site A*) fails to confer TCDD responsiveness upon the CYP1A1 gene (63). Taken out of context, however, an oligonucleotide containing this DRE can bind the transformed AhR complex (in vitro) and confer TCDD responsiveness on a heterologous promoter and gene (Denison, unpublished results). Therefore, investigators examining the functionality of a newly identified DRE sequence need to demonstrate its functionality with respect to its normal context rather than simply providing evidence that an oligonucleotide containing this sequence can both bind AhR in vitro and confer TCDD responsiveness in a heterologous reporter construct.

Utilizing gel retardation analysis we also have examined the ability of the TCDD: AhR complex from a wide variety of species to transform and bind to the DRE (18, 83). These studies have revealed that cytosolic AhR from numerous species can transform in vitro and bind to the DRE, demonstrating that all of the factors necessary for AhR transformation and DNA binding are present in the cytosol after homogenization (ARNT, which is loosely bound within the nucleus, is released into the cytosol during cell disruption). Although species-specific differences in the rate and degree of AhR transformation have been noted (83), DNA-binding analyses using a series of mutant DRE oligonucleotides have indicated no apparent species- or ligand-dependent, nucleotide-specific difference in AhR binding to the DRE (18). These studies not only reveal the highly conserved nature of the DRE and AhR (at least in DNA binding) but also imply that sequences closely related to the murine consensus DRE sequence are responsible for conferring AhR-dependent TCDD responsiveness in each of these species. These data also suggest that factors other than DRE binding by the AhR likely account for the observed species- and tissue-specific differences in TCDD responsiveness.

UV crosslinking studies (84, 85) demonstrated that transformed rat hepatic TCDD: AhR complex binds to DRE-containing DNA as a heterodimer, consisting of ligand- (AhR) and non–ligand-binding (110-kDa) subunit. Comparable studies with the mouse receptor demonstrated that the AhR contacted nucleotides immediately 5′-ward of the CGTG core, whereas the ARNT protein bound at the core sequence (74). The interaction between ARNT and the GTG motif (i.e., an E-box half site) is consistent with the presence in ARNT of a bHLH domain and its homology to other E-box binding bHLH transcription factors (7, 10, 81). In contrast, the absence of an E-box consensus at the AhR binding sequence underscores the lack of homology between the AhR bHLH domain and that of other bHLH proteins.

More recent UV cross-linking experiments using guinea pig hepatic cytosolic AhR revealed three protein–DRE complexes, with molecular weights of approximately 100, 105, and 115 kDa, which covalently crosslinked to DRE-containing DNA in a TCDD-dependent manner (72). The 105-kDa protein was identified as the ligand-binding subunit of the AhR complex by covalent crosslinking with an [^{125}I]-azido TCDD agonist (72) and by use of anti-AhR antibodies (Tullis et al., submitted). In these experiments, the 100-kDa protein also was immunochemically identified as ARNT; the identity and function of the 115-kDa protein remain undetermined. In addition to UV cross-linking experiments supporting the presence of multiple DNA-binding subunits of the AhR complex, multiple DNA-binding forms of the AhR complex also have been identified using gel retardation analysis (86–90). Although the subunit composition and the functionality of these complexes remains to be determined, these results are consistent with the idea that proteins in addition to AhR and ARNT are required to mediate the actions of TCDD and related chemicals. Although initial studies using rat hepatic TCDD:AhR complexes (84, 85) revealed only two inducibly cross-linked protein–DNA complexes, purification of the AhR–DNA bound complex from rat liver by DNA affinity chromatography identified the AhR, ARNT, and 110-kDa proteins (90). The discrepancy in the number of complexes among these studies appears to be due to the fact that the AhR– and ARNT–DNA UV cross-linked products comigrated in the denaturing gels and were not resolved as distinct TCDD-inducible protein–DNA complexes. Whether the 110-kDa rat and 115-kDa guinea pig proteins are homologs is not known. Although the identity of this protein needs to be confirmed by cDNA cloning, partial peptide sequence of the rat 110-kDa protein reveals significant homology with the rat retinoblastoma (Rb) gene product (Elferink, unpublished data). The ramifications of an AhR–Rb association are significant and are addressed later.

ACTIVATION OF TRANSCRIPTION

Recombinant DNA and transfection experiments have revealed the presence of multiple TCDD-inducible, AhR- and ARNT-dependent transcriptional enhancers, the DREs, several hundred base pairs upstream of the transcriptional promoter of the CYP1A1 gene (63–69). In vivo footprinting experiments confirm that the DREs are targets for inducible, AhR- and ARNT-dependent protein–DNA interactions, and that prior to AhR binding, the CYP1A1 gene promoter is relatively inaccessible to DNA–binding proteins in vivo (12, 91–93). Indeed, the inactive enhancer/promoter region assumes a nucleosomal configuration, with a nucleosome specifically positioned at the promoter. Thus, loss of nucleosomes at the promoter may account for the increase in CYP1A1 transcription that occurs in induced cells. Binding of the TCDD:AhR:ARNT complex to a DRE is in the major groove of DNA, resulting in a distortion of the DNA and a localized discontinuous disruption of the nucleosomal organization of the regulatory region (inclusive of the promoter and other regulatory elements). These data support a model that AhR/ARNT–DRE interactions activate transcription by a mechanism that opens the chromatin and thus facilitates the binding of the transcription initiation machinery to the promoter (12). It still is not clear exactly how the AhR:ARNT complex disrupts nucleosomal structure and activates gene transcription, nor whether the AhR or additional factors is responsible for stabilizing the preinitiation complex at the promoter, as has been proposed for steroid hormone receptors. It is clear from deletion studies on the protein, however, that the carboxyl-terminal transactivation domain of the AhR is essential in this event and likely provides the critical contacts with other protein factors necessary for transcriptional activation.

PROTEIN PHOSPHORYLATION IN AhR ACTION

Whether protein phosphorylation is involved in AhR function remains an area of considerable confusion. A role for AhR/ARNT phosphorylation in TCDD responsiveness is supported by studies using phosphatase treatment and protein kinase inhibitors (94, 95). The AhR complex, however, appears to be phosphorylated prior to ligand binding, and there appears to be no significant ligand-induced phosphorylation of AhR necessary to get the complex to bind DNA (96). It also is unclear which protein kinases are responsible for AhR phosphorylation. Extended phorbol ester treatment of human keratinocytes or mice, which is known to suppress protein kinase C (PKC) activity, blocks TCDD-induced AhR transformation and DNA binding (97), implying that PKC activity is required for AhR function. In-vitro studies with PKC inhibitors, however, failed to demonstrate a PKC requirement for AhR–DNA binding (98). Evidence for the involvement of a tyrosine kinase comes from a study in human keratinocytes that failed to induce CYP1A1 in response to AhR agonists when treated with genistein (99). Gradin et al. (99) speculated that tyrosine kinase activity is required for ligand-induced release of the receptor from hsp90. This complements recent biochemical evidence for AhR ligand binding promoting c-src activation and subsequent release of the kinase from the complex (100). The identity of the c-src kinase substrate remains unclear, because examination of AhR phosphorylation reveals that ligand binding and transformation does not alter the phosphorylation pattern (96). Possibly the ARNT protein is a substrate. c-Raf, MEK, or mitogen activated protein kinase (MAPK) also may play a role in AhR function. Using human breast epithelial cell lines or transgenic mice, loss of AhR function was correlated with ras kinase activation (101). Ras signaling involves activation of the downstream kinases c-raf, MEK, and MAPK. Activation of ras, or a downstream kinase, abolished AhR–DNA binding in vitro with no effect on AhR nuclear translocation. The data did not identify the kinase involved, nor could they distinguish between a steady-state phosphorylation incompatible with DNA binding by the AhR complex and ligand-induced phosphorylation during AhR transformation. The latter seems unlikely, because AhR phosphorylation does not appear to change following ligand binding. Instead, the evidence suggests that phosphorylation of the AhR or other components of the DNA-binding complex, although important for function, is dictated by the specific intracellular environment affecting kinase and phosphatase activity.

OVERALL MECHANISM OF AhR ACTION

The overall mechanism of induction of CYP1A1 gene expression is presented in Fig. 4. Briefly, HAHs and PAHs diffuse through the plasma membrane, and those that are good inducers bind with high affinity to the cytoplasmic AhR complex. Following ligand binding, the ligand:AhR complex undergoes transformation, during which at least two molecules of hsp90 and other bound proteins (possibly AIP) dissociate from the complex, the liganded AhR accumulates within the nucleus, and it is converted into its high affinity DNA binding form following its association with ARNT and possibly other proteins. The binding of these transformed heteromeric TCDD:AhR complexes to DREs upstream of the CYP1A1 gene leads to DNA bending, chromatin and nucleosome disruption, increased promoter accessibility, increased rates of transcription initiation of the CYP1A1 gene, the subsequent accumulation of CYP1A1-specific mRNA and increased synthesis of microsomal cytochrome P4501A1 (10–12). The presence of the AhR complex in a wide variety of species and tissues and its ability to act as a ligand-dependent DNA-binding transactivator of gene expression suggests that many of the toxic and biologic effects of HAHs result from differential alteration in expression of genes in addition to that of CYP1A1 in susceptible cells.

In addition to CYP1A1, a large number of other genes and gene products are known to be regulated or influenced by Ah ligands (Table 3). Although roles for the AhR and DREs in the regulation of expression have been documented for a number of these

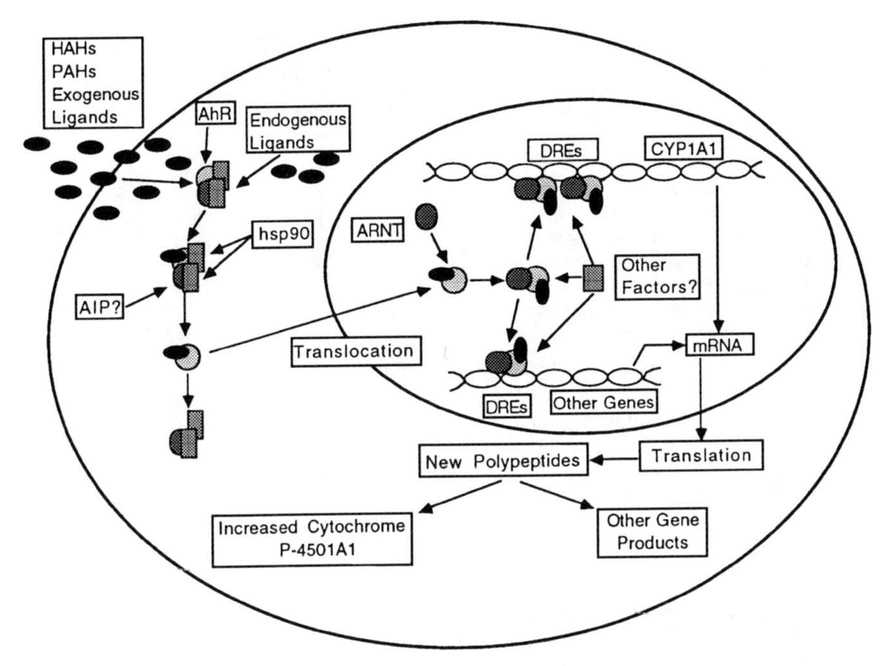

FIG. 4. Ah receptor-mediated mechanism of induction of gene expression by exogenous and endogenous chemicals. See text for additional details.

genes (including CYP1A1/1A2/1B1, glutathione S-transferase Ya, NAD(P)H:quinone oxidoreductase, UDP-glucuronosyltransferase, aldehyde dehydrogenase 3, γ-aminolevulinic acid synthase, and prostaglandin endoperoxide H synthase 2), the mechanism by which TCDD modulates the amount of the other gene products presented in Table 3 and the role if any, of the AhR in these responses are unknown. In addition, the observation that the levels of many of these gene products are up- or down-regulated in species-, tissue-, and development-specific manners argues that the mechanism of action of TCDD is significantly more complex than currently understood. We envision that the differential effects of TCDD can be modulated by numerous protein factors to which TCDD or the TCDD:AhR complex interact. Consequently, it is conceivable that TCDD also can have an action that is independent of the AhR or activation of expression of DRE-regulated genes. In support of this hypothesis, it has been reported that treatment of cells with TCDD can produce a transient influx of calcium into Hepa1 cells which occurs within 1 minute following TCDD exposure and independently of functional intracellular AhR complexes (102). The studies of Hanneman et al. (103) also demonstrate that TCDD treatment can increase intracellular calcium influx into primary cultures of rat hippocampal neuronal cells; however, the inability of the weak AhR agonist 1,2,3,4-TCDD to stimulate calcium influx indirectly supports a role for the AhR in this effect. Given the recent identification of cell-surface binding sites for progesterone that mediate calcium

TABLE 3. *Gene products regulated by AhR ligands and Ah inducers*

Gene product	Regulation	References
Gene products under AhR regulation		
Cytochrome P4501A1	+	12
Cytochrome P4501A2	+	157
Cytochrome P4501B1	+	158, 159
Glutathione S-transferase Ya	+	70
NAD(P)H:Quinone oxidoreductase 1	+	71
UDP-glucuronosyltransferase 1*06	+	160
Aldehyde dehydrogenase 3	+	161
γ-Aminolevulinic acid synthase	+	162
Prostaglandin endoperoxide H Synthase 2	+	163
Gene products possibly under AhR regulation		
Interleukin 1 b	+	164
Plasminogen activator inhibitor 2	+	164
Transforming growth factor $\beta 2$	+	165
Tumor necrosis factor α	+	166
Ornithine decarboxylase	+	167
Malic enzyme	+	168
Tyrosine kinase	+	169
Keratin 17	+	170
Lipoprotein lipase	+	171
25-DX (unknown gene product)	+	172
Glyceraldehye 3-phosphate dehydrogenase	+	171
Adipose type II 5′-deiodinase	+	173
Choline kinase	+	174
c-fos	+	175
Jun-B	+	175
c-jun	+	175
Jun-D	+	175
Keratinocyte transglutaminase	+/−	176, 177
AP2	−	178
Phosphenolpyruvate carboxykinase	−	179
c-erb A	−	180
Transforming growth factor α	−	165
Transforming growth factor β_1	−	181
Cytochrome P4502C11	−	182
Liver type I 5′-deiodinase	−	173
Peroxisome preoliferator activated receptor γ	−	179
Estrogen receptor	−	183
Glucocorticoid receptor	−	184
Epidermal growth factor receptor	−	185
Low density lipoprotein receptor	−	186
Ah receptor	−	187

uptake in human sperm (104), its interesting to speculate that the effect of TCDD on calcium influx is due to its interaction with an analogous cell-surface binding site that might exhibit the same ligand-binding specificity as that of the intracellular AhR. Although the mechanism by which TCDD stimulates calcium influx in these cells remains to be determined, these data suggest that TCDD can exert an effect distinct from that which requires an alteration in gene expression. In addition, although it is currently accepted that the mechanism by which TCDD and the AhR regulate gene expression is their specific interaction of transformed TCDD:AhR complex with a DRE adjacent to the responsive gene, it is conceivable that the AhR also can bind to DNA elements that are distinct from the consensus DRE. Vasiliou et al. (105) recently presented evidence suggesting that the AhR can interact with proteins bound to an electrophile response element involved in murine phase II gene expression. In addition, given that the past couple of years have seen the isolation of several new dimerization partners for the ARNT protein (7, 106–108), it seems plausible that additional AhR partners also exist, to which AhR interactions with hsp90 and AIP attest. Moreover, these new partners need not be bHLH-PAS–containing proteins. In this context, Per, another bHLH-PAS domain protein, was recently shown to interact with Tim, a protein encoded by the *Drosophila* circadian rhythm gene, *timeless*. Tim is not a PAS-containing protein, nor does it contain any identifiable protein-interface domain (109). Thus, the identification and characterization of AhR- and ARNT-associated proteins and their role in the action of TCDD, as well as further analysis of alternative mechanisms of action of TCDD, are exciting areas for future research.

AhR ACTIVITY IN THE ABSENCE OF EXOGENOUS LIGAND

Several recent reports have described the activation of AhR-dependent processes in the absence of exogenous ligand. If we assume that AhR activation absolutely requires ligand, these observations can be interpreted as evidence for the existence of an endogenous AhR ligand or ligands. Furthermore, a number of these observations involve events unrelated to xenobiotic metabolism, implicating a role for the AhR in processes distinct from its better-defined effects on metabolic enzymes.

Spontaneous AhR Activation

Singh and coworkers (110) report that in the absence of exogenous ligand, a significant fraction (16%) of the AhR pool is found in the nucleus of HeLa cells, with biochemical and functional properties characteristic of a liganded AhR. In addition, if monolayers of human keratinocytes or mouse Hepa1 cells (28) are grown in suspension, so as to lose cell–cell and cell–matrix contact, CYP1A1 and other members of the TCDD-inducible gene battery are induced spontaneously in an AhR-dependent mechanism. Both sets of observations are consistent with the notion that an endogenous AhR ligand is responsible for these events. The latter example involves an AhR response to anchorage deprivation, a condition known to trigger cell cycle arrest involving an Rb protein–mediated mechanism (111).

Development Considerations

AhR "knockout" mice possess livers with below normal weights and portal fibrosis (112, 113). A more dramatic consequence is that in one of the knockout lines generated nearly half of the mice die soon after birth and the remainder, although reaching maturity, show immune-system impairment (112). Specifically, these mice reveal a decreased accumulation of lymphocytes in the spleen and lymph nodes. Hence, the AhR plays an important role in liver and possibly immune-system development, independent of exogenous ligand. Additional evidence for AhR involvement in early mouse embryogenesis comes from a study in which disruption of AhR expression was accompanied by a concomitant decrease in blastocyst development (114). This work also suggests endogenous AhR activation and function during embryonic cell differentiation, independent of exogenous ligands.

Cell Cycle Effects

By examining differences in the growth rates of wild-type (Hepa1) and AhR-defective (AhR-D) mouse hepatoma cell lines, Ma and Whitlock (30) determined that the AhR influences G1 cell cycle progression and liver-cell differentiation. AhR-D cells that contain only 10% of wild-type AhR levels exhibit a prolonged transition through the G1 phase, while AhR-D cells transfected with AhR expression constructs demonstrate normal doubling times. Not only does the effect on G1 progression occur independently of TCDD treatment, consistent with an endogenous signal affecting AhR function, but these studies also demonstrate that the AhR functions in some capacity to influence cell cycle progression. Weiss and coworkers (29), using wild-type (5L) and AhR-defective (BP8) rat hepatoma cells, observed TCDD-dependent growth inhibition caused by G1 arrest only in the wild-type cells. Furthermore, stable expression of the AhR in the BP8 mutant line, which ordinarily is refractory to TCDD-induced growth inhibition, reconstituted the inhibitory phenotype (29). These data provide further compelling evidence that ligand-activated AhR can influence cell cycling. The relationship between AhR function and cell-cycle progression becomes even more intriguing with the discovery that the AhR and Rb protein interact (115, 116).

AhR AND THE RETINOBLASTOMA (Rb) PROTEIN

The Rb gene encodes a nuclear phosphoprotein then undergoes cyclic phosphorylation and dephosphorylation during the cell cycle (117–119). Rb is dephosphorylated during early G1 phase and then becomes phosphorylated at several sites by members of the cyclin-dependent kinase family prior to S phase and remains phosphorylated until late mitosis. Dephosphorylated Rb protein arrests cells in the G1 phase; phosphorylation of Rb relieves this inhibition, permitting cell entry into S-phase. Apart from controlling the G1-S transition, Rb also appears to influence cell differentiation and apoptotic events (120), but this property is probably cell type–specific and involves the expression of different Rb-associated proteins as well as the presence of additional proteins that complement Rb function.

Rb forms complexes with many different proteins, and formation of Rb complexes are a prerequisite for its growth-suppressive activity. Rb binds to target proteins by several different mechanisms. Viral oncoproteins (e.g., SV40 large T antigen) and certain cellular proteins (e.g., D-type cyclins and elf-1) contain a LXCXE motif (X denoting unconserved residues) that is involved in binding to the A/B pocket (i.e., short pocket) of Rb (121). E2F and MyoD do not contain the LXCXE motif; instead, their binding requires the A/B pocket plus C-terminal amino acids of Rb, also called the "large pocket" (122), while the bHLH domain of MyoD is involved with its interaction with Rb. Rb also can bind to other bHLH E-box binding proteins with varying affinities (123). Although efforts to identify a Rb–MyoD DNA binding complex in vitro have proven unsuccessful, the presence of Rb appears to greatly stimulate protein–DNA complex formation between MyoD and its binding partners (121). Rb also has been reported to possess DNA-binding properties (124, 125), and it can be found associated with an E2F–DNA complex (117). Given the presence of bHLH domains within the AhR and ARNT proteins as well as the presence of a conserved LXCXE pentapeptide Rb recognition motif only in the AhR protein (which is conserved across species), it is possible that AhR or ARNT interacts with Rb.

Coimmunoprecipitation experiments in rat 5L cells using Rb antibodies precipitated the AhR and ARNT proteins only after ligand-activated AhR transformation (115). The Rb protein in this precipitate was in the active dephosphorylated form. Direct Rb–ARNT interaction did not appear to occur, because Rb antibodies failed to immunoprecipitate ARNT in untreated cells even though both are nuclear proteins. Additional immunoprecipitation experiments using untreated 5L cells demonstrated that the unliganded AhR complex interacts with cyclin-dependent kinase 4 (cdk4), the very kinase that, in association with D-type cyclins, phosphorylates the Rb protein (115). These observations were not unique to cell-culture systems; studies with guinea pig liver cytosol generated similar results (116). Therefore, the available data suggest the following: 1) The AhR also may affect phosphorylation of Rb. 2) The AhR functions in the regulation of G1 cell-cycle progression, probably in concert with the Rb protein. Although the exact mechanism by which this occurs remains unclear, it may involve the transcription factor SP-1. Kobayashi and coworkers (126) presented evidence for a cooperative interaction between the AhR complex and SP-1 binding to the DRE and the basal transcription element respectively, which further stimulated TCDD-inducible CYP1A1 expression. Given that the Rb protein can facilitate SP-1–mediated expression of several cellular genes (127), it is tempting to speculate that Rb can contribute to AhR-mediated gene expression by facilitating AhR and SP-1 interactions and DNA binding. 3) Rb binding to the AhR occurs through the LXCXE motif. Because this pentapeptide motif is located within the ligand binding domain of the AhR, it is possible that the ligand-dependent nature of the Rb–AhR interaction may reflect a ligand-induced conformational change in the AhR protein that exposes this motif and allows the subsequent binding of Rb. Elucidation of the interactions between AhR and Rb and their relationship to alterations in cell-cycle progression is an exciting area of future research.

AhR KNOCKOUT MICE

A powerful tool to uncover protein function is to inactivate specific target genes. Two independent laboratories recently described the generation of AhR knockout mice by deleting either exon 1 (112) or exon 2 (113). The mice possess some common traits such as decreased liver weights and portal fibrosis but also display significant phenotypic differences. Major ones include the severely depressed immune system and 50% mortality rate associated with disruption of exon 1 (112). Schmidt and co-workers (114) attempted to reconcile the differences by suggesting that the exon 1–deleted mouse line is a partial knockout. Whereas the absence of a functional AhR was demonstrated in the line carrying the exon 2 deletion, loss of AhR function in the other line was implied by the inability of chemicals to induce toxicity and expression of several AhR-responsive genes (112, 128). Although the loss of inducible AhR-dependent gene expression and TCDD toxicity is compelling evidence for AhR inactivation, it does not exclude the possibility that an AhR with residual activity is expressed. In fact, amino acids encoded by exon 1 contribute much less to DNA binding than amino acids 34 to 42 encoded by exon 2, which appear critical for DNA binding (7, 129, 131). Whether the mouse line containing the exon 1 deletion expresses a functional AhR protein needs to be addressed experimentally. The possibility that the phenotypic differences reflect differences in genetic background appears unlikely because both mouse lines came from 129x C57BL/6 backgrounds. Likewise, environmental factors seem not to offer a satisfactory explanation because the exon 2–deleted line exposed to various environments continues to express the same phenotype (7). The apparent inability of TCDD to induce its spectrum of toxicologic and biochemical effects in exon-1 AhR knockout mice (128) provide strong evidence that the AhR mediates the toxic and biologic effects of TCDD and related chemicals. Utilization of these AhR knockout lines will greatly facilitate mechanistic studies into the role of the AhR in TCDD toxicity and will provide avenues for future studies into its role in endogenous processes.

CONCLUSIONS AND FUTURE PERSPECTIVES

Much has been learned about AhR biology in the past few years. With the advent of AhR and ARNT clones and, more recently, genetically engineered cell and animal models, there is now an appreciation that AhR function extends well beyond controlling the expression of a few genes involved in drug metabolism. The AhR seems to exhibit a dichotomy of function, regulating drug metabolism in response to xenobiotics on the one hand and influencing cell growth and differentiation on the other. These two processes probably reflect different signaling pathways responding to distinct cues (e.g., exogenous versus endogenous ligands), although there may be some causal overlap between them. This dichotomy appears central to understanding the role of the AhR in TCDD toxicity. By inducing the capacity for xenobiotic metabolism as well as normal cellular processes under AhR control, TCDD is a persistent stimulus that imbalances cellular homeostasis. Although this complicates the effort to understand mechanistically the basis of TCDD toxicity, recognizing that the AhR influences such events as cell-cycle progression may improve overall knowledge of cell growth and differentiation along with its impact on organogenesis and tumorogenensis.

ACKNOWLEDGMENTS

We thank Drs. Robert Rice, Carol Jones, and Michael Ziccardi for their critical review of this chapter. Research in our laboratories has been supported by the National Institutes of Health (ES07072, ES07685, ES07800), the Superfund Basic Research Program (ES04699), the Dutch Technology Foundation, and the California Agriculture Experiment Station.

REFERENCES

1. Safe S. Polychlorinated biphenyls (PCBs), dibenzo-p-dioxins (PCDDs), dibenzofurans (PCDFs), and related compounds: environmental and mechanistic considerations which support the development of toxic equivalency factors (TEFs). *Crit Rev Toxicol* 1990;21:51–88.
2. Zook DR, Rappe C. Environmental sources, distribution and fate of polychlorinated dibenzodixoins, dibenzofurans and related organochlorines. In: Schecter A, ed. *Dioxins and Health.* New York:Plenum Press, 1994;79–113.
3. Safe SH. Polychlorinated biphenyls (PCBs): environmental impact, biochemical and toxic responses and implications for risk assessment. *Crit Rev Toxicol* 1994;24:87–149.
4. Devito MJ, Birnbaum LS. Toxicology of dioxins and related chemicals. In: Schecter A, ed. *Dioxins and Health.* New York: Plenum Press, 1994;139–162.
5. Poland A, Knutson JC. 2,3,7,8-tetrachlorodibenzo-p-dioxin and related halogenated aromatic hydrocarbons: examination of the mechanism of toxicity. *Annu Rev Pharmacol Toxicol* 1982;22:517–542.
6. Safe S. Modulation of gene expression and endocrine response pathways by 2,3,7,8-tetrachloro-dibenzo-p-dioxin and related compounds. *Pharmacol Ther* 1995;67:247–281.
7. Schmidt JV, Bradfield CA. Ah receptor signaling pathways. *Annu Rev Cell Dev Biol* 1996;12:55–89.
8. Brouwer A, Ahlborg UG, van den Berg M, Birnbaum LS, Boersma RE, Bosveld B, Denison MS, Hagmar L, Holene E, Huisman M, Jacobson SW, Jacobson JL, Koopman-Esseboom C, Koppe JG, Kulig BM, Morse DC, Muckle G, Peterson, RE, Sauer PJJ, Seegal R F, Smit-van Prooije AE, Touwen BCL, Weisglas-Kuperus N, Winneke G. Functional aspects of developmental toxicity of polyhalogenated aromatic hydrocarbons in experimental animals and human infants. *Eur J Pharmacol Environ Toxicol Pharmacol* 1994;293:1–40.
9. Hardell L, Eriksson M, Axelson O, Hoar Zahm S. Cancer epidemiology. In: Schecter A, ed. *Dioxins and Health.* New York: Plenum Press, 1994;525–547.
10. Hankinson O. The aryl hydrocarbon receptor complex. *Annu Rev Pharmacol Toxicol* 1995;35:307–340.
11. Denison MS, Whitlock JP, Jr. Xenobiotic-inducible transcription of cytochrome P450 genes. *J Biol Chem* 1995;270:18175–18178.
12. Whitlock JP, Jr, Okino S, Dong L, Ko H, Clarke-Katzenberg R, Ma Q, Li H. Induction of cytochrome P4501A1: a model for analyzing mammalian gene transcription. *FASEB J* 1996;10:809–818.
13. Safe S. Comparative toxicology and mechanism of action of polychlorinated dibenzo-p-dioxins and dibenzofurans. *Annu Rev Pharmacol Toxicol* 1986;26:371–399.
14. Poland A, Glover E. 2,3,7,8-tetrachlorodibenzo-p-dioxin: segregation of toxicity with the Ah locus. *Mol Pharmacol* 1980;17:86–94.
15. Poland A, Glover E, Kende AS. Stereospecific, high affinity binding of 2,3,7,8-tetrachlorodibenzo-p-dioxin by hepatic cytosol: evidence that the binding species is the receptor for induction of aryl hydrocarbon hydroxylase. *J Biol Chem* 1976;251:4936–4946.

16. Denison MS, Wilkinson CF. Identification of the Ah receptor in selected mammalian species and its role in the induction of aryl hydrocarbon hydroxylase. *Eur J Biochem* 1985;147:429–435.
17. Gasiewicz TA, and Rucci G, Cytosolic receptor for 2,3,7,8-tetrachlorodibenzo-p-dioxin: evidence for a homologous nature among various mammalian species. *Mol Pharmacol* 1984;26:90–98.
18. Bank PA, Yao EF, Phelps CL, Harper PA, Denison MS. Species-specific binding of transformed Ah receptor to a dioxin responsive transcriptional enhancer. *Eur J Pharmacol Environ Toxicol Pharmacol* 1990;228:85–94.
19. Denison MS, Wilkinson CL, Okey AB. Ah receptor for 2,3,7,8-tetrachlorodibenzo-p-dioxin: comparative studies in mammalian and nonmammalian species. *Chemosphere* 1986;15:1665–1672.
20. Poellinger L. Mechanism of signal transduction by the basic helix-loop-helix dioxin receptor, In: Baeuerle PA, ed. *Inducible Gene Expression, Environmental Stresses and Nutrients*, Vol. 1. Boston: Birkhäuser, 1995;1177–205.
21. Grady AW, Fabacher DL, Frame G, Steadman BL. Morphological deformities in brown bullheads administered dietary β-naphthoflavone. *Journal of Aquatic Animal Health* 1992;4:7–16.
22. Waller CL, McKinney JD. Three-dimensional quantitative structure-activity relationships of dioxins and dioxin-like compounds: model validation and Ah receptor characterization. *Chem Res Toxicol* 1995;8:847–858.
23. Gillner M, Bergman J, Cambillau C, Alexandersson M, Fernstrom B, Gustafsson J-A. Interactions of indolo[3,2-b]carbazoles and related polycyclic aromatic hydrocarbons with specific binding sites for 2,3,7,8-tetrachlorodibenzo-p-dioxin in rat liver. *Mol Pharmacol* 1993;44:336–345
24. Kafafi SA, Afeefy HY, Said HK, Kafafi AG. Relationship between aryl hydrocarbon receptor binding, induction of aryl hydrocarbon hydroxylase and 7-ethyoxyresorufin o-deethylase enzymes and toxic activities of aromatic xenobiotics in animals: a new model. *Chem Res Toxicol* 1993;6:328–334.
25. Lesca P, Peryt B, Larrieu G, Alvinerie M, Galtier P, Daujat M, Maurel P, Hoogenboom L. Evidence for the ligand-independent activation of the Ah receptor. *Biochem Biophys Res Commun* 1995;209:474–482.
26. Daujat M, Charrasse S, Fabre I, Lesca P, Jounaidi Y, Larroque C, Poellinger L, Maurel P. Induction of CYP1A1 gene by benzimidazole derivatives during Caco-2 differentiation. *Eur J Biochem* 1996;237:642–652.
27. Quattrochi LC, Tukey RH. Nuclear uptake of the Ah (dioxin) receptor in response to omeprazole: transcriptional activation of the human CYP1A1 gene. *Mol Pharmacol* 1993;43:504–508.
28. Sadek CM, Allen-Hoffman BL. Suspension-mediated induction of hepa1c1c7 CYP1a-1 expression is dependent on the Ah receptor signal transduction pathway. *J Biol Chem* 1994;269:31505–31509.
29. Weiss C, Kolluri SK, Kiefer F, Gottlicher M. Complementation of Ah receptor deficiency in hepatoma cells: negative feedback regulation and cell cycle control by the Ah receptor. *Exp Cell Res* 1996;226:154–163.
30. Ma Q, Whitlock JP, Jr. The aromatic hydrocarbon receptor modulates the Hepa 1c1c7 cell cycle and differentiated state independently of dioxin. *Mol Cell Biol* 1996;16:2144–2150.
31. Bjeldanes LF, Kim J-L, Grose KR, Bartholomew JC, Bradfield CA. Aromatic hydrocarbon responsiveness-receptor agonists generated from indole-3-carbinol *in vitro* and *in vivo*: comparisons with 2,3,7,8-tetrachlorodibenzo-p-dioxin. *Proc Natl Acad Sci USA* 1991;88:9543–9547.
32. Gradelet S, Astorg P, Leclerc J, Chevalier J, Vernevaut M-F, Siess M-H. Effects of canthaxanthin, astaxanthin, lycopene and lutein on liver xenobiotic-metabolizing enzymes in the rat. *Xenobiotica* 1996;6:49–63.
33. Perdew GH, Babbs CF. Production of Ah receptor ligands in rat fecal suspensions containing tryptophan or indole-3-carbinol. *Nutr Cancer* 1991;16:209–218.
34. Rannug A, Rannug U, Rosenkranz HS, Winqvist L, Westerholm R, Agurell E, Grafstrom A-K. Certain photooxidized derivatives of tryptophan bind with very high affinity to the Ah receptor and are likely to be endogenous signal substances. *J Biol Chem* 1987;262:15422–15427.
35. Helferich W, Denison MS. Photooxidized products of tryptophan can act as dioxin agonists. *Mol Pharmacol* 1991;40:674–678.
36. Rannug U, Rannug A, Sjoberg U, Li H, Westerhold R, Bergman A. Structure elucidation of two tryptophan-derived, high affinity Ah receptor ligands. *Chem Biol* 1995;2:841–845.
37. Goerz G, Merk H, Bolse K, Tsambaos D, Berger H. Influence of chronic UV-light on hepatic and cutaneous monooxygenases. *Experientia* 1983;39:385–386.
38. Denison MS, Tullis K, Rogers WJ, Heath-Pagliuso S. Tryptamine and indole acetic acid are endogenous water soluble metabolites of tryptamine that are weak AhR ligands. *Toxicologist* 1997;36: 128.

39. El-Fouly MH, Richter C, Giesy JP, Denison, MS. Production of a novel recombinant cell line for use as a bioassay for detection of 2,3,7,8-tetrachlorodibenzo-p-dioxin-like chemicals. *Environ Toxicol Chem* 1994;13:1581–1588.
40. Garrison PM, Aarts JMMJG, Brouwer A, Giesy JP, Denison MS. Ah receptor-mediated gene expression: production of a recombinant mouse hepatoma cell bioassay system for detection of 2,3,7,8-tetrachlorodibenzo-p-dioxin-like chemicals. *Fundam Appl Toxicol* 1996;30:194–203.
41. Denison MS, Vella LM, Okey AB. Structure and function of the Ah receptor for 2,3,7,8-tetrachlorodibenzo-p-dioxin: species differences in molecular properties of the receptor from mouse and rat hepatic cytosol. *J Biol Chem* 1986;261:3987–3995.
42. Denison MS, Garrison PM, Aarts JMMJG, Tullis K, Schalk JAC, Cox MA, Brouwer A. Species-specific differences in Ah receptor ligand binding and transcriptional activation: implications for bioassays for the detection of dioxin-like chemicals. *Organohalogen Compounds* 1995;23:225–229.
43. Aarts JMMJG, Denison MS, Cox MA, Schalk, MAC, Garrison PM, Tullis K, deHaan LHJ, Brouwer A. Species-specific antagonism of Ah receptor action by 2,2′,5,5′-tetrachloro- and 2,2′,3,3′,4,4′-hexachlorobiphenyl. *Eur J Pharmacol* 1996;293:463-474.
44. Kikuchi H, Kato H, Mizuno M, Hossain A, Ikawa S, Miyazaki J, Watanabe M. Differences in inducibility of CYP1A1-mRNA by benzimidazole compounds between human and mouse cells: evidences of a human-specific signal transduction pathway for CYP1A1 induction. *Arch Biochem Biophys* 1996;334:235–240.
45. Chen H-S, Perdew GH. Subunit composition of the heteromeric cytosolic aryl hydrocarbon receptor complex. *J Biol Chem* 1994;269:27554–27558.
46. Carver LA, Bradfield CA. Ligand-dependent interaction of the aryl hydrocarbon receptor with a novel immunophilin homolog *in vivo*. *J Biol Chem* 1977;272:11452–11456.
47. Ma Q, Whitlock JP, Jr. A novel cytoplasmic protein that interacts with the Ah receptor, contains a tetratricopeptide repeat motif and augments the transcriptional response to 2,3,7,8-tetrachlorodibenzo-p-dioxin. *J Biol Chem* 1977;27:8878–8884.
48. Owens-Grillo JK, Czar MJ, Hutchison KA, Hoffman K, Perdew GH, Pratt WB. A model of protein targeting mediated by immunophilins and other proteins that bind to Hsp90 via tetratricopeptide repeat domains. *J Biol Chem* 1996;271:13468–13475.
49. Denison MS, Vella LM. The hepatic Ah receptor for 2,3,7,8-tetrachlorodibenzo-p-dioxin: species differences in subunit dissociation. *Arch Biochem Biophys* 1990;277:382–388.
50. Poland A, Glover E. Variation in the molecular mass of the Ah receptor among vertebrate species and strains of rats. *Biochem Biophys Res Commun* 1987;146:1439–1449.
51. Hahn ME, Poland A, Glover E, Stegeman JJ. Photoaffinity labeling of the Ah receptor: phylogenetic survey of diverse vertebrate and invertebrate species. *Arch Biochem Biophys* 1994;310:218–228.
52. Pratt WB. Heat shock proteins and glucocorticoid receptor function. *Bioessays* 1992;14:841–848.
53. Carver LA, Jackiw V, Bradfield CA. The 90-kDa heat shock protein is essential for Ah receptor signaling in a yeast expression system. *J Biol Chem* 1994;269:1–4.
54. Pongratz I, Mason GGF, Poellinger L. Dual roles of the 90-kDa heat shock protein Hsp90 in modulating functional activities of the dioxin receptor. *J Biol Chem* 1992;267:13728–13734.
55. Bresnick EH, Dalman MK, Sanchez ER, Pratt WB. Evidence that the 90-kDa heat shock protein is necessary for the steroid binding conformation of the L cell glucocorticoid receptor. *J Biol Chem* 1989;264:4992–4997.
56. Bohan S. Hsp90 mutants disrupt glucocorticoid receptor ligand binding and destabilize aporeceptor complexes. *J Biol Chem* 1995;270:29433–29438.
57. Denison MS. Heterogeneity of rat hepatic Ah receptor: identification of two receptor forms which differ in their biochemical properties. *J Biochem Toxicol* 1992;4:249–256.
58. Ko HP, Okino ST, Ma Q, Whitlock JP, Jr. Dioxin-induced CYP1A1 transcription *in vivo*: the aromatic hydrocarbon receptor mediates transactivation, enhancer-promoter communication, and changes in chromatin structure. *Mol Cell Biol* 1996;16:430–436.
59. Dolwick KM, Swanson HI, Bradfield CA. *In vitro* analysis of Ah receptor domains involved in ligand-activated DNA recognition. *Proc Natl Acad Sci USA* 1993;90:8566–8570.
60. Hord NG, Perdew GH. Physiochemical and immunochemical analysis of aryl hydrocarbon receptor nuclear translocator: characterization of two monoclonal antibodies to the aryl hydrocarbon receptor nuclear translocator. *Mol Pharmacol* 1994;46:618–624.
61. Pollenz RS, Sattler, CA, Poland A. The aryl hydrocarbon receptor and aryl hydrocarbon receptor nuclear translocator protein show distinct subcellular localizations in Hepa 1c1c7 cells by immunfluorescence microscopy. *Mol Pharmacol* 1994;45:428–438.

62. Prokipcak RD, Okey AB. Physiochemical characterization of the nuclear form of Ah receptor from mouse hepatoma cells exposed in culture to 2,3,7,8-tetrachlorodibenzo-p-dioxin. *Arch Biochem Biophys* 1988;267:811–828.
63. Jones PBC, Galeazzi DR, Fisher JM, Whitlock JP, Jr. Control of cytochrome P_1-450 gene expression by dioxin. *Science* 1985;227:1499–1502.
64. Jones PBC, Durrin LK, Galeazzi DR, Whitlock JP, Jr. Control of cytochrome P_1-450 gene expression: analysis of a dioxin-responsive enhancer system. *Proc Natl Acad Sci USA* 1986;83:2803–2806.
65. Jones PBC, Durrin LK, Fisher JM, Whitlock JP, Jr. Control of gene expression by 2,3,7,8-tetrachlorodibenzo-p-dioxin: multiple dioxin-responsive domains 5′-ward of the cytochrome P_1-450 gene. *J Biol Chem* 1986;261:6647–6650.
66. Denison MS, Fisher JM, Whitlock JP, Jr. Inducible, receptor-dependent protein-DNA interactions at a dioxin-responsive transcriptional enhancer. *Proc Natl Acad Sci USA* 1988;85:2528–2532.
67. Fisher JM. Wu L, Denison MS, Whitlock JP, Jr. Organization and function of a dioxin responsive enhancer. *J Biol Chem* 1990;265:9676–9681.
68. Whitlock JP, Jr. Mechanistic aspects of dioxin action. *Chem Res Toxicol* 1993;6:754–763.
69. Denison MS, Fisher JM, Whitlock JP, Jr. The DNA recognition site for the dioxin-Ah receptor complex: nucleotide sequence and functional analysis. *J Biol Chem* 1988;263:17721–17724.
70. Rushmore TH, Pickett CB. Glutathione S-transferases, structure, regulation, and therapeutic implications. *J Biol Chem* 1993;268:11475–11478.
71. Favreau LV, Pickett CB. Transcriptional regulation of the rat NAH(P)H: quinone reductase gene: identification of regulatory elements controlling basal level expression and inducible expression by planar aromatic compounds and phenolic antioxidants. *J Biol Chem* 1991;266:4556–4561.
72. Swanson HI, Tullis K, Denison MS. Binding of transformed Ah receptor complex to a dioxin responsive transcriptional enhancer: evidence for two distinct heteromic DNA-binding forms *Biochem* 1993;32:12841–12849.
73. Reyes H, Reisz-Porszasz S, Hankinson O. Identification of the Ah receptor nuclear translocator protein (Arnt) as a component of the DNA binding form of the Ah receptor. *Science* 1992;256:1193–1195.
74. Basci SG, Reisz-Porszasz S, Hankinson O. Orientation of the heteromeric aryl hydrocarbon (dioxin) receptor complex on its assymetrical DNA recognition sequence. *Mol Pharmacol* 1995;47:432–438.
75. Denison MS, Fisher JM, Whitlock JP, Jr. Protein-DNA interactions at recognition sites for the dioxin-Ah receptor complex. *J Biol Chem* 1989;264:16478–16482.
76. Denison MS, Yao EF. Characterization of the interaction of transformed rat hepatic cytosolic Ah receptor with a dioxin-responsive transcriptional enhancer. *Arch Biochem Biophys* 1991;284:158–166.
77. Yao EF, Denison MS. DNA Sequence determinants for binding of transformed Ah receptor to a dioxin-responsive enhancer *Biochemistry* 1992;31:5060–5067.
78. Shen ES, Whitlock JP, Jr. The potential role of DNA methylation in the response to 2,3,7,8-tetrachlorodibenzo-p-dioxin. *J Biol Chem* 1989;264:17754–17758.
79. Saatcioglu F, Perry DJ, Pasco DS, Fagan JB. Aryl hydrocarbon (Ah) receptor DNA-binding activity: sequence specificity and Zn^{2+} requirement. *J Biol Chem* 1990;265:9251.
80. Shen ES, Whitlock JP, Jr. Protein-DNA interactions at a dioxin-responsive enhancer: mutational analysis of the DNA binding site for the liganded Ah receptor. *J Biol Chem* 1992;267:6815–6819.
81. Swanson HI, Chan, WK, Bradfield CA. DNA binding specificities and pairing rules of the Ah receptor, Arnt, and SIM proteins. *J Biol Chem* 1995;270:26292–26302.
82. Luo B, Perry DJ, Zhang L, Krarat I, Basic M, Fagan JB. Mapping sequence specific DNA-protein interactions: a versatile quantitative method and its application to transcription factor XF1. *J Mol Biol* 1997;266:479-492.
83. Denison MS, Phelps CL, DeHoog J, Kim HJ, Bank PA, Harper PA, Yao EF. Species variation in Ah receptor transformation and DNA binding. In: Gallo MA, Scheuplein RJ, Van Der Heijden KA, eds. *Biological Basis of Risk Assessment of Dioxins and Related Compounds*. Banbury Report No. 35. Cold Spring Harbor, NY: Cold Spring Harbor Press, 1991;337–350.
84. Elferink CJ, Gasiewicz TA, Whitlock JP, Jr. Protein-DNA interactions at a dioxin-responsive enhancer: evidence that the transformed Ah receptor is heteromeric. *J Biol Chem* 1990;265:20708–20712.
85. Gasiewicz TA, Elferink CJ, Henry EC. Characterization of multiple forms of the Ah receptor: recognition of a dioxin-responsive enhancer involves heteromer formation. *Biochem* 1991;30:2909–2916.
86. Okino ST, Pendurthi UR, Tukey RH. 2,3,7,8-Tetrachlorodibenzo-p-dioxin induces the nuclear translocation of two XRE binding proteins in mice. *Pharmacogenetics* 1993;3:101–109.
87. Bank PA, Yao EF, Swanson HI, Tullis K, Denison MS. DNA binding of the transformed guinea pig hepatic Ah receptor complex: identification and partial characterization of two high-affinity DNA-binding forms. *Arch Biochem Biophys* 1995;317:439–448.

88. Perdew GH, Hollenback CE. Evidence for two functionally distinct forms of the human Ah receptor. *J Biochem Toxicol* 1995;10:95–102.
89. Cary PA, Martin PA, Dougherty JJ. ATP promotes the appearance of two DNA-binding forms of the AhR. *Arch Biochem Biophys* 1991;288:287–292.
90. Elferink CJ, Whitlock JP, Jr. Dioxin-dependent, DNA sequence-specific binding of a multiprotein complex containing the Ah receptor. *Receptor* 1994;4:157–163.
91. Wu L, Whitlock JP, Jr. Mechanism of dioxin action: Ah receptor-mediated increase in promoter accessibility in vivo. *Proc Natl Acad Sci USA* 1992;89:4811–4815.
92. Morgan JE, Whitlock JP, Jr. Transcriptional-dependent and transcription-independent nucleosome disruption induced by dioxin. *Proc Natl Acad Sci USA* 1992;89:11622–11626.
93. Okino ST, Whitlock JP, Jr. Dioxin induces localized, graded changes in chromatin structure: implications for *Cyp1A1* gene transcription. *Mol Cell Biol* 1995;15:3714–2721.
94. Pongratz I, Strömstedt PE, Mason GG, Poellinger L. Inhibition of specific DNA binding activity of the dioxin receptor by phosphatase treatment. *J Biol Chem* 1991;266:16813–16817.
95. Carrier F, Owens RA, Nebert DW, Puga A. Dioxin-dependent activation of murine Cyp1a-1 gene transcription requires protein kinase C-dependent phosphorylation. *Mol Cell Biol* 1992;12:1856–1863.
96. Mahon MJ, Gasiewicz TA. Ah receptor phosphorylation: Localization of phosphorylation sites to the C-terminal half of the protein. *Arch Biochem Biophys* 1995;318:166–174.
97. Berghard A, Gradin K, Pongratz I, Whitelaw M, Poellinger L. Cross-coupling of signal transduction pathways: the dioxin receptor mediates induction of cytochrome P-4501A1 expression via a protein kinase C-dependent mechanism. *Mol Cell Biol* 1993;13:677–689.
98. Schafer MW, Madhukar BV, Swanson HI, Tullis K, Denison MS. Protein kinase C is not involved in Ah receptor transformation and DNA binding. *Arch Biochem Biophys* 1993;307:267–271.
99. Gradin K, Whitelaw ML, Toftgård R, Poellinger L, Berghard A. A tyrosine kinase-dependent pathway regulates ligand-dependent activation of the dioxin receptor in human keratinocytes. *J Biol Chem* 1994;269:23800–23807.
100. Enan E, Matsumura F. Identification of c-Src as the integral component of the cytosolic Ah receptor complex, transducing the signal of 2,3,7,8-tetrachlorodibenzo-p-dioxin (TCDD) through the protein phosphorylation pathway. *Biochem Pharmacol* 1996;52:1599–1612.
101. Reiners JJ, Jr, Jones CL, Hong N, Clift R, Elferink CJ. Down regulation of aryl hydrocarbon receptor function by expression of Ha-*Ras* oncogenes. *Mol Carcinogenesis* 1997;19:91–100. In press.
102. Puga A, Nebert DW, Carrier F. Dioxin induces expression of c-fos and c-jun proto-oncogenes and a large increase in transcription factor AP-1. *DNA Cell Biol* 1992;11:269–281.
103. Hanneman WH, Legare ME, Barhoumi R, Burghardt RC, Safe S, Tiffany-Castiglioni E. Stimulation of calcium uptake in cultured rat hippocampal neurons by 2,3,7,8-tetrachlorodibenzo-p-dioxin. *Toxicology* 1996;112:19–28.
104. Blackmore PF, Neulen J, Lattanzio F, Beebe SJ. Cell surface-binding sites for progesterone mediated calcium uptake in human sperm. *J Biol Chem* 1991;266:18655–18659.
105. Vasiliou V, Puga A, Chang C-Y, Tabor M, Nebert DW. Interaction between the Ah receptor and proteins binding to the AP-1-like electrophile response element (EpRE) during murine phase II [Ah] battery gene expression. *Biochem Pharmacol* 1995;50:2057–2068.
106. Hogenesch J, Chan W, Jackiw V, Brown R, Gu Y-Z, Pray-Grant R, Perdew GH, Bradfield CA. Characterization of a subset of the basic-helix-loop-helix-PAS superfamily that interacts with components of the dioxin signaling pathway. *J Biol Chem* 1997;272:8581–8593.
107. Zhou Y-D, Barnard M, Tian H, Li X, Ring H, Francke U, Shelton J, Richardson J, Russell D, McKnight S. Molecular characterization of two mammalian bHLH-PAS domain proteins selectively expressed in the central nervous system. *Proc Natl Acad Sci USA* 1997;94:713–718.
108. Li H, Ko H, Whitlock JP, Jr. Induction of phosphoglycerate kinase 1 gene expression by hypoxia. *J Biol Chem* 1996;271:21262–21267.
109. Gekakis N, Saez L, Delahaye-Brown A, Myer MP, Sehgal A, Young MW, Weitz CJ. Isolation of *timeless* by PER protein interaction: defective interaction between *timeless* protein and long-period mutant PER^L. *Science* 1995;270:811–815.
110. Singh S, Hord N, Perdew GH. Characterization of the activated form of the aryl hydrocarbon receptor in the nucleus of Hela cells in the absence of exogenous ligand. *Arch Biochem Biophys* 1996;329:47–55.
111. Assoian RK. Anchorage-dependent cell cycle progression. *J Cell Biol* 1997;136:1–4.
112. Fernandez-Salguero P, Pineau T, Hilbert DM, McPhail T, Lee SST, Kimura S, Nebert DW, Rudikoff S, Ward JM, Gonzalez FJ. Immune system impairment and hepatic fibrosis in mice lacking the dioxin-binding Ah receptor. *Science* 1995;268:722–726.

113. Schmidt J, Su G, Reddy J, Simon M, Bradfield CA. Characterization of a murine AhR null allele: involvement of the Ah receptor in hepatic growth and development. *Proc Nat Acad Sci USA* 1996;93:6731–6736.
114. Peters JM, Wiley LM. Evidence that murine preimplantation embryos express aryl hydrocarbon receptor. *Toxicol Appl Pharmacol* 1995;134:214–221.
115. Hushka DR, Greenlee WF. Ligand-activated Ah receptor induces G1 cell cycle arrest. *Am Assoc Cancer Res Proc* 1996;37:4.
116. Ma X, Wheelock GD, Dong H, Babish JG. Treatment of guinea pig hepatic cytosol with TCDD results in Ah receptor complexing with mitogen-activated protein kinase and rearranging cell cycle subunits. *Toxicologist* 1997;36:128.
117. Goodrich DW, Wang NP, Qian Y-W, Lee EY-HP, Lee W-H. The retinoblastoma gene product regulates progression through the G1 phase of the cell cycle. *Cell* 1991;67:293–302.
118. Riley DJ, Lee EY-HP, Lee W-H. The retinoblastoma protein: more than a tumor suppressor. *Annu Rev Cell Biol* 1994;10:1–29.
119. Weinberg RA. The retinoblastoma protein and cell cycle control. *Cell* 1995;81:323–330.
120. Zacksenhaus E, Jiang Z, Chung D, Marth JD, Phillips RA, Gallie BL. pRb controls proliferation, differentiation, and death of skeletal muscle cells and other lineages during embryogenesis. *Genes Dev* 1996;10:3051–3064.
121. Ewen ME, Sluss HK, Sherr JC, Matsushime H, Kato J-Y, Livingston DM. Functional interactions of the retinoblastoma protein with mammalian D-type cyclins. *Cell* 1993;7:487–497.
122. Qian X-Q, Chittenden D, Livingston DM, Kaelin WG, Jr. Identification of a growth suppression domain within the retinoblastoma gene product. *Genes Dev* 1992;6:953–974.
123. Gu W, Schneider JW, Condorelli G, Kaushal S, Mahdavi V, Nadal-Ginard B. Interaction of myogenic factors and the retinoblastoma protein mediates muscle cell commitment and differentiation. *Cell* 1993;72:309–324.
124. Mittnacht S, Weinberg RA. Phosphorylation of the retinoblastoma protein is associated with an altered affinity for the nuclear compartment. *Cell* 1991;65:381–393.
125. Templeton DJ. Nuclear binding of purified retinoblastoma gene product is determined by cell cycle-regulated phosphorylation. *Mol Cell Biol* 1992;12:435-443.
126. Kobayashi A, Sogawa K, Fujii-Kuriyama Y. Cooperative interaction between AhR-Arnt and Sp1 for the drug-inducible expression of CYP1A1 gene. *J Biol Chem* 1996;271:12310–12316.
127. Udvadia AJ, Rogers KT, Higgins PDR, Murata Y, Martin KH, Humphrey PA, Horowitz JM. Sp-1 binds promoter elements regulated by the RB protein and Sp-1-mediated transcription is stimulated by RB coexpression. *Proc Natl Acad Sci USA* 1993;90:3265–3269.
128. Fernandez-Salguero P, Hilbert D, Rudikoff S, Ward J, Gonzalez FJ. Aryl-hydrocarbon receptor-deficient mice are resistant to 2,3,7,8-tetrachlorodibenzo-p-dioxin-induced toxicity. *Toxicol Appl Pharmacol* 1996;140:173–179.
129. Bacsi SG, Hankinson O. Functional characterization of DNA-binding domains of the subunits of the heterodimeric aryl hydrocarbon receptor complex imputing novel and canonical basic helix-loop-helix protein-DNA interactions. *J Biol Chem* 1996;271:8843–8850.
130. Dong L, Ma Q, Whitlock JP, Jr. DNA binding by the heterodimeric Ah receptor. *J Biol Chem* 1996;271:7942–7948.
131. Swanson HI, Yang J-H. Mapping the protein/DNA contact sites of the Ah receptor and Ah receptor nuclear translocator. *J Biol Chem* 1996;271:31657–31665.
132. Becker M, Phillips T, Safe S. Polychlorinated diphenyl ethers-a review. *Toxicol Environ Chem* 1991;33:189–200.
133. Cheung, ENY, McKinney JD. Polybrominated napthalene and diidobenzene interactions with specific binding sites for 2,3,7,8-tetrachlorodibenzo-p-dioxin in rat liver cytosol. *Mol Toxicol* 1989;2:39–52.
134. Bunce NJ, Landers JP, Schneider UA, Safe SH, Zacharewski TR. Chlorinated trans stilbenes: competitive binding to the Ah receptor, induction of cytochrome P-450 monooxygenase activity and partial 2,3,7,8-TCDD antagonism. *Toxicol Environ Chem* 1989;28:217–229.
135. Greibrokk, T, Lofroth, G, Nilsson, L, Toftgard, R, Carlstedt-Duke, JC and Gustafsson, J-A. Nitroarenes: mutagenicity in the Ames Salmonella/microsome assay and affinity to the TCDD-receptor protein. In: Rickert DE, ed. *Toxicity of Nitroaromatic Compounds*. New York: McGraw-Hill, 1985;167–183.
136. Gasiewicz TA, Kende AS, Rucci G, Whitney B, Willey JJ. Analysis of structural requirements for Ah receptor antagonist activity: ellipticines, flavones, and related compounds. *Biochem Pharmacol* 1996;52:1787–1803.

137. Lu Y-F, Santostefano M, Cunningham BDM, Threadgill MD, Safe S. Substituted flavones as aryl hydrocarbon (Ah) receptor agonists and antagonists. *Biochem Pharmacol* 1996;51:1077–1087.
138. Liu H, Santostefano M, Safe S. 6-Substituted 3,4-benzocoumarins: a new structural class of inducers and inhibitors of CYP1A1-dependent activity. *Arch Biochem Biophys* 1993;306:223–231.
139. Marcus CB, Wilson NM, Jefcoate CR, Wilkinson CF, Omiecinski CJ. Selective induction of cytochrome P450 isozymes in rat liver by 4-*n*-alkyl-methylenedioxybenzenes. *Arch Biochem Biophys* 1990;277:8–16.
140. Jellinck PH, Forket PG, Riddick DS, Okey AB, Michnovicz JJ, Bradlow HL. Ah receptor binding properties of indole carbinols and induction of hepatic estradiol hydroxylation. *Biochem Pharmacol* 1993:45:1129–1136.
141. Gillner M, Bergman J, Cambillau C, Gustafsson J-A. Interactions of rutecarpine alkaloids with specific binding sites for 2,3,7,8-tetrachlorodibenzo-p-dioxin in rat liver. *Carcinogenesis* 1989;10:651–654.
142. Karenlampi SO, Tuome K, Korkalainen M, Raunio H. 2-(4′-Chlorophenyl)benzothiazole is a potent inducer of cytochrome P4501A1 in a human and a mouse cell line. *J Biochem* 1989;181:143–148.
143. Sweatlock JA, Gasiewicz TA. The interaction of 1,3-diaryltriazenes with the Ah receptor. *Chemosphere* 1986;15:1687–1690.
144. Lee IJ, Jeong KS, Roberts BJ, Kallarakal AT. Fernandex-Salguero P, Gonzalez FJ, Song BJ. Transcriptional induction of the cytochrome P4501A1 gene by a thiazolium compound, YH439. *Mol Pharmacol* 1996;49:980–988.
145. Cheung Y-L, Snelling J, Mohammed NND, Gray TJB, Ioannides C. Interaction with the aromatic hydrocarbon receptor, CYP1A induction, and mutagenicity of a series of diaminotoluenes: implications for their carcinogenicity. *Toxicol Appl Pharmacol* 1996;139:203–211.
146. Schafer MW, Madhukar BV, Swanson HI, Tullis K, Denison MS. Protein kinase C is not involved in Ah receptor transformation and DNA binding. *Arch Biochem Biophys* 1993;307:267–271.
147. Kobayashi Y, Matsuura Y, Kotani E, Fukuda T, Aoyagi T, Tobinaga S, Yoshida T, Kuroiwa Y. Structural requirements of the induction of hepatic microsomal cytochrome P450 by imidazole- and pyridine-containing compounds in rats. *J Biochem* 1993;114:697–701.
148. Franklin MR. Induction of rat liver drug-metabolizing enzymes by hetereocycle-containing mono-, di-, tri-, and tetra-arylmethanes. *Biochem Pharmacol* 1993;46:683–689.
149. Karenlampi SO, Tuomi K, Korkalainen M, Raunio H. Induction of cytochrome P4501A1 in mouse hepatoma cells by several chemicals. *Biochem Pharmacol* 1989;38:1517–1525.
150. Gleizes-Escala C, Lesca P, Larrieu G, Dupuy J, Pineau T, Galtier P. Effect of exposure of rabbit hepatocytes to sulfur-containing anthelmintics (oxfendazole and fenbendazole) on cytochrome P4501A1 expression. *Toxicol in Vitro* 1996;10:129–139.
151. Gradelet S, Leclerc J, Siess M-H, Astorg PO. β-Apo-8′-carotenal, but not β-carotene, is a strong inducer of liver cytochromes P4501A1 and 1A2 in rat. *Xenobiotica* 1996;26:909–919.
152. Kleman MI, Overvik E, Mason GGF, Gustafsson J-A. *In vitro* activation of the dioxin receptor to a DNA-binding form by food-borne heterocyclic amines. *Carcinogenesis* 1992;13:1619–1624.
153. Viswalingam A, Caldwell J. Cinnamyl anthranilate causes coinduction of hepatic microsomal and eproxisomal enzymes in mouse but not rat. *Toxicol Appl Pharmacol* 1997;142:338–347.
154. Martinez-Larranaga MR, Anadon A, Diaz MJ, Fernandez R, Sevil B, Fernandez-Cruz ML, Fernandez MC, Martinez MA, Anton R. Induction of cytochrome P4501A1 and P4504A1 activities and peroxisomal proliferation by fumonisin B1. *Toxicol Appl Pharmacol* 1996;141:185–194.
155. Sadar MD, Ash R, Andersson TB. Picrotoxin is a CYP1A1 inducer in rainbow trout hepatocytes. *Biochem Biophys Res Commun* 1995;214:1060–1066.
156. Sadar MD, Westlind A, Blomstrand F, Andersson TB. Induction of CYP1A1 by GABA receptor ligands. *Biochem Biophys Res Commun* 1996;229:231–237.
157. Quattrochi LC, Vu T, Tukey RH. The human CYP1A2 gene and induction by 3-methylcholanthrene. *J Biol Chem* 1994;269:6949–6954.
158. Sutter TR, Tang YM, Hayes CL, Wo Y-YP, Jabs EW, Li X, Yin H, Cody CW, Greenlee WF. Complete cdna sequence of a human dioxin-inducible mrna identifies a new gene subfamily of cytochrome P450 that maps to chromosome 2. *J Biol Chem* 1994;269:13092–13099.
159. Savas U, Bhattacharyya KK, Christou M, Alexander DL, Jefcoate CR. Mouse cytochrome P-450EF, representative of a new 1B subfamily of cytochrome P-450s. *J Biol Chem* 1994;269:14905–14911.
160. Munzel PA, Bookjans G, Mehner G, Lehmkoster T, Bock KW. Tissue-specific 2,3,7,8-tetrachlorodibenzo-p-dioxin-inducible expression of human UDP-glucuronosyltransferase UGT1A6. *Arch Biochem Biophys* 1996;335:205–210.

161. Takimoto K, Lindahl R, Pitot HC. Superinduction of 2,3,7,8-tetrachlorodibenzo-p-dioxin-inducible expression of aldehyde dehydrogenase by the inhibition of protein synthesis. *Biochem Biophys Res Commun* 1991;180:953–959.
162. Poland A, Glover E. Chlorinated dibenzo-p-dioxins: potent inducers of γ-aminolevulinic acid synthase and aryl hydrocarbon hydroxylase. *Mol Pharmacol* 1973;9:736–747.
163. Kraemer S, Arthur K, Denison MS, Smith WL, DeWitt DL. Regulation of prostaglandin endoperoxide H synthase-2 expression by 2,3,7,8-tetrachlorodibenzo-p-dioxin. *Arch Biochem Biophys* 1996;330:319–328.
164. Sutter TR, Guzman R, Dold KM, Greenlee WF. Targets for dioxin: genes for plasminogen activator inhibitor-2 and interleukin-1B. *Science* 1991;254:415–418.
165. Gaido KW, Maness SC, Leonard LS, Greenlee WF. 2,3,7,8-Tetrachlorodibenzo-p-dioxin-dependent regulation of transforming growth factors-α and -β_2 expression in a human keratinocyte cell line involves both transcriptional and posttranscriptional control. *J Biol Chem* 1992;267:24591–24595.
166. Clark GC, Taylor MJ, Trischer AM, Lucier GW. Tumor necrosis factor involvement in 2,3,7,8-tetrachlorodibenzo-p-dioxin-mediated endotoxin hypersensitivity in C57BL/6J mice congenic at the Ah locus. *Toxicol Appl Pharmacol* 1991;111:422–431.
167. Raunio H and Pelkonen O. Effect of polycyclic aromatic compounds and phorbol esters on ornithine decarboxylase and aryl hydrocarbon hydroxylase activities in mouse liver. *Cancer Res* 1983;43:782–786.
168. Roth W, Voorman R, Aust SD. Activity of thyroid hormone-inducible enzymes following treatment with 2,3,7,8-tetrachlorodibenzo-p-dioxin. *Toxicol Appl Pharmacol* 1988;92:65–74.
169. DeVito MJ, Ma X, Babish JG, Menache M, Birnbaum LS. Dose-response relationships in mice following subchronic exposure to 2,3,7,8-tetrachlorodibenzo-p-dioxin: CYP1A1, CYP1A2, estrogen receptor, and protein tyrosine phosphorylation. *Toxicol Appl Pharmacol* 1994;124:82–90.
170. Panteleyev AA, Thiel R, Wanner R, Zhang J, Roumak VS, Paus R, Neubert D, Henz BM, Rosenbach T. 2,3,7,8-tetrachlorodibenzo-p-dioxin (TCCD) affects keratin 1 and keratin 17 gene expression and differentially induces keratinization in hairless mouse skin. *J Invest Dermatol* 1997;108:330–335.
171. Brodie AE, Azarenko VA, Hu CY. 2,3,7,8-Tetrachlorodibenzo-p-dioxin (TCDD) inhibition of fat cell differentiation. *Toxicol Lett* 1996;84:55–59.
172. Selmin O, Lucier GW, Clark GC, Tritscher AM, Vanden Heuvel JP, Gastel JA, Walker NJ, Sutter TR, Bell DA. Isolation and characterization of a novel gene induced by 2,3,7,8-tetrachlorodibenzo-p-dioxin in rat liver. *Carcinogenesis* 1996;17:2609–2615.
173. Raasmaja A, Viluksela M, Rozman KK. Decreased liver type I 5′-deiodinase (−) and increased brown adipose tissue type II 5′-deiodinase activity (+) in 2,3,7,8-tetrachlorobibenzo-p-dioxin (TCDD)-treated Long-Evans rats. *Toxicology* 1996;114:199–205.
174. Paulson TJ, Kent C. The effect of polycyclic aromatic hydrocarbons on choline kinase activity in mouse hepatoma cells. *Biochim Biophys Acta* 1989;1004:274–277.
175. Puga A, Nebert DW, Carrier F. Dioxin induces expression of c-fos and c-jun proto-oncogenes and a large increase in transcription factor AP-1. *DNA Cell Biol* 1992;11:269–281.
176. Puhvel SM, Ertl DC, Lynberg CA. Increased epidermal transglutaminase activity following 2,3,7,8-tetrachlorodibenzo-p-dioxin: *in vivo* and *in vitro* studies with mouse skin. *Toxicol Appl Pharmacol* 1984;73:42–47.
177. Rubin AL, Rice RH. 2,3,7,8-tetrachlorodibenzo-p-dioxin and polycyclic aromatic hydrocarbons suppress retinoid-induced tissue transglutaminase in SCC-4 cultured human squamous carcinoma cells. *Carcinogenesis* 1988;9:1067–1070.
178. Brodie AE, Azarenko VA, Hu CY. Inhibition of increases of transcription factor mRNAs during differentiation of primary rat adipocytes by *in vivo* 2,3,7,8-tetrachlorodibenzo-p-dioxin (TCDD) treatment. *Toxicol Lett* 1997;90:91–95.
179. Stahl B. 2,3,7,8-Tetrachlorodibenzo-p-dioxin blocks the physiological regulation of hepatic phosphoenolpyruvate carboxykinase activity in primary rat hepatocytes. *Toxicology* 1995;103:45–52.
180. Bombick DW, Jankun J, Tullis K, Matsumura F. 2,3,7,8-tetrachlorodibenzo-p-dioxin causes increases in expression of c-*erb* -A and levels of protein-tyrosine kinases in selected tissues of responsive mouse strains. *Proc Natl Acad Sci USA* 1988;85:4128–4132.
181. Abbott BD, Birnbaum LS. TCDD-induced altered expression of growth factors may have a role in producing cleft palate and enhancing the incidence of clefts after coadministration of retinoic acid and TCDD. *Toxicol Appl Pharmacol* 1990;106:418–432.
182. Safa B, Lee C, Riddick D. Role of the aromatic hydrocarbon receptor in the suppression of cytochrome P-4502C11 by polycyclic aromatic hydrocarbons. *Toxicol Lett* 1997;90:163–175.

183. DeVito MJ, Thomas T, Martin E, Umbreit TH, Gallo MA. Antiestrogenic action of 2,3,7,8-tetrachlorodibenzo-p-dioxin: tissue-specific regulation of estrogen receptor in CD1 mice. *Toxicol Appl Pharmacol* 1992;113:284–292.
184. Sunahara GI, Lucier GW, McCoy Z, Bresnick EH, Sanchez ER, Nelson KG. Characterization of 2,3,7,8-tetrachlorodibenzo-p-dioxin-mediated decreases in dexamethasone binding to rat hepatic cytosolic glucocorticoid receptor. *Mol Pharmacol* 1989;36:239–247.
185. Madhukar BV, Brewster DW, Matsumura F. Effects of *in vivo*-administered 2,3,7,8-tetrachlorodibenzo-p-dioxin on receptor binding of epidermal growth factor in the hepatic plasma membrane of rat, guinea pig, mouse, and hamster. *Proc Natl Acad Sci USA* 1984;81:7407–7411.
186. Bombick DW, Matsumura F, Madhukar BV. TCDD (2,3,7,8-tetrachlorodibenzo-p-dioxin) causes reduction in the low density lipoprotein (LDL) receptor activities in the hepatic plasma membrane of the guinea pig and rat. *Biochem Biophys Res Comm* 1984;118:548–554.
187. Prokipcak RD, Okey AB. Downregulation of the Ah receptor in mouse hepatoma cells treated in culture with 2,3,7,8-tetrachlorodibenzo-p-dioxin. *Can J Physiol Pharmacol* 1991;69:1204–1210.

Toxicant–Receptor Interactions
Edited by Michael S. Denison and William G. Helferich

2

The Peroxisome Proliferator–Activated Receptor

An Orphan Receptor in Xenobiotica Response

Martin Göttlicher

Forschungszentrum Karlsruhe GmbH, Institute of Genetics, Karlsruhe, Federal Republic of Germany

---

Peroxisomes are relatively recently discovered cell organelles coated by a single membrane layer. Large numbers of peroxisomes are found in the liver but also in virtually all other eukaryotic cells including yeast [reviewed in (1)]. Peroxisomes contain structural proteins and enzymes that are encoded by nuclear genes and are imported into preexisting organelles (2). Peroxisomes contain a complete fatty acid β-oxidation system in addition to that present in the mitochondria, and they carry enzymes involved in reactive oxygen metabolism such as rich supplies of catalase. The

β-oxidation in peroxisomes is essential for very long–chain fatty acids and bile acids, which are poor substrates for the mitochondrial β-oxidation system (3). Therefore, inheritable defects in peroxisome structure or function lead to severe degenerative diseases with the most prominent manifestation in the central nervous system, such as X-linked adrenoleukodystrophy or Zellweger syndrome [reviewed in (4)]. Deposits of very long–chain fatty acids in neuronal tissues due to a defect in metabolism appear to play an essential role in pathogenesis (3). β-Oxidation by peroxisomes, in contrast to that by mitochondria, generates H_2O_2 during the acyl–coenzymeA (CoA) oxidation reaction that is degraded by rich supplies of catalase in the peroxisomes (1, 5), but it may also give rise to other diffusable reactive oxygen species, DNA damage, and possibly cancer (6, 7).

Peroxisomes are found in virtually all eukaryotic cells, but only in some organs and animal species, such as the rodent liver, the size and number of these organelles drastically increase on exposure to certain compounds, so-called *peroxisome proliferators* (PPs) (8). Interestingly, peroxisomes are highly inducible in the rodent liver but not in human liver or hepatocytes (9), although PPs are bioactive compounds in both species (see later). PPs are a heterogeneous group of about 100 compounds (recently reviewed in [10]) including drugs used for therapeutic purposes (11). Also, products or contaminants of chemical production processes such as the phthalate esters, which are used in large amounts as polymer plastisizers (8), certain herbicides (12–15), fatty acid derivatives (16–18) or simple molecules as trichloroethylene and trichloroacetic acid (19) induce peroxisome proliferation in the rodent liver (for selected structures see Fig. 1). Treatment of rodents or humans with PPs leads to changes in lipid homeostasis, which are exploited for therapy of hyperlipidemia in humans. Clofibric acid is an early, but still-applied representative of therapeutically used PPs (11), and other members of the

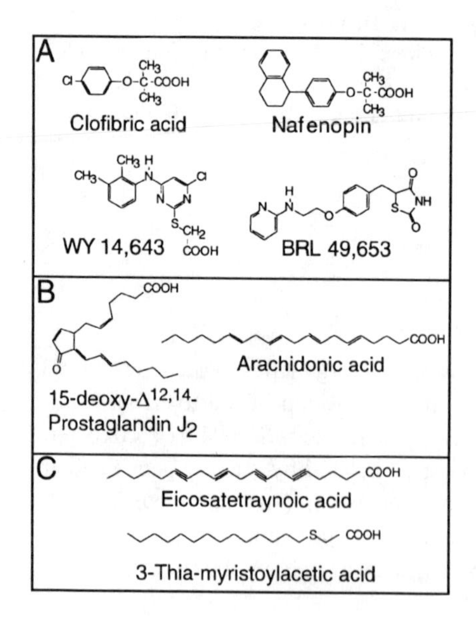

FIG. 1. Chemical structure of selected peroxisome proliferators and activators of PPARs. *A*, Synthetic compounds used (or initially intended) for human therapy. *B*, Arachidonic acid–derived compounds. *C*, Poorly metabolizable fatty acid derivatives.

fibrate class of compounds have been introduced into therapy since. Other nonrelated drugs such as nafenopin (20) and WY14,643 (21) have been developed but have not been introduced into human therapy. Interestingly, PPs lower blood lipid levels in humans, although the morphologic changes of peroxisomal proliferation observed in the rodent liver are not found in humans during therapeutic use of PPs (9, 22, 23).

Whereas defects in peroxisomes cause severe neurodegenerative syndromes, induced levels of peroxisomes also are associated with various diseases. Toxicologic studies in rodents revealed that PPs are liver carcinogens (24) (reviewed in [10, 25–28]) and some are toxic to the developing embryo (29). The striking correlation of peroxisome proliferation and carcinogenicity, as well as the implications of tumor-promoting signals from H_2O_2 production by peroxisomes, suggest on first glance that peroxisome proliferation per se enhances the development of cancer. It is also possible, however, that peroxisome proliferation and carcinogenesis are two distinct biologic responses to a common primary event, and other actions of PPs, distinct from the induction of peroxisomes, mediate the carcinogenic properties of this class of compounds. Indeed, there is evidence suggesting that proliferation of peroxisomes and increased hepatocyte proliferation as well as formation of preneoplastic lesions in response to PPs arc distinct events.

As with rodents, it is not clear whether a risk of PP-induced cancer exists in humans. This risk has been discussed extensively, particularly as a potential side effect of treatment of hyperlipidemic patients. There is no epidemiologic evidence, however, to support an increased cancer rate owing to PP treatment among these patients, and the compounds have not been banned from therapy considering the potential risks from nontreated hyperlipidemia. This debate, however, may have delayed the introduction of novel compounds into clinical use. The main argument against the comparable carcinogenic potential of PPs in humans and rodents is that the same compounds that induce morphologic proliferation of peroxisomes in rodents do not do so in humans. Because PPs are also biologically active compounds with respect to lipid homeostasis in humans, however, one would like to know which of the biologic activities of PPs in rodents, such as the morphologic changes in peroxisomes, the changes in lipid metabolism, or an alternative set of changes, are relevant for carcinogenicity. More specifically, from the perspective of a mechanism-based toxicologic evaluation, it is necessary to know whether the proliferation of peroxisomes is the cause of liver cancer in rodents or whether these morphologic changes and the development of cancer are two noncausally related endpoints of a common initial signaling pathway by PPs. If these two biologic activities follow distinct and separable pathways than it is conceivable that one activity of PPs is conserved between rodents and humans whereas the other, proliferation of peroxisomes, is not.

The existence of a specific signaling molecule, such as a receptor for PPs, has been proposed (30, 31). Subsequently, the peroxisome proliferator–activated receptor (PPAR), which is activated by a wide variety of known PPs (32) and fatty acids (33), was identified. With the receptor we now have a tool to help us determine the primary target of intracellular PP action. A clear picture is evolving of how the PPAR and a family of related receptors regulate target genes coding for lipid-metabolizing enzymes or proteins involved in adipogenesis. Other effects of PPs in addition to those in lipid metabolism may be mediated by different genes that have yet to be identified. The isolation of such genes may finally support (or disprove) the as-yet-correlative evidence that PPARs also mediate the adverse effects of PPs, for example in carcinogenesis. If such signaling pathways are identified they will provide a rational base to predict dose–response and structure–activity relationships and also the putative adverse effects of PPs on human health. Knowledge on the mechanism of action of PPs would provide simple tools to screen for potentially hazardous compounds not only during drug development but also in the environment or in food.

This chapter summarizes what has been learned in less than a decade about the different functions of PPARs in physiology and pharmacology and also the toxicity of PPs. Some of the mechanistic links in the different activities of PPs still are not understood. Recently other compound classes such as eicosanoids (34, 35) and antidiabetic thiazolidinediones (36) have been shown to bind and activate PPARs. This suggests that PPARs function in more physiologic processes than initially thought and that the identification of the first isoform, which is activated by lipid-lowering drugs, is just the beginning.

AN ORPHAN RECEPTOR ENGAGES AN ACTIVATOR AND GETS A NAME: ACTIVATION OF A NUCLEAR RECEPTOR BY PPs

Classical steroid hormone receptors share a common structure with a highly conserved central zinc-finger domain that enables the receptors to bind to specific DNA sequences and, on ligand binding, to activate transcription of target genes [reviewed in (37, 38)]. Subtle changes in a few amino acids in this conserved DNA-binding domain determine which DNA elements are recognized and subsequently which target genes are activated by the receptors. The carboxy-terminus harbors the ligand-binding domain which upon recognition of the receptor's specific steroid ligand undergoes dramatic structural changes and by that enhances transcription of target genes. Another class of receptors shows a closely related structure and includes the receptors for thyroid hormones, retinoids, and vitamin D [reviewed in (38–40)]. Yet these receptors (class II) add variations to the mode of steroid-hormone receptor function because they bind to DNA as heterodimers with a second protein, the so-called *retinoid X receptor*, rather than homodimers, which are preferentially formed by the classic steroid-hormone receptors (class I). The conservation of the backbone in the DNA-binding domain of all these ligand-dependent transcription factors triggered a search for novel genes that code for putative receptors for yet-unidentified ligands, the so-called *orphan receptors* (39). There are several dozen genes for putative receptors known by now which may serve as putative receptors for a manyfold larger number of low–molecular weight molecules such as intermediary metabolites or xenobiotica that regulate expression of certain target genes. A major family of such target genes regulated by low molecular weight hormone-like compounds or xenobiotica consists of the cytochrome CYP450 genes [reviewed in (41, 42)]. It is tempting to simplify

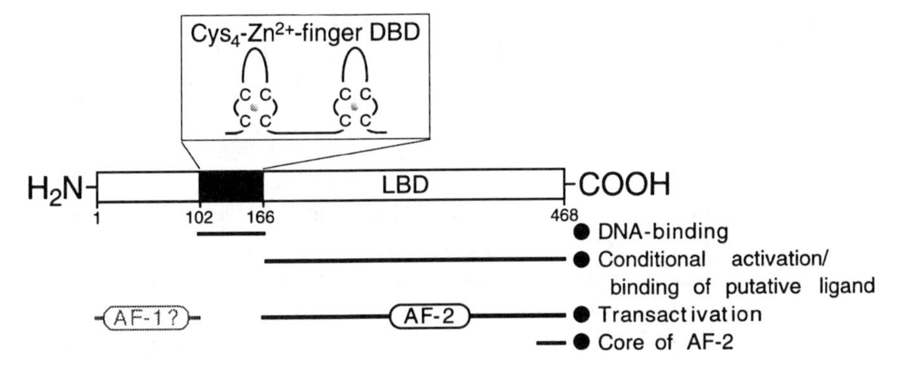

FIG. 2. Domain structure of PPARs. Numbers refer to amino acid positions in rat PPARα. Domains of the receptor that are required for certain receptor functions are indicated. DBD, DNA-binding domain; LBD, ligand-binding domain; AF-n, transcription activating function 1 or 2.

the picture to a scenario in which one class of genes, such as the nuclear receptors, constitute the sensors and another class, such as the cytochromes P450, are important effectors in the response of the organism to xenobiotica. The match of activator or ligand and receptor, however, has succeeded in very few cases only. Most of the receptors have remained orphans in search of their ligands. One of the most prominent matches [besides the identification of 9-cis retinoic acid as ligand to a new class of retinoic acid receptors (43–45)] was pioneered by Issemann and Green (32); they showed that one receptor is activated by a structurally diverse group of PPs and hence called it the peroxisome proliferator activated receptor (PPAR) (Fig. 2).

In cell culture, structurally diverse PPs and even simple peroxisome proliferation–inducing molecules such as trichloroacetic acid activate the PPAR (32). The previously cloned murine PPARα gene is homologous to more recently cloned PPAR genes from other rodent species (33), *Xenopus* (46), and humans (47), suggesting conserved signaling between PPs and its associated receptor in all these species.

Considering the high selectivity of steroid receptors for their respective ligands, it appears unlikely that all PPs will bind to PPAR themselves. Alternatively, formation of an ultimate ligand by metabolism or ligand-independent modes of receptor activation have to be considered. The physiologic role of PPAR discussed subsequently provides cues to molecules in lipid intermediary metabolism that may qualify as ligands to the PPAR. An arachidonic-acid derivative, the leukotriene B4, indeed appears to bind with high affinity to PPARα, and WY14,643 is capable of competing for this binding site (48). It remains to be established whether all other activators bind to the PPAR as well and whether the binding observed in vitro also activates the receptor in vivo. Nevertheless, the identification of "activator-" or "ligand-" dependent transcription factors as the intracellular target for PPs marked a milestone in understanding the biologic activity of this class of compounds. The knowledge of the primary site of action provides a starting point for identifying the target genes regulated by these receptors and understanding their roles in physiologic and pathologic processes.

PPAR FINDS A PLACE IN PHYSIOLOGY: PPAR ACTIVATION BY PHYSIOLOGICALLY OCCURRING FATTY ACIDS, FATTY-ACID DERIVATIVES, AND EICOSANOIDS

PPs are manmade compounds of this century, and they certainly are not the signaling molecules for which evolution has conserved PPARs. The effects of PPs on lipid metabolism by induced fatty acid β-oxidation or fatty acid ω-hydroxylating cytochrome P450 4A enzymes (49, 50) suggest a role for the PPAR in lipid homeostasis. Therefore, intermediates in fatty-acid and cholesterol metabolism were tested for activation of PPAR and, indeed, physiologically occurring fatty acids activate PPARα (33, 51). Poorly degradable fatty acid derivatives, such as those owing to sulfur substitution of the β-carbon (16, 17) or perfluorination (18), are even more potent activators of the PPAR in cell culture (52) and act as PPs in vivo. Yet, as discussed previously this does not necessarily mean that fatty acids are the ligands for the PPAR nor how synthetic PPs could substitute for physiologically occurring fatty acids. Assuming that all compounds that activate PPAR also are ligands to PPAR would require both high-affinity binding to the PPAR and, equally importantly, transformation of the receptor into a transcriptionally active conformation similar to the ligand-dependent conformational changes that other nuclear receptors undergo on engagement of ligands (53, 54). If PPARs accepted such a broad spectrum of ligands with an apparently low specificity these were unusual in the nuclear receptor superfamily since the other members are characterized by high specificity of each receptor for a well-defined subset of ligands. Fatty acids added to a cell-culture medium undergo rapid metabolism, and any metabolite which occurs even as a transient intermediate may qualify as a ligand to the PPAR. Accumulation of fatty acids or their metabolites is essential for activation of the PPAR. High concentrations, close to toxic levels, of regular fatty acids are required for efficient activation of the PPAR (33), and fatty-acid derivatives, which are poor substrates for β-oxidation, usually are more potent activators of PPAR than their degradable homologs (16–18, 52, 55). These fatty-acid derivatives and many of the synthetic PPs all are carboxylates (or esters that are hydrolyzed to carboxylates) that are poorly degradable by β-oxidation. This common feature supports the hypothesis that the carboxylates themselves or their activated forms, such as CoA esters, or further derivatives thereof are ligands to PPAR. For some of the fibrates (56) and several fatty acid–derived PPs (57), but not for other fatty acid–derived PPs (58), the formation of CoA esters has been shown directly. Certainly the CoA esters contain a sufficiently large conserved part of the molecule, the CoA moiety bound to a hydrophobic residue, to allow specific binding to the receptor. The formation of CoA esters, however, could not been shown for all PPs. It also must be realized that at least some free acids could be ligands for PPARs, leaving the PPAR with an unprecedented high level of promiscuity in choice of ligands. Leukotriene B4 may qualify as one such arachidonic acid–derived ligand which has recently been shown to bind to PPAR directly (48). Alternatively, the poorly degradable carboxylates may inhibit cellular metabolic pathways, and endogenous molecules could accumulate that finally activate

PPAR. Since the initial isolation of activators or ligands for PPARα a number of synthetic compounds as well as physiologically occurring arachidonic-acid derivatives have been identified as ligands for a different isoform of PPARs, the PPARγ (34–36).

The identification of manmade carboxylates and fatty-acid derivatives as activators of the PPAR suggests that the physiologically occurring fatty acids regulate the activity of PPARs also under physiologic conditions. Thus, unlike the aromatic hydrocarbon and dioxin receptor, for which an endogenous ligand is still not known (see chapter 1), the PPAR has a plausible role in homeostasis of lipids. Manmade exogenous carboxylates apparently do not provide completely novel signals but may mimic signals in preexisting physiologic regulatory loops. They differ, however, from endogenous compounds by their poor degradability. Thus, endogenous fatty acids and related compounds may activate PPARs, and the products of the subsequently induced target genes, such as those involved in peroxisomal β-oxidation or cytochrome P450-dependent ω-hydroxylation, can metabolize the PPAR-activating fatty acid moieties. Most of the exogenous PPs, however, can activate PPARs but are poor substrates for the induced enzyme systems. Therefore, they are expected to lead to a much longer persisting signal compared with that induced by the endogenously occurring activators.

FINDING A PARTNER FOR PPAR AND TARGET GENES FOR THE HETERODIMER: PPAR AND THE RETINOID X RECEPTOR ACT AS A HETERODIMER TO ACTIVATE GENES INVOLVED IN LIPID STORAGE AND METABOLISM

The PPAR and its isoforms belong to the class II nuclear receptors. They bind to DNA target elements in regulated genes as heterodimers with another nuclear receptor, the retinoid X receptor (RXR) (39, 59, 60). The search for target genes that are regulated by the PPAR was initiated with the genes known to be regulated by PPs and to participate in lipid metabolism. The first PPAR-responsive element was mapped in the gene coding for the key enzyme of peroxisomal β-oxidation, acyl-CoA oxidase (61). Subsequently, similar response elements were found in other genes of the peroxisomal β-oxidation pathway (62, 63), the CYP450 4 genes that code for cytocrome P450s with fatty-acid ω-hydroxylase activity (64–66), and also genes for mitochondrial β-oxidation (67) as well as other mitochondrial proteins (68). Also, genes that affect lipid homeostasis by forming an intracellular pool of protein-bound lipids such as the fatty acid–binding proteins are regulated through the PPAR (69).

In summary, the target genes of the PPAR identified so far code for proteins that are able to reduce free fatty-acid levels either through metabolism or storage in protein complexes. Together with the finding that fatty acids can activate the PPAR, this gives rise to the speculation that the PPAR and PPAR-regulated target genes constitutes a feedback loop in fatty-acid homeostasis (schematically drawn in Fig. 3). A key role of PPARα in peroxisomal proliferation and in PP-inducible fatty-acid degradation

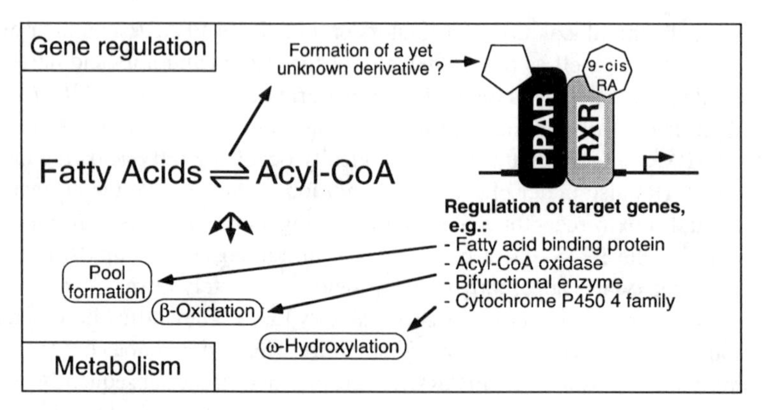

FIG. 3. Fatty acids as transcriptional inducers and substrates of enzymes and storage proteins in lipid homeostasis.

has been confirmed by inactivation of this gene by homologous recombination in the germ line of mice. In these animals the PPs clofibric acid or WY14,643 do not induce the pleiotropic response induced in normal mice, including hepatomegalie, proliferation of peroxisomes, or induction of PPAR target genes coding for fatty-acid metabolizing enzymes (70).

So far, PPAR-regulated genes have been found mostly by specific searches in the regulatory regions of genes related to lipid metabolism. These genes, however, are not necessarily relevant to all the biologic activities of PPs, and particularly they do not account easily for the carcinogenic activity of PPs. Therefore, a complete understanding requires a broad and unbiased search for PPAR-regulated genes.

To add even more complexity to the PPAR-dependent regulatory gene network, the PPAR-responsive DNA element that is recognized by PPAR/RXR heterodimers also is the target of a number of other transcription factors. Thus, the RXR heterodimer, the retinoic acid receptors (71), the chicken ovalbumin upstream promoter (COUP) transcription factor (72), the Arp-1 protein that was identified as a regulator of apolipoprotein genes (73), and the hepatocyte nuclear factor 4, which plays a role in tissue-specific differentiation of hepatocytes (74), all bind to similar DNA-recognition motifs that are collectively called the *DR-1 element* [reviewed in (38, 39)]. The target sites are similar, although subtle changes in bases adjacent to the core DR-1 motif affect preference for certain transcription factors, such as Arp-1 over the PPAR/RXR heterodimer (66, 75). In other cases several factors compete for the same binding site (76, 77).

In summary, PPs act through PPARs as the primary mechanism to regulate gene expression. The mechanism by which this primary interaction with a receptor leads to altered expression of genes in lipid homeostasis is well understood. PPARs, however, also may regulate an additional set of genes which is relevant for PP toxicity but so far no such gene is known.

PPAR FINDS ITS RELATIVES: PPARs COMPOSE A FAMILY OF SIMILAR PROTEINS, BUT ACTIVATORS/LIGANDS AS WELL AS TARGET GENES ARE DISTINCT

A screen for steroid hormone receptor–like proteins during embryonic development of *Xenopus* revealed that PPARs form a subfamily with at least three members (α, β, γ) in this species (46). Similarly, three isoforms are known in mouse, and by sequence homology they are called PPARα, PPARγ, and PPARδ (32, 78), of which PPARδ also has been cloned independently as a fatty acid–activated receptor (79). From sequence comparison, however, it is not clear whether the rodent PPARδ is homologous to *Xenopus* PPARβ, possibly functional similarities, such as in ligand specificity, may clarify this issue in the future. In humans the homologous genes also are known [α (47), γ (80), and δ [also called *NUC1*] (81, 82)]. PPARs are characterized as a subfamily of nuclear receptors by two properties, sequence similarities, between PPARs and activation at least to some degree by fatty acids. Beyond these common properties, the receptors are distinct with regard to their expression pattern in the organism (78), their response to different ligands, and their putative physiologic functions (78, 82).

Studies in cell culture and gene inactivation in mice suggest that the PPARα form is a key regulator of PP-induced genes involved in fatty-acid metabolism (70). PPARγ plays an interesting role in the differentiation of adipocytes. Thus, overexpression of PPARγ (in combination with a PPAR activator) is sufficient to differentiate cultured fibroblasts into lipid-storing adipocytes. This function cannot be substituted by the PPARα form (83–85). PPARγ has been shown by biochemical methods to bind ligands with high specificity and affinity. The binding of an antidiabetic drug, the thiazolidinedione BRL49653 (see Fig. 1), to PPARγ may play a role in the activity of this drug on intermediary metabolism (36). The drug, although binding with high affinity (K_d = 40 nmol), is not a physiologic compound and may substitute for an endogenous activator of PPARγ, in the process of adipocyte differentiation. Possibly the prostaglandin derivative 15-deoxy-$\Delta^{12,14}$-prostaglandin J2 (34, 35) is such a physiologic ligand. This finding suggests exciting links between eicosanoid metabolism, PPARs, and cell differentiation.

PPARs in *Xenopus* initially have been identified from a cDNA library of developing embryos (46). The exciting proposition that PPARs play an important role during embryonic development, although lacking experimental proof, gains support from an expression analysis in the development of mice. PPARα and PPARγ are expressed relatively late during embryogenesis, and their known functions are capabilities of the relatively matured embryo, such as in inducible fatty acid metabolism and adipogenesis. PPARδ is the only known isoform that is present in mammals from early embryonic development (78). If one accepts that evolution has not conserved a redundant protein, PPARδ is likely to play a role from early embryonic development. This putative role, however, is still elusive, and targeted disruption of the gene by the gene knockout technology will provide experimental evidence.

Whatever the research of the near future shows, the functions of PPARs appear much more profound than the induction of the morphologic changes called peroxisome proliferation. Therefore, comparing humans and rodents only by the fact that the former lack the changes in liver morphology on PP treatment certainly does not cover the complete picture, considering the diverse effects of PPARs. Rather, the conservation of the genes for all the PPAR isoforms through the mammalian species supports the idea that their main functions (e.g. in lipid metabolism and cell differentiation) are conserved. Certain features such as the morphologic changes of peroxisomes apparently depend on additional factors that differ between species.

PPARs FROM THE TOXICOLOGIST'S POINT OF VIEW: PPARs IN REGULATION OF CELL DIFFERENTIATION, PROLIFERATION, AND CANCER

The previously identified target genes of PPARs explain to a reasonable degree how PPs act on peroxisome function and cytochrome-P450 induction, which is to be expected from earlier biochemical data on PP action in lipid metabolism. PP action in a whole organism, however, is more complex, including proliferation and morphologic changes in the rodent liver [reviewed in (1)], liver carcinogenicity in rodents (24–26), and embryotoxicity in the mouse (albeit at high doses that may exceed those for which one might expect a receptor-dependent mode of action [29]). These are the biologic activities that incriminate PPs as toxins. PPARs are intracellular targets of PPs, and some functions of these receptors in lipid homeostasis are understood in detail as described previously. It is tempting to assume that similarly the PPARs also are key mediators of the more complex activities of PPs such as carcinogenicity or liver hyperplasia. Interestingly, there is correlative evidence to support this proposition in rodents. The potency of a number of PPs as carcinogens in the rodent liver roughly correlates with their potency as activators of the PPAR and inducers of PPAR target genes (28). If PPARs are assigned a causal role in the PP-induced development of cancer, then genes have to be identified that are regulated by PPARs in the process of carcinogenesis. Certainly, PPs appear to induce cancer without substantial mutagenesis of the DNA, and therefore they have to be considered nongenotoxic carcinogens, which in the multistage model of carcinogenesis act on later steps, such as tumor promotion and progression rather than initiation. The current knowledge allows two major hypotheses to explain how altered gene expression by PPARs may relate to liver carcinogenicity (for a schematic presentation of the discussed relationships see Fig. 4). First, the induced enzyme activities such as peroxisomal β-oxidation, that produce reactive oxygen species (6, 7, 86), and cytochrome P450 4A–dependent ω-hydroxylation of fatty acids (64, 65), which affects for example eicosanoid homeostasis (48, 87), may provide signals that lead to increased cell damage or altered cell signaling, and these changes finally may play a role in cell proliferation and carcinogenesis. Alternatively, PPARs may regulate an additional set of genes that is not related to lipid metabolism but affects the cell's program in proliferation or differentiation, much like other steroid hormone receptors exert trophic, differentiation,

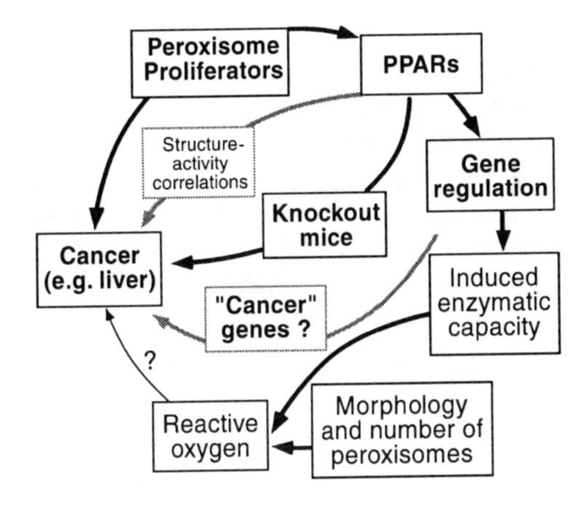

FIG. 4. Schematic presentation of causal, correlative, and speculative relations between peroxisome proliferators, PPARs, lipid metabolism, and PP-induced cancer. Relationships that are supported by direct experimental evidence are shown by bold black arrows. Grey arrows indicate evidence from correlations in structure–activity relationships. Thin arrows and question marks identify concepts that are controversially discussed but not proven or disproven yet.

or apoptotic signals in the immune system or hormone-dependent cancers. A specific example of a role of PPARs in the cell's differentiation and proliferation program is the induction of adipocyte differentiation in certain fibroblast cell ines [reviewed in (88)]. To what degree other PPARs play similar roles in the differentiation of other tissues is unclear. Targeted inactivation of PPAR genes in mice by gene knockout technology has started to make mutant mice available in which the role of PPARs can be evaluated with respect to the different biologic effects of PPs including carcinogenesis (70, 96).

Despite the suspected role of PPARs in mediating adverse effects of synthetic PPs, these compounds nevertheless trigger preexisting signaling pathways presumably by mimicking endogenous ligands rather than generating qualitatively new signals. A major difference, however, between endogenous activators and exogenous PPs resides in the time course of the signal. Endogenously occurring activators such as fatty acids are readily degraded by enzymes such as in peroxisomal β-oxidation, which they induce through the activation of PPARs. Most of the exogenously applied PPs activate PPARs but are resistant to degradation by the induced enzyme activities and therefore probably create a much longer-lasting signal. Also, activation of PPARs at a wrong time during development or adult life or in an inappropriate cell population may lead to the adverse effects of this group of compounds. Finding the relevant genes regulated by PPs and PPARs and understanding the corresponding physiologic endogenous signals will finally allow us to estimate the doses at which an exogenous compound will override physiologic signaling and by doing so exert its toxicity.

GUILTY OR NOT? THE DIFFICULTIES IN ESTIMATING HUMAN RISKS FROM RODENT AND IN-VITRO STUDIES

Carcinogenicity and embryotoxicity of peroxisome proliferators in rodents foster the concern that similar adverse effects also occur in humans. Assessing the risks to humans, assuming a common principle of action for all PPs through PPARs, relies on two presuppositions and an open question. First, PPARs mediate the adverse effects of PPs. Second, adverse actions of PPARs require that doses of xenobiotica are employed that are sufficient for PPAR activation and that the levels of these doses also are reached in the human body under the conditions of interest. These presuppositions are probably true although not strictly proven yet. The essential question not answered yet is however to what extent the relevant target genes for adverse PPAR activities are conserved between rodents and humans.

It is highly probable that PPARs mediate adverse effects such as liver carcinogenesis in rodents, and the evidence for this is based on correlations between PPAR activation and carcinogenesis (28) or carcinogenesis and peroxisome proliferation [reviewed in (10)] by a large number of compounds. Studies in genetically manipulated mice have more recently provided direct evidence for a key role of PPARs in liver carcinogenesis (70, 96).

The conservation of relevant signaling pathways between humans and rodents is much more difficult to judge. Pharmacologic activities of PPAR-activating compounds, such as those in lipid homeostasis, are established both in humans and laboratory animals. Adverse effects are shown in rodents, and the risk to humans is subject to debate. Therefore, assessment of the risk to humans poses the problem of whether PP signaling is conserved between humans and rodents, including the adverse effects of PPs. The pharmacologic activities of lipid-lowering drugs in both rodent and human populations suggest that a substantial part of PP-induced signaling indeed is conserved. The morphologic phenomenon of peroxisomal proliferation, however, certainly is not conserved in humans (9, 89). Yet this does not argue against potential adverse effects of PPs in humans, because there is evidence, albeit controversial, that the proliferation of peroxisomes per se and the subsequent production of reactive oxygen intermediates is not directly linked to carcinogenicity even in the rodent. First, proliferation of peroxisomes on one hand and increased hepatocyte proliferation and formation of preneoplastic lesions on the other hand in response to PP treatment of rodents occur in distinct areas of the liver (90). The second, less convincing, line of evidence is based on the observation that nafenopin is a much more potent carcinogen in old compared with young rats, although indicators of oxidative stress or lipid peroxidation do not change with age. This argues against a simple and direct role of oxidative stress in rodent liver carcinogenesis but also might reflect simply the accumulation of cells with age that are susceptible to carcinogenic transformation (promotion) by peroxisome proliferators (91). Therefore, PPAR signaling itself and the subsequent events that are relevant to PP toxicity and carcinogenicity may be well conserved between rodent species and humans, although the particular event of peroxisomal proliferation is not conserved.

With respect to PPAR signaling, rodents and humans appear similar because all PPAR isoforms found in the mouse or rat have their homologs in humans that differ only slightly in agonist profiles (92). Subsequent PPAR-triggered events, however, that are relevant to the adverse PP activities are not known, even in rodents, and on their conservation in human one may only speculate. Certainly, species-specific differences contribute to complex biologic phenomena such as the development of cancer or birth defects. For example, a different compound, phenobarbital, is a tumor promoter in the rodent liver, but there is no indication of substantial carcinogenicity in humans even during long-term barbiturate therapy of epilepsy. Thus we have to understand the differences between human and rodent carcinogenesis, before the relevance of the rodent PP-carcinogenicity studies for human health can be assesssed from a mechanistic point of view. More specifically, the events that lead from PPAR activation as the putative primary target of PPs to the final development of cancer have to be analyzed. After all, liver carcinogenicity is both common and spontaneous in rodents but rare in humans, and thus the liver may be one of the least sensitive organs in humans. Possibly other tumor locations and the effects on the developing embryo (although there is no epidemiologic evidence in humans yet) may be more pronounced and relevant in humans than liver carcinogenicity.

PPAR-mediated adverse effects on humans also depend on the concentrations that putative PPAR agonists reach in the body. Two major types of exposure have to be considered. First, PPAR-activating drugs, such as those used in the therapy of hyperlipidemia, are taken at biologically active doses. Lipid-lowering drugs are taken over long periods, and as yet there is no convincing epidemiologic evidence to support a carcinogenic potential of these drugs in humans. The lack of epidemiological evidence, however, does not prove that there is no risk. Prolonged exposure may not result in liver cancer in humans but, although there currently are no data to support this assumption, PPs might promote the occurrence of other tumors. Should such tumors be present in the nontreated human population, the mathematical tools of epidemiology would not suffice to detect a marginal or even moderate increase from PP treatment.

The second major route of exposure, usually at low doses, occurs through the environment and food with contaminants such as polymer plasticizers or herbicides for which the risks to human health are not very well known. A substantial risk from the PPAR-activating property of these compounds would require that they reach receptor-activating concentrations in the body. PPAR activation in cell culture systems at least by the hitherto known agonists requires relatively high concentrations (micromolar to millimolar range) that may occur in specific (accidental or occupational) situations but are unlikely to be reached from uptake through the environment or as food contaminants on a general basis. The initial interaction of a xenobiotic with the receptor may follow, as in the case of the Ah receptor, a linear relationship down to very low doses (93, 94). This implies that a very low dose of a receptor-activating compound also has a biologic effect on the receptor. Whether such an increment in receptor activity is relevant to a toxic endpoint depends on the response curves of subsequent events, and unfortunately different receptor-dependent events may follow distinct dose–activity relationships as shown for example in the case of the glucocorticoid receptor (95). In the case of PPARs a substantial effect from low doses of receptor-activating molecules seems unlikely because the known PPAR isoforms, under normal conditions, probably also are activated by endogenously occurring compounds such as fatty acids or prostaglandin derivatives. Thus, a small increment by exogenous compounds is likely to be marginal compared with activation occurring through endogenous activators. As long as we have not reached a reasonably complete understanding of the physiology and the target genes of all the PPARs, however, it cannot be excluded that there exist certain tissues or stages of development in which endogenous activation of PPARs is extremely low and an increment by exogenous compounds in the wrong cell type at the wrong time may trigger wrong decisions in the fate of a cell.

CONCLUSIONS AND FUTURE PERSPECTIVES

In summary, PPs are bioactive molecules, and relevant exposures occur in human at least during use of pharmaceutically active drugs. The identification of ligand-dependent transcription factors as receptors for this class of compounds has opened a new chapter in understanding the biologic effects of these compounds in molecular terms. Genetic manipulation in mice and studies in cell models allow to show decisively which receptor is relevant to which biologic phenomenon. Learning to know the molecular targets downstream of PPARs will provide the rationale to understand in mechanistic terms, the therapeutically desired and adverse toxic effects of PPs, to assess dose–response relationships particularly for low doses of potentially toxic contaminants in the environment and food chain, and finally to design novel compounds that may avoid adverse and toxic activities.

ACKNOWLEDGMENTS

The author thanks Hans J. Rahmsdorf for critical reading of the manuscript and the Deutsche Forschungsgemeinschaft for financial support (Go473-2).

REFERENCES

1. vanden Bosch H, Schutgens RBH, Wanders RJA, Tager JM. Biochemistry of peroxisomes. *Annu Rev Biochem* 1992;61:157–197.
2. Rachubinski RA, Subramani S. How proteins penetrate peroxisomes. *Cell* 1995;83:525–528.
3. Singh I, Moser AE, Goldfischer S, Moser HW. Lignoceric acid is oxidized in the peroxisome: implications for the Zellweger cerebro-hepato-renal syndrome and adrenoleukodystrophy. *Proc Natl Acad Sci USA* 1984;81:4203–4207.
4. Aubourg P. Adrenoleukodystrophy and other peroxisomal diseases. *Curr Opin Genet Dev* 1994;4:407–411.
5. Reddy JK. Hepatic and renal effects of peroxisome proliferators: biological implications. *Ann N Y Acad Sci* 1982;386:81–110.
6. Lake BG, Gray TJ, Smith AG, Evans JG. Hepatic peroxisome proliferation and oxidative stress. *Biochem Soc Trans* 1990;18:94–97.
7. Fahl WE, Lalwani ND, Watanabe T, Goel SK, Reddy JK. DNA damage related to increased hydrogen peroxide generation by hypolipidemic drug-induced liver peroxisomes. *Proc Natl Acad Sci USA* 1984;81:7827–7830.
8. Reubsaet FA, Veerkamp JH, Dirven HA, Bruckwilder ML, Hashimoto T, Trijbels JM, Monnens LA. The effect of di(ethylhexyl)phthalate on fatty acid oxidation and carnitine palmitoyltransferase in various rat tissues. *Biochim Biophys Acta* 1990;1047:264–270.
9. Blaauboer BJ, van Holsteijn CW, Bleumink R, Mennes WC, van Pelt FN, Yap SH, van Pelt JF, van Iersel AA, Timmerman A, Schmid BP. The effect of beclobric acid and clofibric acid on peroxisomal beta-

oxidation and peroxisome proliferation in primary cultures of rat, monkey and human hepatocytes. *Biochem Pharmacol* 1990;40:521–528.
10. Ashby J, Brady A, Elcombe CR, Elliott BM, Ishmael J, Odum J, Tugwood JD, Kettle S, Purchase IF. Mechanistically-based human hazard assessment of peroxisome proliferator-induced hepatocarcinogenesis. *Hum Exp Toxicol* 1994;3:S1–S117.
11. Lazarow PB, De Duve C. A fatty acyl-CoA oxidizing system in rat liver peroxisomes: enhancement by clofibrate, a hypolipidemic drug. *Proc Natl Acad Sci USA* 1976;73:2043–2046.
12. Moody DE, Hammock BD. The effect of tridiphane (2-(3,5-dichlorophenyl)-2-(2,2,2-trichloroethyl)oxirane) on hepatic epoxide-metabolizing enzymes: indications of peroxisome proliferation. *Toxicol Appl Pharmacol* 1987;9:37–48.
13. Moody DE, Narloch BA, Shull LR, Hammock BD. The effect of structurally divergent herbicides on mouse liver xenobiotic-metabolizing enzymes (P-450-dependent mono-oxygenases, epoxide hydrolases and glutathione S-transferases) and carnitine acetyltransferase. *Toxicol Lett* 1991;59:175–185.
14. Espandiari P, Thomas VA, Glauert HP, O'Brien M, Noonan D, Robertson LW. The herbicide dicamba (2-methoxy-3,6-dichlorobenzoic acid) is a peroxisome proliferator in rats. *Fundam Appl Toxicol* 1995;26:85–90.
15. Hietanen E, Ahotupa M, Heinonen T, Hamalainen H, Kunnas T, Linnainmaa K, Mantyla E, Vainio H. Enhanced peroxisomal beta-oxidation of fatty acids and glutathione metabolism in rats exposed to phenoxyacetic acids. *Toxicology* 1985;34:103–111.
16. Hertz R, Bar-Tana J, Sujatta M, Pill J, Schmidt FH, Fahimi HD. The induction of liver peroxisomal proliferation by beta,beta'-methyl-substituted hexadecanedioic acid (MEDICA 16). *Biochem Pharmacol* 1988;37:3571–3577.
17. Berge RK, Aarsland A, Kryvi H, Bremer J, Aarsaether N. Alkylthioacetic acid (3-thia fatty acids): a new group of non-beta-oxidizable, peroxisome-inducing fatty acid analogues. I. A study on the structural requirements for proliferation of peroxisomes and mitochondria in rat liver. *Biochim Biophys Acta* 1989;1004:345–356.
18. Intrasuksri U, Feller DR. Comparison of the effects of selected monocarboxylic, dicarboxylic and perfluorinated fatty acids on peroxisome proliferation in primary cultured rat hepatocytes. *Biochem Pharmacol* 1991;42:184–188.
19. Mitchell AM, Bridges JW, Elcombe CR. Factors influencing peroxisome proliferation in cultured rat hepatocytes. *Arch Toxicol* 1984;55:239–246.
20. Reddy JK, Azarnoff DL, Svoboda DJ, Prasad JD. Nafenopin-induced hepatic microbody (peroxisome) proliferation and catalase synthesis in rats and mice: absence of sex difference in response. *J Cell Biol* 1974;61:344–358.
21. Reddy JK, Krishnakantha TP. Hepatic peroxisome proliferation: induction by two novel compounds structurally unrelated to clofibrate. *Science* 1975;190:787–789.
22. Gariot P, Barrat, E, Drouin P, Genton P, Pointel JP, Foliguet B, Kolopp M, Debry G. Morphometric study of human hepatic cell modifications induced by fenofibrate. *Metabolism* 1987;36:203–210.
23. Blümcke S, Schwartzkopff W, Lobeck H, Edmondson NA, Prentice DE, Blane GF. Influence of fenofibrate on cellular and subcellular liver structure in hyperlipidemic patients. *Atherosclerosis* 1983;46:105–116.
24. Reddy JK, Azarnoff DL, Hignite CE. Hypolipidaemic hepatic peroxisome proliferators form a novel class of chemical carcinogens. *Nature* 1980;283:397–398.
25. Reddy JK. Carcinogenicity of peroxisome proliferators: evaluation and mechanisms. *Biochem Soc Trans* 1990;18:92–94.
26. Rao MS, Reddy JK. The relevance of peroxisome proliferation and cell proliferation in peroxisome proliferator-induced hepatocarcinogenesis. *Drug Metab Rev* 1989;21:103–110.
27. Popp JA, Cattley RC, Miller RT, Marsman DS. Relationship of peroxisome proliferator-induced cellular effects to hepatocarcinogenesis. *Prog Clin Biol Res* 1994;387:193–207.
28. Green S. Peroxisome proliferators: a model for receptor mediated carcinogenesis. *Cancer Surv* 1992;14:221–232.
29. Ujhazy E, Onderova E, Horakova M, Bencova E, Durisova M, Nosal R, Balonova T, Zeljenkova D. Teratological study of the hypolipidaemic drugs etofylline clofibrate (VULM) and fenofibrate in Swiss mice. *Pharmacol Toxicol* 1989;64:286–290.
30. Lalwani ND, Fahl WE, Reddy JK. Detection of a nafenopin-binding protein in rat liver cytosol associated with the induction of peroxisome proliferation by hypolipidemic compounds. *Biochem Biophys Res Commun* 1983;116:388–393.

31. Huang Q, Alvares K, Chu R, Bradfield CA, Reddy JK. Association of peroxisome proliferator-activated receptor and Hsp72. *J Biol Chem* 1994;269:8493–8497.
32. Issemann I, Green S. Activation of a member of the steroid hormone receptor superfamily by peroxisome proliferators. *Nature* 1990;347:645–650.
33. Göttlicher M, Widmark E, Li Q, Gustafsson J-Å. Fatty acids activate a chimera of the clofibric acid-activated receptor and the glucocorticoid receptor. *Proc Natl Acad Sci USA* 1992;89:4653–4657.
34. Kliewer SA, Lenhard JM, Willson TM, Patel I, Morris DC, Lehmann JM. A prostaglandin J2 metabolite binds peroxisome proliferator-activated receptor gamma and promotes adipocyte differentiation. *Cell* 1995;83:813–819.
35. Forman BM, Tontonoz P, Chen J, Brun RP, Spiegelman BM, Evans RM. 15-Deoxy-delta 12, 14-prostaglandin J2 is a ligand for the adipocyte determination factor PPAR gamma. *Cell* 1995;83:803–812.
36. Lehmann JM, Moore LB, Smith-Oliver TA, Wilkison WO, Willson TM, Kliewer SA. An antidiabetic thiazolidinedione is a high affinity ligand for peroxisome proliferator-activated receptor gamma (PPAR gamma). *J Biol Chem* 1995;270:12953–12956.
37. Beato M, Herrlich P, Schütz G. Steroid hormone receptors: many actors in search of a plot. *Cell* 1995;83:851–857.
38. Mangelsdorf DJ, Thummel C, Beato M, Herrlich P, Schütz G, Umesono K, Blumberg B, Kastner P, Mark M, Chambon P. The nuclear receptor superfamily: the second decade. *Cell* 1995;83:835–839.
39. Mangelsdorf DJ, Evans RM. The RXR heterodimers and orphan receptors. *Cell* 1995;83:841–850.
40. Kastner P, Mark M, Chambon P. Nonsteroid nuclear receptors: what are genetic studies telling us about their role in real life? *Cell* 1995;83:859–869.
41. Estabrook RW. The remarkable P450s: a historical overview of these versatile hemeprotein catalysts. *FASEB J* 1996;10:202–204.
42. Nelson DR, Koymans L, Kamataki T, Stegeman JJ, Feyereisen R, Waxman DJ, Waterman MR, Gotoh O, Coon, MJ, Estabrook RW, Gunsalus IC, Nebert DW. P450 Superfamily: update on new sequences, gene mapping, accession numbers and nomenclature. *Pharmacogenetics* 1996;6:1–42.
43. Mangelsdorf DJ, Ong ES, Dyck JA, Evans RM. Nuclear receptor that identifies a novel retinoic acid response pathway. *Nature* 1990;345:224–229.
44. Levin AA, Sturzenbecker LJ, Kazmer S, Bosakowski T, Huselton C, Allenby G, Speck J, Kratzeisen C, Rosenberger M, Lovey A. 9-cis Retinoic acid stereoisomer binds and activates the nuclear receptor RXR alpha. *Nature* 1992;355:359–361.
45. Heyman RA, Mangelsdorf DJ, Dyck JA, Stein RB, Eichele G, Evans RM, Thaller C. 9-cis Retinoic acid is a high affinity ligand for the retinoid X receptor. *Cell* 1992;68:397–406.
46. Dreyer C, Krey G, Keller H, Givel F, Helftenbein G, Wahli W. Control of the peroxisomal beta-oxidation pathway by a novel family of nuclear hormone receptors. *Cell* 1992;68:879–887.
47. Sher T, Yi HF, McBride OW, Gonzalez FJ. cDNA Cloning, chromosomal mapping, and functional characterization of the human peroxisome proliferator activated receptor. *Biochem* 1993;32:5598–5604.
48. Devchand PR, Keller H, Peters JM, Vazquez M, Gonzalez FJ, Wahli W. The PPARalpha-leukotriene B4 pathway to inflammation control. *Nature* 1996;384:39–43.
49. Aoyama T, Hardwick JP, Imaoka S, Funae Y, Gelboin HV, Gonzalez FJ. Clofibrate-inducible rat hepatic P450s IVA1 and IVA3 catalyze the omega- and (omega-1)-hydroxylation of fatty acids and the omega-hydroxylation of prostaglandins E1 and F2 alpha. *J Lipid Res* 1990;31:1477–1482.
50. Kimura S, Hanioka N, Matsunaga E, Gonzalez FJ. The rat clofibrate-inducible CYP4A gene subfamily: I. Complete intron and exon sequence of the CYP4A1 and CYP4A2 genes, unique exon organization, and identification of a conserved 19-bp upstream element. *DNA* 1989;8:503–516.
51. Issemann I, Prince RA, Tugwood JD, Green S. The peroxisome proliferator-activated receptor:retinoid X receptor heterodimer is activated by fatty acids and fibrate hypolipidaemic drugs. *J Mol Endocrinol* 1993;11:37–47.
52. Göttlicher M, Demoz A, Svensson D, Tollet P, Berge RK, Gustafsson J-Å. Structural and metabolic requirements for activators of the peroxisome proliferator-activated receptor. *Biochem Pharmacol* 1993;46:2177–2184.
53. Bourguet W, Ruff M, Chambon P, Gronemeyer H, Moras D. Crystal structure of the ligand-binding domain of the human nuclear receptor RXR-alpha. *Nature* 1995;375:377–382.
54. Renaud JP, Rochel N, Ruff M, Vivat V, Chambon P, Gronemeyer H, Moras D. Crystal structure of the RAR-gamma ligand-binding domain bound to all-trans retinoic acid. *Nature* 1995;378:681–689.

55. Dreyer C, Keller H, Mahfoudi A, Laudet V, Krey G, Wahli W. Positive regulation of the peroxisomal beta-oxidation pathway by fatty acids through activation of peroxisome proliferator-activated receptors (PPAR). *Biol Cell* 1993;77:67–76.
56. Lygre T, Aarsaether N, Stensland E, Aarsland A, Berge RK. Separation and measurement of clofibroyl coenzyme A and clofibric acid in rat liver after clofibrate adminstration by reversed-phase high-performance liquid chromatography with photodiode array detection. *J Chromatogr* 1986;381:95–105.
57. Aarsland A, Berge RK. Peroxisome proliferating sulphur- and oxy-substituted fatty acid analogues are activated to acyl coenzyme A thioesters. *Biochem Pharmacol* 1991;41:53–61.
58. Kuslikis BI, Vanden Heuvel J, JP, Peterson RE. Lack of evidence for perfluorodecanoyl- or perfluorooctanoyl-coenzyme A formation in male and female rats. *J Biochem Toxicol* 1992;7:25–29.
59. Kliewer SA, Umesono K, Noonan DJ, Heyman RA, Evans RM. Convergence of 9-cis retinoic acid and peroxisome proliferator signalling pathways through heterodimer formation of their receptors. *Nature* 1992;358:771–774.
60. Gearing KL, Göttlicher M, Teboul M, Widmark E, Gustafsson, J-Å. Interaction of the peroxisome-proliferator-activated receptor and retinoid X receptor. *Proc Natl Acad Sci USA* 1993;90:1440–1444.
61. Tugwood JD, Issemann I, Anderson RG, Bundell KR, McPheat WL, Green S. The mouse peroxisome proliferator activated receptor recognizes a response element in the 5′ flanking sequence of the rat acyl CoA oxidase gene. *EMBO J* 1992;11:433–439.
62. Schoonjans K, Watanabe M, Suzuki H, Mahfoudi A, Krey G, Wahli W, Grimaldi P, Staels B, Yamamoto T, Auwerx J. Induction of the acyl-coenzyme A synthetase gene by fibrates and fatty acids is mediated by a peroxisome proliferator response element in the C promoter. *J Biol Chem* 1995;270:19269–19276.
63. Bardot O, Aldridge TC, Latruffe N, Green S. PPAR-RXR heterodimer activates a peroxisome proliferator response element upstream of the bifunctional enzyme gene. *Biochem Biophys Res Commun* 1993;192:37–45.
64. Aldridge TC, Tugwood JD, Green S. Identification and characterization of DNA elements implicated in the regulation of CYP4A1 transcription. *Biochem J* 1995;306:473–479.
65. Muerhoff AS, Griffin KJ, Johnson EF. The peroxisome proliferator-activated receptor mediates the induction of CYP4A6, a cytochrome P450 fatty acid omega-hydroxylase, by clofibric acid. *J Biol Chem* 1994;267:19051–19053.
66. Palmer CN, Hsu MH, Muerhoff AS, Griffin KJ, Johnson EF. Interaction of the peroxisome proliferator-activated receptor alpha with the retinoid X receptor alpha unmasks a cryptic peroxisome proliferator response element that overlaps an ARP-1-binding site in the CYP4A6 promoter. *J Biol Chem* 1994;269:18083–18089.
67. Gulick T, Cresci S, Caira T, Moore DD, Kelly DP. The peroxisome proliferator-activated receptor regulates mitochondrial fatty acid oxidative enzyme gene expression. *Proc Natl Acad Sci USA* 1994;91:11012–11016.
68. Rodriguez JC, Gil-Gomez G, Hegardt FG, Haro D. Peroxisome proliferator-activated receptor mediates induction of the mitochondrial 3-hydroxy-3-methylglutaryl-CoA synthase gene by fatty acids. *J Biol Chem* 1994;269:18767–18772.
69. Kaikaus RM, Chan WK, Ortiz de Montellano PR, Bass NM. Mechanisms of regulation of liver fatty acid-binding protein. *Mol Cell Biochem* 1993;123:93–100.
70. Lee SS, Pineau T, Drago J, Lee EJ, Owens JW, Kroetz DL, Fernandez-Salguero PM, Westphal H, Gonzalez FJ. Targeted disruption of the alpha isoform of the peroxisome proliferator-activated receptor gene in mice results in abolishment of the pleiotropic effects of peroxisome proliferators. *Mol Cell Biol* 1995;15:3012–3022.
71. Durand B, Saunders M, Leroy P, Leid M, Chambon P. All-trans and 9-cis retinoic acid induction of CRABPII transcription is mediated by RAR-RXR heterodimers bound to DR1 and DR2 repeated motifs. *Cell* 1992;71:73–85.
72. Wang LH, Tsai SY, Cook RG, Beattie WG, Tsai MJ, O'Malley BW. COUP Transcription factor is a member of the steroid receptor superfamily. *Nature* 1989;340:163–166.
73. Ladias JA, Karathanasis SK. Regulation of the apolipoprotein AI gene by ARP-1, a novel member of the steroid receptor superfamily. *Science* 1991;251:561–565.
74. Sladek FM, Zhong WM, Lai E, Darnell, JE, Jr. Liver-enriched transcription factor HNF-4 is a novel member of the steroid hormone receptor superfamily. *Genes Dev* 1990;4:2353–2365.
75. Palmer CN, Hsu MH, Griffin HJ, Johnson EF. Novel sequence determinants in peroxisome proliferator signaling. *J Biol Chem* 1995;270:16114–16121.

76. Juge-Aubry CE, Gorla-Bajszczak A, Pernin A, Lemberger T, Wahli W, Burger AG, Meier CA. Peroxisome proliferator-activated receptor mediates cross-talk with thyroid hormone receptor by competition for retinoid X receptor: possible role of a leucine zipper-like heptad repeat. *J Biol Chem* 1995;270:18117–18122.
77. Baes M, Castelein H, Desmet L, Declercq PE. Antagonism of COUP-TF and PPAR alpha/RXR alpha on the activation of the malic enzyme gene promoter: modulation by 9-cis RA. *Biochem Biophys Res Commun* 1995;215:338–345.
78. Kliewer SA, Forman BM, Blumberg B, Ong ES, Borgmeyer U, Mangelsdorf DJ, Umesono K, Evans RM. Differential expression and activation of a family of murine peroxisome proliferator-activated receptors. *Proc Natl Acad Sci USA* 1994;91:7355–7359.
79. Amri EZ, Bonino F, Ailhaud G, Abumrad NA, Grimaldi PA. Cloning of a protein that mediates transcriptional effects of fatty acids in preadipocytes: homology to peroxisome proliferator-activated receptors. *J Biol Chem* 1995;270:2367–2371.
80. Greene ME, Blumberg B, McBride OW, Yi HF, Kronquist K, Kwan K, Hsieh L, Greene G, Nimer SD. Isolation of the human peroxisome proliferator activated receptor gamma cDNA: expression in hematopoietic cells and chromosomal mapping. *Gene Expression* 1995;4:281–299.
81. Schmidt A, Endo N, Rutledge SJ, Vogel R, Shinar D, Rodan GA. Identification of a new member of the steroid hormone receptor superfamily that is activated by a peroxisome proliferator and fatty acids. *Mol Endocrinol* 1992;6:1634–1641.
82. Jow L, Mukherjee R. The human peroxisome proliferator-activated receptor (PPAR) subtype NUC1 represses the activation of hPPAR alpha and thyroid hormone receptors. *J Biol Chem* 1995;270:3836–3840.
83. MacDougald OA, Lane MD. Transcriptional regulation of gene expression during adipocyte differentiation. *Annu Rev Biochem* 1995;64:345–373.
84. Tontonoz P, Hu E, Spiegelman BM. Stimulation of adipogenesis in fibroblasts by PPAR gamma 2, a lipid-activated transcription factor. *Cell* 1994;79:1147–1156.
85. Tontonoz P, Graves RA, Budavari AI, Erdjument-Bromage H, Lui M, Hu E, Tempst P, Spiegelman BM. Adipocyte-specific transcription factor ARF6 is a heterodimeric complex of two nuclear hormone receptors, PPAR gamma and RXR alpha. *Nucleic Acids Res* 1994;22:5628–5634.
86. Reddy JK, Rao MS. Oxidative DNA damage caused by persistent peroxisome proliferation: its role in hepatocarcinogenesis. *Mutat Res* 1989;214:63–68.
87. Tanaka S, Imaoka S, Kusunose E, Kusunose M, Maekawa M, Funae Y. Omega- and (omega-1)-hydroxylation of arachidonic acid, lauric acid and prostaglandin A1 by multiple forms of cytochrome P-450 purified from rat hepatic microsomes. *Biochim Biophys Acta* 1990;1043:177–181.
88. Tontonoz P, Hu E, Spiegelman BM. Regulation of adipocyte gene expression and differentiation by peroxisome proliferator activated receptor gamma. *Curr Opin Genet Dev* 1995;5:571–576.
89. Elcombe CR, Mitchell AM. Peroxisome proliferation due to di(2-ethylhexyl) phthalate (DEHP): species differences and possible mechanisms. *Environ Health Perspect* 1986;70:211–219.
90. Kraupp-Grasl B, Huber W, Timmermann-Trosiener I, Schulte-Hermann R. Peroxisomal enzyme induction uncoupled from enhanced DNA synthesis in putative preneoplastic liver foci of rats treated with a single dose of the peroxisome proliferator nafenopin. *Carcinogenesis* 1993;14:2435–2437.
91. Huber W, Kraupp-Grasl B, Esterbauer H, Schulte-Hermann R. Role of oxidative stress in age dependent hepatocarcinogenesis by the peroxisome proliferator nafenopin in the rat. *Cancer Res* 1993;51:1789–1792.
92. Mukherjee R, Jow L, Noonan D, McDonnell DP. Human and rat peroxisome proliferator activated receptors (PPARs) demonstrate similar tissue distribution but different responsiveness to PPAR activators. *J Steroid Biochem Mol Biol* 1994;51:157–166.
93. Maronpot RR, Foley JF, Takahashi K, Goldsworthy T, Clark G, Tritscher A, Portier C, Lucier G. Dose response for TCDD promotion of hepatocarcinogenesis in rats initiated with DEN: histologic, biochemical, and cell proliferation endpoints. *Environ Health Perspect* 1993;101:634–642.
94. Vanden Heuvel JP, Clark GC, Kohn MC, Tritscher AM, Greenlee WF, Lucier GW, Bell DA. Dioxin-responsive genes: examination of dose-response relationships using quantitative reverse transcriptase-polymerase chain reaction. *Cancer Res* 1994;54:62–68.
95. Jonat C, Rahmsdorf HJ, Park KK, Cato, AC, Gebel S, Ponta H, Herrlich P. Antitumor promotion and antiinflammation: down-modulation of AP-1 (Fos/Jun) activity by glucocorticoid hormone. *Cell* 1990;62:1189–1204.
96. Peters JM, Cattley RC, Gonzalez FJ. Role of PPAR alpha in the mechanism of action of the nongenotoxic carcinogen and peroxisome proliferator Wy-14,643. *Carcinogenesis* 1997 Nov;18:2029–2033.

Toxicant–Receptor Interactions
Edited by Michael S. Denison and William G. Helferich

3

Effects of Xenobiotics on Hormone Receptors

Debie J. Hoivik and Stephen H. Safe

Department of Veterinary Physiology and Pharmacology, Texas A&M University, College Station, Texas, USA

Kevin W. Gaido

Chemical Industry Institute of Toxicology, Research Triangle Park, North Carolina, USA

There has been considerable recent interest in endocrine-active chemicals and their potential role in wildlife reproductive problems (1). Several studies have demonstrated the adverse effects of the organochlorine pesticides 2,2-bis(p-chlorophenyl)-1,1,1-trichloroethane (p,p′-DDT) and its degradation product, 2,2-bis(p-chlorophenyl)-1,1-dichloroethylene (p,p′-DDE) on reproduction in birds. A recent study (2) has identified p,p′-DDE as an antiandrogen, and this activity may contribute to the observed effects in some highly exposed wildlife populations. It also has been hypothesized that environmental endocrine disruptors, with estrogenic (xenoestrogens) or antiandrogenic activity may be associated with decreased male reproductive capacity in humans, which includes increased testicular cancer and decreased sperm counts (3, 4). Moreover, a second hypothesis has proposed that xenoestrogens also may play a role in the development of breast cancer in women (5). The validity of both hypotheses have been questioned (6, 7) and therefore require further investigation; however, the levels of public, scientific, and regulatory concerns regarding endocrine-active environmental contaminants are very high.

There are multiple mechanisms associated with xenobiotic-induced disruption of endocrine response pathways; however, this review primarily focuses on chemicals that interact directly with hormone receptors and chemicals that modify endocrine pathways via aryl hydrocarbon receptor (AhR)–mediated pathways.

HORMONE RECEPTORS

The steroid/thyroid hormone–retinoic acid receptor superfamily is a rapidly expanding group of nuclear proteins that includes the estrogen receptor (ER), androgen receptor (AR), mineralocorticoid receptor, progesterone receptor (PR), thyroid hormone receptor, vitamin D receptor, retinoic acid receptor family, and orphan receptors for which no specific endogenous or exogenous ligands have been identified (8–10). Proteins of the steroid receptor superfamily exhibit several common structural features that include a variable N-terminal region (A/B), a highly conserved DNA binding domain (C) that interacts with specific DNA-binding sequences via zinc fingers, a hinge region (D) that is highly variable, a conserved ligand-binding domain (E), and a highly variable C-terminal domain (F) (11). A model for steroid receptor function is depicted in Fig. 1. Steroid hormone–induced gene expression requires direct or indirect interaction of hormone-receptor dimers with hormone receptor–responsive elements (HREs) typified by the classic palindromic estrogen responsive element (ERE). Formation of a hormone receptor–HRE complex is accompanied by interaction with components of the general transcriptional machinery as well as other coactivators resulting in transcriptional activation of target genes. This simplified model for steroid hormone–induced transactivation does not adequately describe the multiplicity of pathways that have been described recently (9, 12–14). For example, it is recognized now that estrogen-induced responses are dependent on a number of factors including ligand structure, variable DNA recognition sequences, cell- and promoter-specific responses, protein phosphorylation status, and ERE-dependent and -independent interactions with the ER with other nuclear coactivators and transcriptional factors. Tamoxifen is a widely used antiestrogenic drug used in treatment of breast cancer; however, the mechanistic pathways associated with tamoxifen-induced responses are highly complex (13, 15, 16). Tamoxifen binds to the ER, and

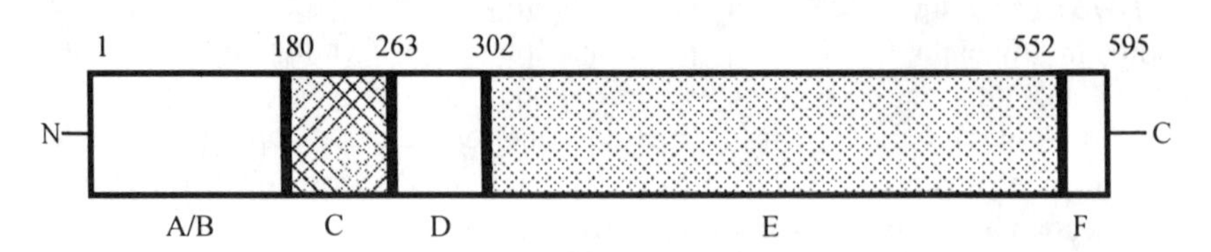

FIG. 1. Mechanism of hormone-induced gene expression.

in breast cancer cells the nuclear ER complex liganded with tamoxifen is relatively inactive and thereby acts as a competitive inhibitor of ER-induced transactivation. In contrast, tamoxifen acts as an ER agonist in the uterus and endometrium (17). One explanation of this ER agonist/antagonist action of tamoxifen is the preferential inactivation of the C-terminal activation function (AF) 2 of the ER, which is dominant in breast cancer cells. In contrast, the N-terminal AF1 domain of the ER plays a major role in the endometrium, and tamoxifen activates this function. The class II antiestrogens, ICI164,384 and ICI182,780, which have been defined as "*pure anitiestrogens*," inactivate both AF1 and AF2 (13). Presumably the various activities of estrogens and antiestrogens are related to ligand-induced conformational changes in the ER that modulate interactions with promoter sequences, coactivators, and other nuclear factors. Ligand-induced conformational changes also play a role in modulating the activities of other steroid receptors.

NATURAL AND XENOESTROGENS

It has long been recognized that various plant species consumed by domestic animals and humans contain relatively high levels of estrogenic compounds (18–20). For example, after short-term exposure to phytoestrogen-rich pasture, irregular cycling and decreased lambing rates were observed in sheep (21–23). In addition, after prolonged grazing the animals exhibited a broad spectrum of reproductive tract problems (including infertility) consistent with overexposure to estrogenic compounds. Humans also consume fruits, vegetables, and nuts that contain diverse ER ligands (Fig. 2). The potential human health impacts of estrogen-rich food such as soy-based infant formula is being discussed in terms of both risks and benefits (18, 24). Isoflavonoids and other dietary phytoestrogens exhibit both estrogenic and antiestrogenic activities, and most studies show that populations that consume large quantities of phytoestrogens (e.g.,

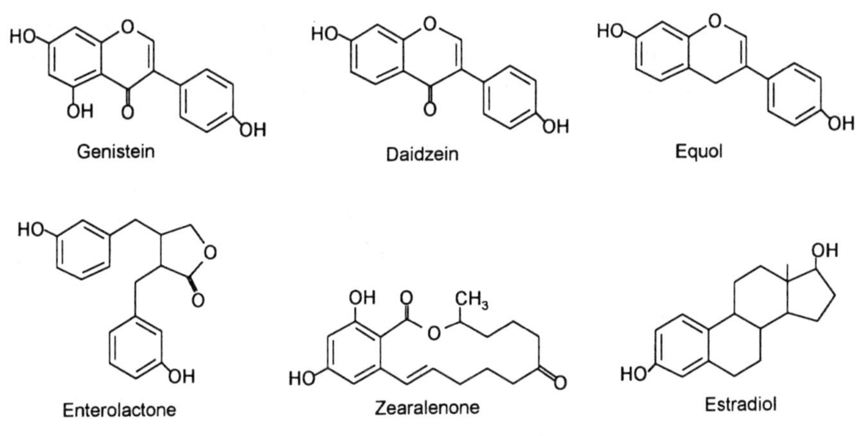

FIG. 2. Structures of naturally occurring estrogens.

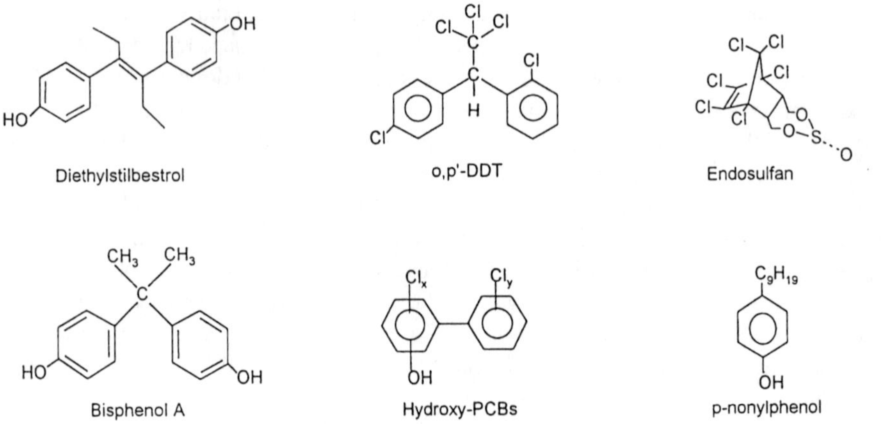

Diethylstilbestrol o,p'-DDT Endosulfan

Bisphenol A Hydroxy-PCBs p-nonylphenol

FIG. 3. Structures of xenoestrogens.

far-eastern countries) generally exhibit lower incidences of several hormone-related diseases including breast cancer. The health benefits derived from phytoestrogen-enriched diets also may be due to many of the other estrogen-independent activities of these compounds (e.g., protein kinase C inhibition, P450 inhibition, cancer cell growth inhibition).

Prior to recent concerns regarding the potential adverse health effects of xenoestrogens, several studies identified industrial compounds as estrogenic including o,p′-DDT and other o,p′-substituted analogs, methoxychlor, kepone, polychlorinated biphenyl (PCB) mixtures and congeners, hydroxylated PCBs, and other phenolics (25–29). Subsequent studies in several laboratories have significantly extended the list of industrially derived estrogenic compounds and this includes widely used chemicals such as nonylphenol, nonylphenyl ethoxylates, dibenzyl and benzylbutyl phthalate, and bisphenol A (30–36) (Fig. 3).

There has been considerable emphasis on development of new in-vivo and in-vitro screening methods for xenoestrogens (37). Many of the standard in-vivo methods still are used widely and include measurement of estrogen-induced responses in the immature and ovariectomized rat and mouse models (uterine hypertrophy, vaginal cornification, and gene expression). Some of the in-vitro assays that are used routinely in screening for estrogenic compounds include proliferation of MCF-7 human breast cancer cells, ER binding assays, induction of estrogen-regulated genes in estrogen (E_2)-responsive cell lines, induction of reporter gene expression in various transiently or stably transfected cell lines (30–33, 35, 36). The results illustrated in Fig. 4 summarize the effects of various estrogenic compounds in an estrogen-responsive yeast assay system (38). The yeast cells have been transformed with an expression plasmid for the human ER cDNA as described by (39). The yeast assay system is both sensitive and robust, and dose–response curves for estrogenic compounds can be readily obtained. The results also show that antiestrogens such as ICI164,384 also induce β-galactosidase gene expression, because yeast cells do not express repressor proteins

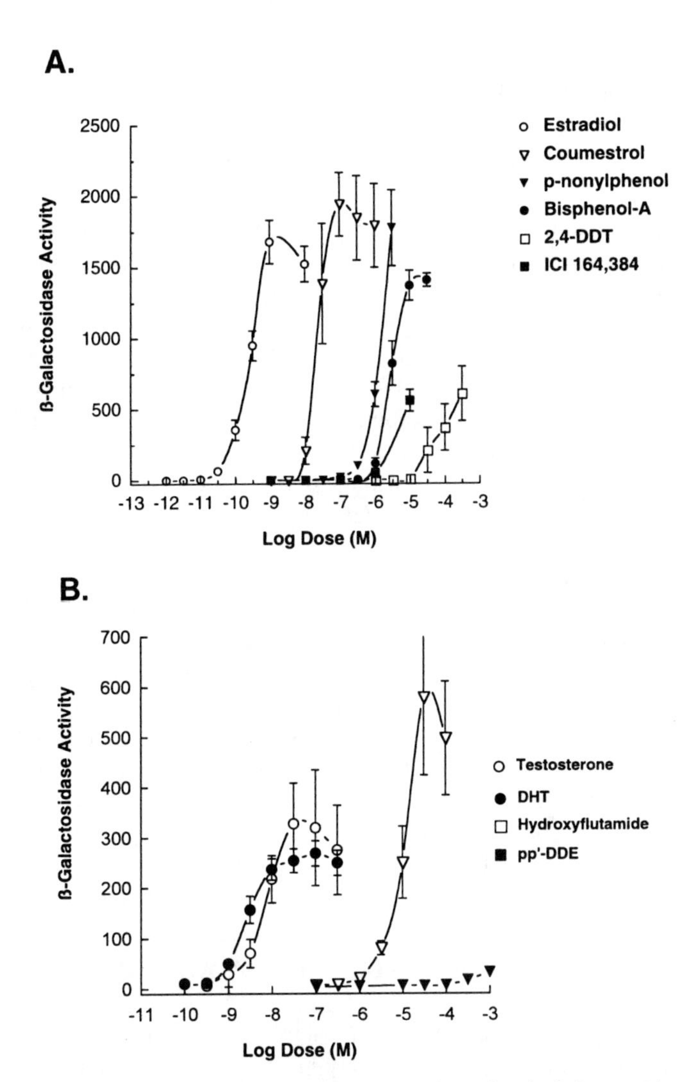

FIG. 4. Estrogenic (*A*) and androgenic (*B*) activity of various chemicals in yeast-based systems that express the estrogen receptor (*A*) and androgen receptor (*B*) receptors and the appropriate hormone-responsive promoter–reporter constructs (38).

associated with ER antagonism. Thus the assay system dose not distinguish between ER agononists and antagonists but provides a rapid method for screening individual compounds or mixtures that bind the ER.

There are both advantages and disadvantages associated with the in-vitro and in-vivo bioassays for estrogenicity. Most of these assays can determine estrogenic potency for a response at the gene or cellular level. Other important determinants for hazard and risk assessment, however, including organ- or cell- and promoter-specific responses as well as important pharmacokinetic parameters and effects at critical periods of exposure, are not readily determined in any single bioassay. Therefore, a tiered screening approach has been recommended (37).

ANDROGENS AND ANTIANDROGENS

The systemic fungicide, 3-(3,5-dichlorophenyl)-5-methyl-5-vinyl-oxazolidine-2,4-dione (vinclozolin) is used as an antifungal agent for treatment of fruits and vegetables. Vinclozolin exhibits weak binding affinity to the AR; however, two of its metabolites (M1 and M2) bind to the AR with higher affinity and exhibit antiandrogenic activity in both in-vivo and in-vitro models (40, 41). More recent studies by Kelce and coworkers (2) have reported that p,p′-DDE, a major persistent environmental contaminant in wildlife and humans, also binds to the AR and exhibits antiandrogenic activity. It has been suggested that the antiandrogenic activity of p,p′-DDE may be responsible for reproductive problems in alligators in Lake Apopka, Florida (42), and also may contribute to the hypothesized decrease in male reproductive activity (4). Although it is clear that p,p′-DDE is an antiandrogen, the potency of this compound in most bioassays is relatively low compared to testosterone or dihydrotestosterone. For example, in an androgen-responsive yeast assay (see Fig. 4*B*) it was estimated that p,p′-DDE was 2,500,000 times less potent than dihydrotestosterone. Waller and coworkers (43) recently reported quantitative structure–activity relationships for AR ligands and showed that overall structure and electrostatic properties were sufficient for predicting the AR-binding affinities for structurally diverse compounds. It is evident from their AR-binding data and quantitative structure–activity relationships that the AR (like the ER) is highly promiscuous and binds with moderate to low affinity to diverse steroids, organochlorine compounds, chloracetanilides, and other chemical classes. The androgenic or antiandrogenic activities of these compounds and their potential adverse impacts as endocrine disruptors have not been determined.

OTHER HORMONALLY ACTIVE AGENTS

An overall hazard and risk assessment of hormonally active compounds requires initial hazard identification and information on daily intakes of these chemicals. Although some information is available on a limited number of naturally occurring and xenoestrogens only minimal data on serum levels of these compounds are available. Exposure to endocrine disruptors has focused primarily on compounds that bind to the AR, ER, or AhR and exhibit agonist and antagonist activities. Hydroxy-PCBs, which exhibit both estrogenic and antiestrogenic activity (29, 33), also bind to transthyretin (44), a thyroid hormone transport protein, and it is possible that these compounds may interfere with multiple hormone-dependent responses. There are significant dietary intakes of many other "hormone-acting" molecules, however, that may affect ER- or AR-mediated signaling pathways via cross-talk. For example, humans consume large quantities of chlorophyll, which is metabolized to various diterpenoid-derived compounds including phytol, phytenic acid, phytanic acid, and pristanic acid; levels of phytanic and phytenic acid in human serum are 6 and 2 μmol, respectively (45). These compounds have been identified as ligands for the retinoid acid X receptor (46, 47), which is a highly versatile member of the nuclear receptor superfamily (9). Other fatty acids that also are abundant in the diet activate the peroxisome proliferator–activator receptor (48). The retinoic acid X receptor and peroxisome proliferator–activator receptor modulate multiple response pathways and interact directly or indirectly with other hormone receptors and transcription factors (9, 49). Thus the effects of xenoendocrine disruptors must be evaluated in the presence of high background levels of natural endocrine disruptors that modulate multiple hormonally regulated pathways. Research in our laboratory has focused on cross-talk between the AhR and the ER in the rodent uterus and mammary and in human breast-cancer cells (7). These studies show that AhR agonists inhibit multiple E_2-induced responses including the development and growth of mammary tumors. In contrast, male adult rats exposed in utero to AhR agonists or estrogens exhibit effects associated with demasculinization suggesting that both signaling pathways induce similar responses in this model. Thus, cross-talk between ligand-induced receptors is dependent on the response, and these complex interactions further complicate hazard and risk assessment of endocrine disruptors.

AhR–MEDIATED ANTIESTROGENICITY

2,3,7,8-Tetrachlorodibenzo-p-dioxin (TCDD) modulates diverse endocrine responses in exposed animals and humans; however, the mechanisms of interaction between AhR–mediated pathways and most endocrine effects have not been elucidated clearly (7). TCDD inhibits several E_2-induced responses in the rodent uterus and mammary gland and in human breast-cancer cells, and the antiestrogenic effects resulting from cross-talk between the AhR and ER signaling pathways can serve as a model for understanding modulation of other endocrine responses by these compounds.

In the Sprague-Dawley rat uterus, TCDD caused dose-dependent decreases in basal and E_2-induced binding to cytosolic and nuclear ER and PR (50–52), epidermal growth factor (EGF) receptor binding and mRNA levels (53), peroxidase activity (54), and c-fos protooncogene mRNA levels (55). Inhibition of E_2-induced activities by TCDD was similar to that observed for progesterone; however, TCDD was over 30 times more potent and the effects were longer-lived (51). TCDD also decreases E_2-induced uterine wet weight and cytosolic and nuclear ER levels in mice (56, 57). For several polychlorinated dibenzo-p-dioxin and dibenzofuran congeners, there was a correlation between their potencies as antiestrogens and their rank-order binding affinities for the AhR suggesting that inhibition of E_2-induced responses by TCDD was mediated by the AhR (50, 58–60). Similarly, the polycyclic aromatic hydrocarbon (PAH) 3-methylcholanthrene (61) and the coplanar PCB congeners 3,3′,4,4′-tetraCB (62) and 3,3′,4,4′,5,5′-hexaCB (63) bind to the AhR and also exhibit antiestrogenic activity. Consistent with these observations, age-dependent expression of the AhR correlates with susceptibility of rodents to the antiestrogenic effects of TCDD (61). For example, TCDD does not influence estrogen-induced uterine wet weight in weanling rats (61, 64), which express lower levels of the AhR than 21-day-old or mature rodents, which are responsive to the TCDD-induced downregulation of uterine estrogenic responses (65).

Female Sprague-Dawley rats spontaneously develop benign neoplasms of the mammary gland and benign uterine tumors, and it is likely that formation of these tumors is estrogen-dependent. Chronic dietary administration of TCDD over 2 years significantly decreased spontaneous mammary and uterine tumor incidence in rats (66). TCDD also delayed carcinogen (7,12-dimethylbenzanthracene)–induced mammary tumors in Sprague-Dawley rats (67) and caused a dose-dependent decrease in mammary tumor growth in B6D2F1 mice transplanted with MCF-7 cells (68). In agreement with results of animal experiments, epidemiologic studies indicate that the accidental release of TCDD in to the environment in Seveso, Italy, resulted in decreased incidence of endometrial and breast cancer in women exposed to TCDD (69). Exposure to TCDD-like compounds that have been identified in food also can have antiestrogenic effects. For example, cruciferous vegetables contain relatively high levels of indole 3-carbinol (I3C), which can form I3C derivatives such as indolo-[3,2-b]carbazole (ICZ) by acidic rearrangement in the gut. ICZ is a pentacyclic heteroaromatic hydrocarbon that binds to the AhR with relatively high affinity (70, 71) and induces CYP1A-dependent enzyme activity (72, 73). I3C inhibits endometrial and mammary tumor formation in animals (74–76), and I3C and ICZ exhibit antiestrogenic activities in human breast cancer cells in culture (77, 78).

The antiestrogenic activity of TCDD and the mechanism of this response also have been investigated in breast-cancer cell lines. In MCF-7 or T47D human breast cancer cells, TCDD inhibited E_2-induced cell proliferation (68, 79, 80), postconfluent focus production (81, 82), PR binding and PR mRNA and protein levels (83), glucose-to-lactate conversion (84, 85), secretion of tissue plasminogen activator activity (86), cathepsin D, procathepsin D (79, 87, 88) and a 160-kDa protein (79), heat shock protein (HSP) 27 (89) and c-fos (90) mRNA, and insulin-like growth factor binding proteins (91). Inhibition of insulin-like growth Factor-I (IGF-I) (92), transforming growth Factor-α (TGFα) ((93, 94), and EGF- (93, 94) induced growth also has been observed after treatment of breast-cancer cells with TCDD. For several compounds there was a correlation between their AhR-binding affinities and their activities as inhibitors of E_2-induced procathepsin D secretion (87, 88) and tissue plasminogen activator activity (86), postconfluent focus production (82), and pS2 (95) and PR (83) gene expression. These results support a role for the AhR in mediating the antiestrogenic responses, and similar results were reported for ER downregulation by AhR agonists (96). Additionally, several AhR agonists including I3C and ICZ (77, 78), polycyclic aromatic hydrocarbons (97, 98), and alkyl dibenzofurans (99) exhibit antiestrogenic activity in human breast-cancer cells, and the AhR antagonist, α-naphthoflavone, inhibited several antiestrogenic responses elicited by TCDD (80, 100). Moreover, TCDD is not antiestrogenic in cell lines that do not express the nuclear AhR. For example, TCDD does not decrease nuclear ER levels (101) or inhibit E_2-induced pS2 gene expression in mutant AhR-deficient Hepa 1c1c7 cells (95). The effects of TCDD on E_2-induced cell proliferation and procathepsin D secretion were not observed in benzo[a]pyrene resistant Ah-nonresponsive MCF-7 cells (102).

The pathways by which TCDD downregulates these E_2-induced responses is not clear; however, results of several studies suggest that there are several possible mechanisms including modulation of estrogen metabolism, direct interaction of AhR with cis-acting genomic elements, or induction of trans-acting factors that directly or indirectly block E_2 or growth factor-induced responses.

Treatment of breast cancer cells with TCDD results in an increase in E_2 2-, 4-, 15α-, and 6α-hydroxylase activities (103, 104). Induced E_2-hydroxylase in microsomal fractions from MCF-7 cells were decreased after treatment with CYP1A antibodies (105), suggesting that induction of CYP1A gene expression by TCDD may increase E_2 metabolism. Decreased levels of E_2 owing to enhanced metabolic biotransformation could account for the antiestrogenic activity of TCDD in cell-culture experiments. Structure–activity studies for several halogenated aromatics showed a correlation between their activity as inhibitors of E_2-dependent postconfluent growth and induction of E_2 metabolism (82).

However, in rodents TCDD does not alter serum E_2 levels at doses that are antiestrogenic (57, 106). Moreover, TCDD exhibits antiestrogenic activity in breast-cancer cells at concentrations that do not induce CYP1A1-dependent enzyme activity (77, 78). For example, 1-pmol TCDD, a concentration that does not increase CYP1A activity, inhibits E_2-induced secretion of 34-, 52-, and 160-kDa proteins in MCF-7 cells (79). In addition, AhR agonists such as methylcholanthrene, benzo[a]pyrene, and 7,12-dimethylbenz[a]anthracene decrease ER content and inhibit other E_2-induced responses in MCF-7 cells in the absence of CYP1A induction (83, 97).

The time-course induction of CYP1A activity does not necessarily correspond to the time required for inhibition of some E_2-induced responses, suggesting that the two events are not causally related. Induction of CYP1A enzyme activity is induced by TCDD within 3 hours of treatment of MCF-7 cells (84), whereas downregulation of E_2-induced glycolysis (84), procathepsin D protein and mRNA levels (107), cathepsin D (108), c-fos (90) and hsp27 (109) mRNA, and insulin-like growth factor binding proteins (91) was observed within 2 hours of treatment with TCDD. Thus, enhanced metabolism of E_2 by TCDD is not required for the antiestrogenic activity of AhR agonists.

Structure–activity studies have shown that several alkyl PCDF compounds that competitively bind the AhR, do not induce CYP1A activity and exhibit AhR antagonist activity for this response (110). 6-Methyl-1,3,8-trichlorodibenzofuran (MCDF), a congener that binds with relatively high affinity to the AhR was over 100,000 times less potent than TCDD as an inducer of CYP1A1-dependent activity. In contrast, MCDF and several alkyl-substituted analogs caused a dose-dependent decrease in uterine wet weight and ER and PR content at dosages that minimally increased CYP1A activity (52, 111). MCDF also inhibited E_2-induced cell proliferation and secretion of 34-, 52-, and 160-kDa proteins in MCF-7 human breast cancer cells in the absence of enhanced CYP1A activity (99). Current studies are focused on the development of these relatively nontoxic AhR agonists as drugs for clinical treatment of breast cancer.

Many of the responses induced by TCDD are mediated by the AhR and the antiestrogenic activities appear to involve cross-talk between the ER and AhR signaling pathways. Both cathepsin D and HSP27 are induced by E_2, and TCDD decreases these responses. These genes have been utilized as models to further investigate the mechanism of AhR-mediated antiestrogenicity.

Studies on the 5′ promoter region of the cathepsin D gene have identified an Sp1/ERE half site (GGGCGG(N)23ACGGG) (−199 to −165) that is characterized as an E_2-responsive enhancer sequence (108). Also in the cathepsin D promoter a dioxin responsive element (DRE) (GTGCGTG) core-binding sequence is located at position –175/–181, and it was hypothesized that binding of the AhR complex to this sequence destabilizes formation of the ER/Sp1 complex and is responsible for inhibition of hormone-induced cathepsin D gene expression by TCDD. A comprehensive study by Krishnan and coworkers (108) provided several lines of evidence that support this proposed mechanism. In MCF-7 human breast cancer cells transiently transfected with a cathepsin D promoter–reporter construct, E_2-induced chloramphenicol acetyltransferase activity and TCDD inhibited this response. The inhibitory effect of TCDD was blocked by α-naphthoflavone, an AhR antagonist. Mutation of the DRE within the cathepsin D promoter abolished the antiestrogenic activity of TCDD. The ligand-bound AhR dimerizes with AhR nuclear translocator (ARNT), and this heterodimer is required for induction. In MCF-7 breast cancer cells transfection with antisense ARNT and AhR expression plasmids decreased the antiestrogenic activity of TCDD. Taken together these studies suggest that inhibition of E_2-induced cathepsin D gene expression by TCDD requires a functional AhR–ARNT complex. Electrophoretic mobility shift analysis showed that TCDD inhibited E_2-dependent binding of nuclear extract proteins to a [^{32}P]ERE/Sp1/DRE oligonucleotide whereas TCDD did not affect binding to a [^{32}P]ERE/Sp1/DRE oligonucleotide containing C-A mutations in the DRE sequence. Furthermore, cross-linking studies showed that the AhR binds directly to the ERE/Sp1/DRE oligonucleotide, and this interaction was inhibited by competition with unlabeled wild-type but not mutant ERE/Sp1/DRE oligonucleotides. Collectively, these studies indicate that the antiestrogenic activity of TCDD is mediated by the AhR which interacts with a cis-acting DRE located within the 5′ flanking region of the cathepsin D gene.

Analogous to the cathepsin D gene, inhibition of E_2-inducible HSP27 gene expression by TCDD also is mediated by an inhibitory DRE (iDRE) (89). In transient transfection of MCF-7 human breast-cancer cells with a promoter–reporter construct containing the −105 to +23 region of the HSP27 gene, E_2 induced reporter gene activity, and this response was inhibited by TCDD. A functional Sp1/ERE half site (−105/−84) was identified in the 5′ flanking region of the gene and is required for E_2 responsiveness. An iDRE was identified at the +2 transcription start site, and the antiestrogenic activity of TCDD was not observed in studies utilizing constructs mutated in the iDRE sequence. Moreover, electrophoretic gel mobility shift and cross-linking assays further confirmed that the iDRE sequence was required for inhibition of E_2-induced transactivation.

CONCLUSION AND FUTURE PERSPECTIVES

TCDD inhibits several E_2-induced responses in the rodent uterus and mammary gland and in human breast-cancer cells, and several mechanisms of action have been proposed. Structure–activity studies support a role for the AhR in mediating the antiestrogenic activities of AhR agonists, and for some genes hormone-induced responses are inhibited by direct interaction of the nuclear AhR complex with 5′-iDREs. Additional trans-acting factors also may be involved, but these have not yet been identified. Studies are underway to further delineate the mechanism of TCDD antiestrogenicity in the hope that other AhR agonists that exhibit low toxicity but high antiestrogenic activity can be used for therapeutic treatment of mammary cancer in women.

ACKNOWLEDGMENTS

The financial assistance of the National Institutes of Health (ESO4176; ESO5734) and the Texas Agricultural Experiment Station are gratefully acknowledged. S. Safe is a Sid Kyle Toxicology Scholar.

REFERENCES

1. Colborn T, Vom Saal F, Soto A. Development effects of endocrine-disrupting chemicals in wildlife and humans. *Environ Health Perspect* 1993;101:378–384.
2. Kelce W, Stone C, Laws S, Gray L. Persistent DDT metabolite p,p′-DDE is a potent androgen receptor antagonist. *Nature* 1995;375:581–585.
3. Sharpe R, Skakkebaek N. Are oestrogens involved in falling sperm counts and disorders of the male reproductive tract. *Lancet* 1993;341:1392–1395.
4. Sharpe R. Reproductive biology: another DDT connection. *Nature* 1995; 375:538–539.
5. Davis D, Bradlow H, Wolff M, Woodruf T, Hoel D, Anton-Culver H. Medical hypothesis: xenoestrogens as preventable causes of breast cancer. *Environ Health Perspect* 1993;101:327–377.
6. Ahlborg U, Lipworth L, Titusernstoff L, Hsieh C, Hanberg A, Baron J, Trichopoulos D, Adami H. Organochlorine compounds in relation to breast cancer, endometrial cancer and endometriosis: as assessment of the biological and epidemiological evidence. *Crit Rev Toxicol* 1995;25:463–531.
7. Safe S. Environmental and dietary estrogens and human health: is there a problem? *Environ Health Perspect* 1995;103:346–351.
8. Gronemeyer H. Transcription activation by estrogen and progesterone receptors. *Annu Rev Genet* 1991;25:89–123.
9. Mangelsdorf DJ, Thummel C, Beato M, Herrlich P, Schutz G, Umesono K, Blumberg B, Kastner P, Mark M, Chambon P. The nuclear receptor superfamily: the second decade. *Cell* 1995;83:835–839.
10. Tsai MJ, O'Malley BW. Molecular mechanisms of action of steroid/thyroid receptor superfamily members. *Annu Rev Biochem* 1994;63:451–486.
11. Ing NH, O'Malley BW. The steroid hormone receptor superfamily: molecular mechanisms of action. In: Weintraub BD, ed. *Molecular Endocrinology: Basic Concepts and Clinical Correlations.* New York: Raven Press, 1995;195–215.
12. Katzenellenbogen JA, O'Malley BW, Katzenellenbogen BS. Tripartite steroid hormone receptor pharmacology: Interaction with multiple effector sites as a basis for the cell- and promoter-specific action of these hormones. *Mol Endocrinol* 1996;10:110–131.
13. McDonnell DP, Clemm DL, Hermann T, Goldman ME, Pike JW. Analysis of estrogen receptor function *in vitro* reveals three distinct classes of antiestrogens. *Mol Endocrinol* 1995;9:659–669.
14. Horwitz KB, Jackson TA, Bain DL, Richer JK, Takimoto GS, Tung L. Nuclear receptors coactivators and corepressors. *Mol Endocrinol* 1996;10:1167–1177

15. Tzukerman MT, Esty A, Santiso-Mere D, Danielian P, Parker MG, Stein RB, Pike JW, McDonnell DP. Human estrogen receptor transactivational capacity is determined by both cellular and promoter context and mediated by two functionally distinct intramolecular regions. *Mol Endocrinol* 1994;8: 21–30.
16. Webb P, Lopez GN, Uht RM, Kushner PJ. Tamoxifen activation of the estrogen receptor/AP-1 pathway: potential for the cell-specific estrogen-like effects of antiestrogens. *Mol Endocrinol* 1995;9:443–456.
17. Jordan VC. The development of tamoxifen for breast cancer therapy: a tribute to the late Arthur L. Walpole. *Breast Cancer Res Treatment* 1988;11:197–209.
18. Clarkson TW. Environmental contaminants in the food chain. *Am J Clin Nutr* 1995;61:682S–686S.
19. Setchell KDR. Naturally occurring nonsteroidal estogens of dietary origin. In: McLachlan JA, ed. *Estrogens in the Environment.* New York, Elsevier Science Publishers, 1985;69–85.
20. Aldercreutz H. Western diet and western diseases: some hormonal and biochemical mechanisms and associations. *Scand J Clin Lab Invest* 1990;50:3–23.
21. Moule GR, Braden AWH, Lamond DR. The significance of oestrogens in pasture plants in relation to animal production. *Anim Breed Abstr* 1963;31:139–157.
22. Croker KP, Lightfoot RJ, Johnson TJ, Adams NR, Carrick MJ. The effects of selection for resistance to clover infertility on the reproductive performances of Merino ewes grazed on oestrogenic pastures. *Aust J Agric Res* 1989;40:165–176.
23. Kaldas RS, Hughes CL. Reproductive and general metabolic effects of phytoestrogens in mammals. *Reprod Toxicol* 1989;3:81–89.
24. Aldecreutz H, Yaghoob M, Clark J, Höckerstedt K, Hämäläinen E, Wähälä K, Mäkelä T, Hase T. Dietary phytoestrogens and cancer: *in vitro* and *in vivo* studies. *J Ster Biochem Mol Biol* 1992;41:331–337.
25. Gellert RJ. Uterotrophic activity of polychlorinated biphenyls (PCB) and induction of precocious reproductive aging in neonatally treated female rats. *Environ Res* 1987;16:123–130.
26. Gellert RJ. Kepone, mirex, dieldren, and aldrin: estrogenic activity and the induction of persistent vaginal estrus and anovulation in rats following neonatal treatment. *Environ Res* 1987;16:131–138.
27. Hammond B, Katzenellenbogen BS, Kruthammer N, McConnell J. Estrogenic activity of the insecticide chlorodecone (Kepone) and interaction with uterine estrogen receptor. *Proc Natl Acad Sci USA* 1979;76:6641–6645.
28. Bitman J, Cecil HC, Harris SJ, Fries GF. Estrogenic activity of o,p′-DDT in the mammalian uterus and avian oviduct. *Science* 1968;162:371–372.
29. Korach K, Sarver P, Chae K, McLachlan J, McKinney J. Estrogen receptor-binding activity of polychlorinated hydroxybiphenyls: conformationally restricted structural probes. *Mol Pharmacol* 1988;33:120–126.
30. Soto A, Justicia H, Wray J, Sonnenschein C. P-Nonolphenol: an estrogenic xenobiotic released from "modified" polystyrene. *Environ Health Perspect* 1991;92:67–173.
31. Soto A, Chung K, Sonnenschein C. The pesticides endosulfan, toxaphene, and dieldrin have estrogenic effects on human estrogen-sensitive cells. *Environ Health Perspect* 1994;102:380–383.
32. Soto A, Sonnenschein C, Chung K, Fernandez M, Olea N, Serrano F. The E-screen assay as a tool to identify estrogens: an update on estrogenic environmental pollutants. *Environ Health Perspect* 1995;103:113–122.
33. Moore M, Mustain M, Daniel K, Chen I-C, Safe S, Zacharewski T, Gillesby B, Joyeux A, Balaguer P. Antiestrogenic activity of hydroxylated polychlorinated bipheynyl congeners identified in human serum. *Toxicol Appl Pharmacol* 1997;142:160–168.
34. Krishnan A, Stathis P, Permuth S, Tokes L, Feldman D. Bisphenol-A: an estrogenic substance is released from polycarbonate flasks during autoclaving. *Endocrinol* 1993;132:2279–2286.
35. White R, Jobling S, Hoare S, Sumpter J, Parker M. Environmentally persistent alkylphenolic compounds are estrogenic. *Endocrinology* 1993;135:175–182.
36. Jobling S, Reynolds T, White R, Parker M, Sumpter J. A variety of environmentally persistent chemicals, including some phthalate plasticizers, are weakly estrogenic. *Environ Health Perspect* 1995;3:582–587.
37. Reel J, Lamb IV, J, Neal B. Survey and assessment of mammalian estrogen biological assays for hazard characterization. *Fundam Appl Toxicol* 1996;33:288–305.
38. Gaido KW, Leonard LS, Lovell S, Gould JC, Dariouch B, Portier CJ, McDonnell DP. Evaluation of chemicals with endocrine modulating activity in a yeast-based steroid hormone receptor gene transcription assay. *Toxicol Appl Pharmacol* 1997;143:205–212.

39. Pham TA, Hwung YP, Santiso MD, McDonnell DP, O'Malley BW. Ligand-dependent and -independent function of the transactivation regions of the human estrogen receptor in yeast. *Mol Endocrinol* 1992;6:1043–1050.
40. Gray LE, Ostby JS, Kelce WR. Developmental effects of an environmental antiandrogen: the fungicide vinclozolin alters sex differentiation of the male rat. *Toxicol Appl Pharmacol* 1987;129:46–52.
41. Kelce WR, Monosson E, Gamcsik MP, Laws SC, Gray LE. Environmental hormone disruptors: evidence that vinclozolin developmental toxicity is mediated by antiandrogenic metabolite. *Toxicol Appl Pharmacol* 1994;126:276–285.
42. Guillette LJ, Gross TS, Masson GR, Matter JM, Percival HF, Woodward AR. Developmental abnormalities of the gonad and abnormal sex hormone concentrations in juvenile alligators from contaminated and control lakes in Florida. *Environ Health Perspect* 1994;102:680–688.
43. Waller CL, Juma BW, Gray LE, Kelce WR. Three dimensional quantitative structure-activity relationships for androgen receptor ligands. *Toxicol Appl Pharmacol* 1996;137:219–227.
44. Lans M, Klasson-Wehler E, Willemsen M, Meussen E, Safe S, Brouwer A. Structure-dependent competition interaction of hydroxy-polychlorobiphenyls, -dibenzo-p-dioxins and -dibenzofurans with human transthyretin. *Chem Biol Interact* 1993;88:7–21.
45. Avignan J. The presence of phytanic acid in normal human and animal plasma. *Biochem Biophys Acta* 1966;166:391–294.
46. Lemotte PK, Keidel S, Apfel CM. Phytanic acid is a retinoid X receptor ligand. *Eur J Biochem* 1996;236:328–333.
47. Kitareewan S, Burka LT, Tomer KB, Parker CE, Deterding LJ, Stevens RD, Forman BM, Mais DE, Heyman RA, McMorri T, Weinberger C. Phytol metabolites are circulating dietary factors that activate the nuclear receptor RXR. *Mol Biol Cell* 1996;7:153–1166.
48. Göttlicher M, Widmark E, Li Q, Gustafsson JA. Fatty acids activate a chimera of the clofibric acid-activated receptor and the glucocorticoid receptor. *Proc Natl Acad Sci USA* 1987;89:4653–4657.
49. Mangelsdorf DJ, Evans RM. The RXR heterodimers and orphan receptors. *Cell* 1995;83:841–850.
50. Romkes M, Piskorska-Pliszczynska J, Safe S. Effects of 2,3,7,8-tetrachlorodibenzo-p-dioxin on hepatic and uterine estrogen receptor levels in rats. *Toxicol Appl Pharmacol* 1987;87:306–314.
51. Romkes M, Safe S. Comparative activities of 2,3,7,8-tetrachlorodibenzo-p-dioxin and progesterone on antiestrogens in the female rat uterus. *Toxicol Appl Pharmacol* 1988;92:368–380.
52. Astroff B, Safe S. Comparative antiestrogenic activities of 2,3,7,8-tetrachlorodoibenzo-p-dioxin and 6-methyl-1,3,8-trichlorodibenzofuran in the female rat. *Toxicol Appl Pharmacol* 1988;95:435–443.
53. Astroff B, Rowlands C, Dickerson R, Safe S. 2,3,7,8-tetrachlorodibenzo-p-dioxin inhibition of 17 beta-estradiol-induced increases in rat uterine epidermal growth factoe receptor binding activity and gene expression. *Mol Cell Endocrinol* 1990;72:247–252.
54. Astroff B, Safe S. 2,3,7,8-Tetrachlorodibenzo-p-dioxin as an antiestrogen: effect on rat uterine peroxidase activity. *Biochem Pharmacol* 1990;39:485–488.
55. Astroff B, Eldridge B, Safe S. Inhibition of 17 beta-estradiol-induced and constitutive expression of the cellular protooncogene c-fos by 2,3,7,8-tetrachlorodibenzo-p-dioxin (TCDD) in the female rat uterus. *Toxicol Lett* 1991;56:315–315.
56. Gallo MA, Hesse EJ, MacDonald GJ, Umbreit TH. Interactive effects of estradiol and 2,3,7,8-tetrachlorodibenzo-p-dioxin on hepatic cytochrome P450 and mouse uterus. *Toxicol Lett* 1986;32: 123–132.
57. DeVito MJ, Thomas T, Martin E, Umbreit TH, Gallo MA. Antiestrogenic action of 2,3,7,8-tetrachlorodibenzo-p-dioxin: tissue specific regulation of estrogen receptor in CD1 mice. *Toxicol Appl Pharmacol* 1992;113:284–292.
58. Safe S. Polychlorinated biphenyls (PCBs), dibenzo-p-dioxins (PCDDs), dibenzofurans (PCDFs) and related compounds: environmental and mechanistic considerations which support the development of toxic equivalency factors (TEFs). *Crit Rev Toxicol* 1990;21:51–88.
59. Mason G, Sawyer T, Keys B, Bandiera S, Romkes M, Piskorska-Pliszczynska J, Zmudzka B, Safe S. Polychlorinated dibenzofurans (PCDFs): correlation between *in vivo* and *in vitro* structure activity relationships. *Toxicology* 1985;37:1–12.
60. Mason G, Farrell K, Keys B, Piskorska-Pliszczynska J, Safe L, Safe S. Polychlorinated dibenzo-p-dioxins: quantitative *in vitro* and *in vivo* structure-activity relationships. *Toxicology* 1986;41:21–31.
61. Dickerson R. Antiestrogenic and immunotoxic effects of polychlorinated dibenzo-p-dioxins and dibenzofurans: mechanistic studies. Ph.D. diss., Texas A&M University, 1992.
62. Jansen HT, Cooke PS, Porcelli J, Liu TC, Hansen LG. Estrogenic and antiestrogenic actions of PCBs in the female rat: *in vitro* and *in vivo* studies. *Reprod Toxicol* 1993;7:237–248.

63. Patnode KA, Curtis LR. 2,2′,4,4′,5,5′- and 3,3′,4,4′,5,5′-hexachloro-biphenyl alteration of uterine progesterone and estrogen receptors coincides with embryotoxicity in mink (Mustela vision). *Toxicol Appl Pharmacol* 1994;127:9–18.
64. White TEK, Rucci G, Liu Z, Gasiewicz TA. Weanling female Sprague-Dawley rats are not sensitive to the antiestrogenic effects of 2,3,7,8-tetrachlorodibenzo-p-dioxin (TCDD). *Toxicol Appl Pharmacol* 1995;133:13–320.
65. Dickerson R, Howie L, Safe S. The effect of 6-nitro-1,3,8-trichlorodibenzofuran as a partial estrogen in the female rat uterus. *Toxicol Appl Pharmacol* 1992;113:55–63.
66. Kociba RJ, Keyes DG, Beger JE, Carreon RM, WadeCE, Dittenber DA, Kalnins R, Frauson LE, Park CL, Barnard SD, Hummel RA, Humiston CG. Results of a 2-year chronic toxicity and oncogenicity study of 2,3,7,8-tetrachlorodibenzo-p-dioxin (TCDD) in rats. *Toxicol Appl Pharmacol* 1978;46: 279–303.
67. Holcomb M, Safe S. Inhibition of 7,12-dimethyl benzanthracene-induced rat mammary tumor growth by 2,3,7,8-tetrachlorodibenzo-p-dioxin. *Cancer Lett* 1994;82:43–47.
68. Gierthy JF, Bennett JA, Bradley LM, Cutler DS. Correlation of *in vitro* and *in vivo* growth suppression of MCF-7 human breast cancer by 2,3,7,8-tetrachlorodibenzo-p-dioxin. *Cancer Res* 1993;53: 3149–3153.
69. Bertazzi PA, Pesatori AC, Consonni, D, Tironi A, Landi MT, Zocchetti C. Cancer incidence in a population accidently exposed to 2,3,7,8-tetrachlorodibenzo-p-dioxin. *Epidemiology* 1993;4: 398–406.
70. Bjeldanes LF, Kim JY, Grose KR, Bartholemew CJ, Bradfield CA. Aromatic hydrocarbon responsiveness-receptor agonists generated from indole-3-carbinol *in vitro* and *in vivo*: comparisons with 2,3,7,8-tetrachlorodibenzo-p-dioxin. *Proc Natl Acad Sci USA* 1991;88:9543–9547.
71. Jellinick PH, Forkert PG, Riddick DS, Okey AB, Michnovicz JJ, Bradlow HL. Ah receptor binding properties of indole carbinols and induction of hepatic estradiol hydroxylation. *Biochem Pharmacol* 1993; 43:1129–1136.
72. Vang O, Jensen MB, Autrup H. Induction of cytochrome P4501A1 in rat colon and liver by indole-3-carbinol and 5,6-benzoflavone.*Carcinogenesis* 1990;11:1259–1263.
73. Gillner M, Bergman J, Cambillau C, Alexandersson M, Fernstron B, Gustafsson JA. Interactions of indolo[3,2-b] carbazoles and related polycyclic aromatic hydrocarbons with specific binding sites for 2,3,7,8-tetrachlorodibenzo-p-dioxin in rat liver. *Mol Pharmacol* 1987;44:336–345.
74. Stoewsand GS, Anderson JL, Munson L. Protective effect of dietary brusels sprouts against mammary carcinogenesis in Sprague-Dawley rats. *Cancer Lett* 1988;39:199–207.
75. Kojima T, Tanaka T, Mori H. Chemoprevention of spontaneous endometrial cancer in female Donryu rats by dietary indol-3-carbinol. *Cancer Res* 1994;54:1446–1449.
76. Grubbs CJ, Steele VE, Casebolt T, Juliana MM, Eto I, Whitaker LM, Dragneu KH, Kelloff GJ, Lubet RL. Chemoprevention of chemically-induced mammary carcinogenesis by indole-3-carbinol. *Anticancer Res* 1995;15:709–716.
77. Liu H, Wormke M, Safe S, Bjeldanes LF. Indole[3,2-b]carbazole: a dietary-derived factor that exhibits both antiestrogenic and estrogenic activity. *J Natl Cancer Inst* 1994;86:1758–1765.
78. Tiwari RK, Gou L, Bradlow HL, Telang NT, Osborn MP. Selective responsiveness of breast cancer cells to indole-3-carbinol, a chemopreventive agent. *J Natl Cancer Inst* 1994;86:126–131.
79. Biegel L, Safe S. Effects of 2,3,7,8-tetrachlorodibenzo-p-dioxin (TCDD) on cell growth and secretion of the estrogen-induced 34-, 52- and 160 kDa proteins in human breast cancer cells. *J Steroid Biochem Mol Biol* 1990;37:725–732.
80. Merchant M, Krishnan V, Safe S. Mechanism of action of β-naphthoflavone as an Ah receptor antagonist in MCF-7 human breast cancer cell. *Toxicol Appl Pharmacol* 1993;120:179–185.
81. Gierthy JF, Lincoln DW. Inhibition of postconfluent focus production in cultures of MCF-7 breast cancer cells by 2,3,7,8-tetrachlorodibenzo-p-dioxin. *Breast Cancer Res Treat* 1987;12:227–233.
82. Spink DC, Johnson JA, Conner SP, Aldous KMM Gierthy JF. Stimulation of 17β-estradiol metabolism in MCF-7 cells by bromochloro-and chloromethyl-substituted dibenzo-p-dioxins and dibenzofurans: correlations with antiestrogenic activity. *J Toxicol Environ Health* 1994;41:451–466.
83. Harper N, Wang X, Liu H, Safe S. Inhibition of estrogen-induced progesterone receptor in MCF-7 human breast cancer cells by aryl hydrocarbon (Ah) receptor agonists. *Mol Cell Endocrinol* 1994;104: 47–55.
84. Narasimhan TR, Safe S, Williams HJ, Scott AI. Effects of 2,3,7,8-tetrachlorodibenzo-p-dioxin on 17β-estradiol-induced glucose metabolism in MCF-7 human breast cancer cells: 13C nuclear magnetic resonance spectroscopy studies. *Molec Pharmacol* 1991;40:1029–1035.

85. Moore M, Narasimhan TR, Wang X, Krishnan V, Safe S, Williams HJ, Scott AI. Interaction of 2,3,7,8-tetrachlorodibenzo-p-dioxin, 12-O-tetra-decanoylphorbol-13-acetate (TPA) and 17β-estradiol in MCF-7 human breast cancer cells. *J Steroid Biochem Mol Biol* 1993;44:251–261.
86. Gierthy JF, Lincoln DW, Gillespie MB, Seeger JT, Martinez HL, Dickerson HW, Kumar SA. Suppression of estrogen-regulated extracellular plasminogen activator activity of MCF-7 cells by 2,3,7,8-tetrachlorodibenzo-p-dioxin. *Cancer Res* 1987;47:9198–6203.
87. Krishnan V, Narasimhan TR, Safe S. Development of gel staining techniques for detecting the secretion of procathepsin D (52 kDa protein) in MCF-7 human breast cancer cell. *Anal Biochem* 1992;204: 137–142.
88. Krishnan V, Safe S. Polychlorinated biphenyls (PCBs), dibenzo-p-dioxins (PCDDs) and dibenzofurans (PCDFs) as antiestrogens in MCF-7 human breast cancer cells: quantitative structure-activity relationships. *Toxicol Appl Pharmacol* 1993;120:55–61.
89. Porter W. Regulation of heat shock protein 27 gene expression. Ph.D. diss., Texas A&M University, 1997
90. Duan R, Porter W, Safe S. Inhibition of estrogen-induced c-fos protooncogene mRNA levels by 2,3,7,8-tetrachlorodibenzo-p-dioxin (TCDD) in MCF-7 human breast cancer cells *Toxicologist* 1995;15: 1253.
91. Schrope K, Porter W, Safe S. Effects of 2,3,7,8-tetrachlorodibenzo-p-dioxin on expression of insulin-like growth factor binding proteins in MCF-7 cells. *Toxicologist* 1995;15:1254.
92. Liu H, Biegel L, Narasimhan TR, Rowlands C, Safe S. Inhibition of insulin-like growth factor-I responses in MCF-7 cells bt 2,3,7,8-tetrachlorodibenzo-p-dioxin and related compounds. *Mol Cell Endocrinol* 1992;87:19–28.
93. Fernandez P, Safe SH. Growth inhibitory and antimitogenic activity of 2,3,7,8-tetrachlorodibenzo-p-dioxin (TCDD) in T47D human breast cancer cells. *Toxicol Lett* 1992;61:185–197.
94. Fernandez P, Burghardt R, Smith R, Nodland K, Safe S. High passage T47D human breast cancer cells: altered endocrine and 2,3,7,8-tetrachlorodibenzo-p-dioxin responsiveness. *Eur J Pharmacol* 1994;270:53–66.
95. Zacharewski T, Bondy K, McDonell P, Wu ZF. Antiestrogenic effects of the 2,3,7,8-tetrachlorodibenzo-p-dioxin or 17β-estradiol-induced pS2 expression. *Cancer Res* 1994;54:2707–2713.
96. Harris M, Zacharewski T, Safe S. Effects of 2,3,7,8-tetrachlorodibenzo-p-dioxin and related compounds on the occupied nuclear estrogen receptor in MCF-7 human breast cancer cell. *Cancer Res* 1990;50:3579–3584.
97. Chaloupka K, Krishnan V, Safe S. Polychlorinated aromatic hydrocarbon carcinogens as antiestrogens in MCF-7 human breast cancer cells: role of the Ah receptor. *Carcinogenesis* 1992;13:2233–2239.
98. Chaloupka K, Harper N, Krishman V, Santestefano M, Rodriguez LV, Safe S. Synergistic activity of aromatic hydrocarbon mixtures as aryl hydrocarbon (Ah) receptor agonists. *Chem Biol Interact* 1993;89:141–158.
99. Zacharewski T, Harris M, Biegel L, Morrison V, Merchant M, Safe S. 6-methyl-1,3,8-trichlorodibenzofuran (MCDF) as an antiestrogen in human and rodent cancer cell lines: evidence for the role of the Ah receptor. *Toxicol Appl Pharmacol* 1992;113:311–318.
100. Merchant M, Morrison V, Santostefano M, Safe S. Mechanism of action of aryl hydrocarbon receptor antagonists: inhibition of 2,3,7,8-tetrachlorodibenzo-p-dioxin-induced CYP1A1 gene expression. *Arch Biochem Biophys* 1992;298:389–394.
101. Zacharewski T, Harris M, Safe S. Evidence for a possible mechanism of action of the 2,3,7,8-tertachlorodibenzo-p-dioxin-mediated decrease of nuclear estrogen receptor levels in wild type and mutant Hepa 1c1c7 cells. it Biochem Pharmacol 1991;41:1931–1939.
102. Moore M, Wang X, Lu YF, Wormke M, Craig A, Gerlach JH, Burghardt R, Barhoumi R, Safe S. Benzo[a]pyrene-resistant MCF-7 human breast cancer cells: a unique aryl hydrocarbon-nonresponsive clone. *J Biol Chem* 1994;269:11751–11759.
103. Spink DC, Lincoln DW, Dickerman HW, Gierthy JF. 2,3,7,8-tetrachlorodibenzo-p-dioxin causes an extensive alteration of 17β-estradiol metabolism in MCF-7 breast tumor cells. *Proc Natl Acad Sci USA* 1990;87:6917–6921.
104. Spink DC, Hayes CL, Young NR, Christou M, Sutter TR, Jefcoate CR, Gierthy JF. The effects of 2,3,7,8-tetrachlorodibenzo-p-dioxin on estrogen metabolism in MCF-7 breast cancer cells: evidence for induction of a novel 17β-estradiol 4-hydroxylase. *J Steroid Biochem Molec Biol* 1994;51:251–258.
105. Spink DC, Eugster HP, Lincoln DW, Schuetz JD, Schuetz EG, Johnson JA, Kaminsky L, Gierthy JF. 17β-estradiol hydroxylation catalyzed by human cytochrome P4501A1: a comparison of the

activities induced by 2,3,7,8-tetrachlorodibenzo-p-dioxin in MCF-7 cells with those from heterologous expression of the cDNA. *Arch Bioch Biophys* 1992;293:342–348.
106. Shiverick KT, Muther TF. Effects of 2,3,7,8-tetrachlorodibenzo-p-dioxin on serum concentrations and the uterotrophic action of exogenous estrone in rats. *Toxicol Appl Pharmacol* 1982;65:170–176.
107. Krishnan V, Wang X, Ramamurthy P, Safe S. Effect of 2,3,7,8-tetrachlorodibenzo-p-dioxin (TCDD) in formation of estrogen-induced ER/SP1 complexes on the cathepsin D promoter. *Toxicologist* 1994;14:47.
108. Krishnan V, Porter W, Santostefano M, Wang X, Safe S. Molecular mechanism of inhibition of estrogen-induced cathepsin D gene expression by 2,3,7,8-tetrachlorodibenzo-p-dioxin (TCDD) in MCF-7 cells. *Mol Cell Biol* 1995;15:6710–6719.
109. Porter W, Duan R, Safe S. 2,3,7,8-tetrachlorodibenzo-p-dioxin (TCDD) as an antiestrogen in MCF-7 human breast cancer cells: inhibition of estrogen-induced heat shock protein 27 gene expression. *Toxicologist* 1995;15:1261.
110. Astroff B, Safe S. 6-substituted-1,3,8-trichlorodibenzofurans as 2,3,7,8-tetrachlorodibenzo-p-dioxin antagonists in the rat: structure activity relationships. *Toxicology* 1989;59:285–296.
111. Astroff B, Safe S. 6-alkyl-1,3,8-trichlorodibenzo furans as antiestrogens in female Sprague-Dawley rats. *Toxicology* 1991;69:187–197.

Toxicant–Receptor Interactions
Edited by Michael S. Denison and William G. Helferich

4

Interactions Between Environmental Xenobiotics and Estrogen Receptor–Mediated Responses

Daniel L. Villeneuve, Alan L. Blankenship, and John P. Giesy

Department of Zoology, National Food Safety and Toxicology Center and Institute for Environmental Toxicology, Michigan State University, East Lansing, Michigan, USA

INTRODUCTION

Environmental Estrogens

Endocrine "disruption" by environmental contaminants has become a cause for concern among scientists, environmental advocates, and politicians (1–3). A number of compounds released into the environment by human activities can modulate endogenous hormone activities and have been termed *endocrine-disrupting compounds* (EDCs) (3–5). It has been hypothesized that such compounds may elicit a variety of adverse effects in both humans and wildlife, including promotion of hormone-dependent cancers, reproductive-tract disorders, and a reduction in reproductive fitness (6).

The neuroendocrine system is the primary mechanism by which organisms maintain homeostasis (7). Thus, generalized adaptations to stress, which allow organisms to resist perturbations from normal homeostatic ranges, typically involve a variety of endocrine and physiologic responses (8). As a result, any stressor can be loosely defined as an "endocrine disruptor." Furthermore, there are a number of receptor-mediated hormonal responses to toxicity. These include xenobiotic effects on thyroid hormone receptor– (9), epidermal growth-factor (EGF) receptor– (10), aryl hydrocarbon receptor (AhR)–, and estrogen and androgen receptor– (ER, AR) mediated mechanisms. In this chapter we restrict our discussion to direct-acting estrogenic and antiestrogenic compounds. These are defined as those compounds that bind competitively to the ER and cause or inhibit estrogen-like responses in vitro or in vivo. Although our discussion focuses on estrogen agonists and antagonists, compounds that can cause tissue-level responses without ER binding are also covered.

There are a wide variety of compounds in the environment that have been shown to bind to the ER and function as either agonists or antagonists. These include both natural products (11–14) and synthetic compounds (3, 4). The synthetic compounds include both chlorinated and nonchlorinated compounds (3, 4). Some act as xenoestrogens, which either mimic or antagonize the effects of endogenous estrogen (Table 1). Others act as androgens in the case of tributyltin or as antiandrogens in the case of the fungicide vinclozolin (15) and 2,2-bis(*p*-chlorophenyl)-1,1 dichloroethane (p,p'-DDE) (16). Some compounds such as 2,3,7,8-tetrachloro-dibenzo-p-dioxin (TCDD) can modulate a number of hormone functions, by acting as both antiestrogens and antiandrogens (17) as well as affecting thyroid-hormone function (18, 19), EGF (10), insulin/insulin-like growth factor I (20), transforming growth factor α (21), or other signal transduction pathways including protein kinases (22–26). Thus, a wide

TABLE 1. *Compounds implicated as xenoestrogens or antiestrogens*

Compounds	References
Organochlorine pesticides	
DDT/DDE	32, 33
toxaphene	34
dieldrin	34
chlorodecone	35
Polychlorinated biphenyls	36–38
Polycyclic aromatic hydrocarbons	39
Polychlorinated dibenzo dioxins	40, 41
Plasticizers (e.g., bisphenol A)	42
Phthalates	43, 44
Surfactants (e.g., alkylphenol ethoxylates and alkylphenols)	44, 45
Synthetic steroids (e.g., DES; ethynyl estradiol)	47
Phytoestrogens (e.g., genistein, coumestrol)	11–14

variety of types of xenobiotics can exert endocrine-modulating effects through many different mechanisms including altered steroid-receptor function, estrogen–androgen ratios, and changes in concentrations of hormones in specific tissues.

Of all the endocrine-modulating compounds, those that are direct-acting ER agonists (xenoestrogens or estrogen mimics), direct-acting ER antagonists (antiestrogens), or androgen antagonists have received the greatest attention (27, 28). This is due to their importance in embryonic development (29). It also reflects the fact that some xenobiotics that have been released into the environment seem to act through mechanisms that affect the expression of sex steroid hormones (30, 31). Examples of some compounds that have received attention as potential xenoestrogens or antiestrogens are presented in Table 1. The "estrogen hypothesis" stemmed from the observation that some of the effects observed in oviparous wildlife exposed to persistent and bioaccumulative chemicals were similar to those that could be caused by injecting estrogen into eggs. This hypothesis was supported by the fact that naturally occurring exoestrogens, such as phytoestrogens, could cause reproductive dysfunction in animals (47, 48). Further support came from the observation that some xenobiotics that can bind the ER were weak estrogen agonists or antagonists in in-vitro expression assays (6, 27).

Estrogen agonists are compounds that mimic the effects of estrogen. The classic definition of an estrogen is a compound that produces changes in an estrogen-responsive tissue, such as cornification of the vaginal epithelium of rodents (49). Other physiologic endpoints have been used to define estrogenicity, including increased uterine weight, uterine glycogen content, protein induction (50), and cellular proliferation (51). It has been recognized that numerous synthetic as well as naturally occurring compounds fit the classic definition of an estrogen (52). A refined definition of an estrogen, recognizing the role of nuclear hormone receptors in gene expression, is a compound that binds to the estrogen receptor and induces dimerization of the receptors that specifically bind to and activate transcription of genes under the regulatory control of cis-acting estrogen responsive elements (53).

Estrogen antagonists block the action of estrogens by interfering with the normal functioning of the estrogen receptor. Antiestrogenic compounds act by several mechanisms, not necessarily related to estrogen-receptor binding and activation. Classical estrogen antagonists such as ICI 164,384 and tamoxifen bind competitively to the estrogen receptor, displacing the natural ligand E_2 and blocking or reducing the effectiveness of the ligand-bound receptor to enhance gene expression (54–56). AhR agonists, such as 2,3,7,8-TCDD and non-*ortho* chlorinated polychlorinated biphenyls (PCB_s), cause downregulation of the ER and may also interfere with DNA binding (6). Aromatase inhibitors, such as aminoglutethimide, block conversion of testosterone to E_2 and are used in the treatment of metastatic estrogen-dependent breast cancer (57). Inducers of phase I and phase II metabolic detoxification enzymes, such as 2,3,7,8-TCDD and non-*ortho* chlorinated PCBs, reduce the estrogen-dependent expression of proteins by enhancing the enzymatic conversion of latent E_2, thus eliciting an antiestrogenic response (37).

Another important factor in determining a compound's ability to modulate ER function is bioavailability. Three of the most important characteristics for determining bioavailability of ER ligands are lipid solubility, biologic half-life, and amount of protein binding. Comparisons of these properties for natural estrogens, synthetic estrogens, phytoestrogens, and *o,p′*-DDT are presented in Table 2. Some researchers have suggested that xenoestrogens have greater bioavailability than E_2, mostly because of differences in protein binding and biological half-lives (66, 67). It is important to note, however, that this hypothesis has yet to be tested in vivo.

It has been suggested that relative to E_2, *o,p′*-DDT, a weak estrogen agonist, may be more active in vivo because E_2 binds to steroid hormone–binding globulins (SHBGs). Because it does not bind to SHBGs, *o,p′*-DDT is thought to be free in plasma and thus more bioavailable. Except in the third trimester of pregnancy, however, less than 40% of E_2 is bound to SHBGs (see Table 2). Most E_2 is bound to albumin and other serum proteins, with only about 2% being available as free E_2. Although exact values for *o,p′*-DDT distribution in blood are not available, there is a high capacity for binding to serum proteins (48). Furthermore, DDT is more lipophilic than E_2. The octanol–water partitioning coefficient (K_{ow}) for DDT is over 300,000 times that of E_2 (58, 61). Thus, DDT is likely to be bound as tightly as E_2 and, in fact, may be less available. Assuming differences on the order of two orders of magnitude are significant for risk assessment, efforts should be made to determine the binding characteristics of xenobiotics relative to E_2.

Description of the ER

The ER is an intracellular receptor belonging to the steroid hormone–receptor superfamily (68, 69). It functions as a ligand-activated transcription factor mediating the effects of estrogens, which regulate the growth, differentiation, and functioning of diverse target tissues. To date, two isoforms, ER-α and ER-β, have been described (69, 70). ER-α has been well characterized and is a highly-conserved protein of approximately 66 kDa (69). ER-β has been described only recently and has a calculated molecular weight of 54.2 kDa (70). The ligand–binding specificities are similar between the two isoforms; however, there are differences in the distribution and relative binding affinities that could contribute to selective actions of ER agonists and antagonists in different tissues (70).

Mechanism of Action for ER Agonists or Antagonists

A simplified mechanism of action for activation of the ER is shown in Fig. 1. First, an ER ligand must enter the cell and bind to the ER. The non–ligand bound form of the ER is predominantly localized to the nucleus (71) and is part of a complex with associated proteins (72). On ligand binding, the associated proteins dissociate, and then ligand–receptor complexes associate with additional nuclear factors (73) and bind to estrogen-responsive elements (EREs) as homodimers. Binding of the transformed complex to the ERE results in production of mRNA for a number of estrogen-responsive proteins, such as pS2, and cathepsin D (6).

TABLE 2. *Comparison of factors affecting bioavailability of estrogenic compounds*

Compound	Lipid solubility (log K_{ow})	Biological half-life (h)		Amount of protein binding		Amount unbound (%)
				Steroid hormone binding globulin	Albumin and other serum proteins	
17-β-Estradiol (E_2)	1.99 (Ref. 58)	13.5 (Ref. 59)	Men (Ref. 61)	19.6%	78.0%	2.32
			Women (Ref. 61)			
			Follicular	37.3%	60.8%	1.81
			Luteal	37.0%	61.1%	1.82
			3rd trimester	87.8%	11.7%	0.49
Diethylstibestrol	—	24 (Ref. 61)		Slight (Ref. 60)	10- to 20-fold greater affinity than E_2; high-capacity low-affinity binding (Ref. 48)	—
Genistein	—	<12 (Ref. 62)		27% relative binding affinity compared with E_2 (Ref. 63)	—	—
Technical DDT	7.48 (Ref. 64)	—		Unknown	High-capacity, low-affinity binding (Ref. 65)	—

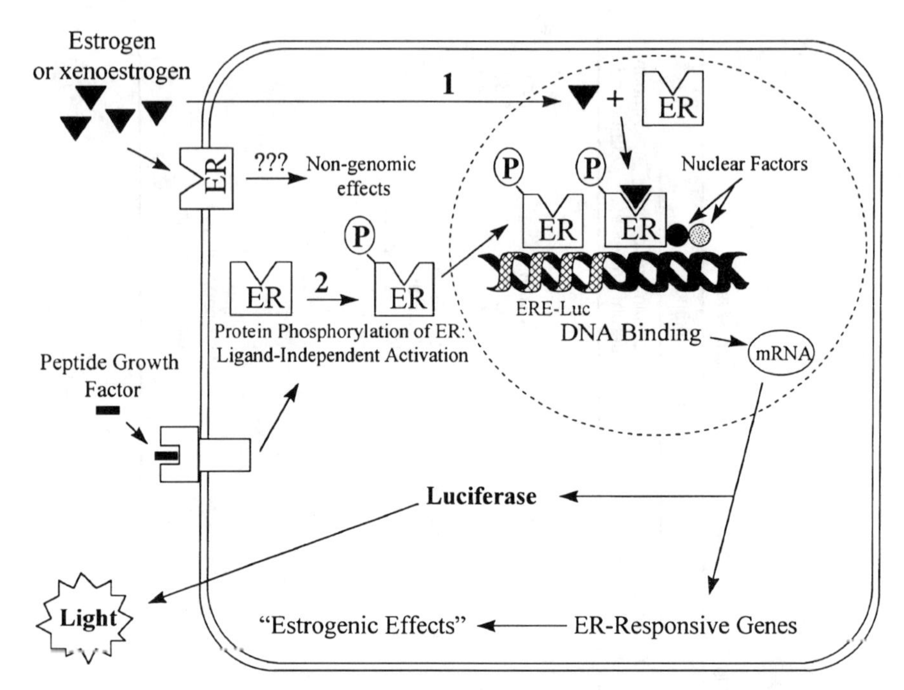

FIG. 1. Model depicting mechanisms for activation of the estrogen receptor (ER) in an MCF-7 cell line stably transfected with an ER-controlled luciferase reporter gene construct (MCF-7-luc). Refer to the text for a full description. In brief, pathway 1 depicts an ER agonist entering a cell and interacting with an intracellular ER, which then binds to estrogen-responsive elements (EREs) in the promoter region of ER-responsive genes. Pathway 2 depicts ligand-independent activation of the ER through a protein phosphorylation pathway.

In addition to endogenous responses, exogenous reporter systems, such as firefly luciferase, under control of EREs can be stably transfected into cells such that exposure to estrogens, luciferase expression is induced (74). Luciferase or other such reporter systems provide a rapid and convenient means for receptor-mediated dose–response assessment as well as allowing for mechanistic investigations into transcriptional activation.

There is considerable evidence for hyperphosphorylation of the ER after ligand binding, supporting the hypothesis that phosphorylation is an important regulatory mechanism for ER function. Phosphorylation of key serines (75–77) and at least one tyrosine residue of the ER, Tyr527 (78), play an important role in regulating the transcriptional activity of the ER. Phosphorylation increases the negative charge and acidity of a region of a protein, thereby modifying interactions with other proteins and DNA. Hypo- and hyperphosphorylation at the same time in different regions of the ER could explain differential transcriptional regulation of certain genes, in addition to tissue- and cell-specific regulatory mechanisms (79). In addition, differential phosphorylation of other nuclear factors with which the ER interacts to mediate transcription potentially plays a significant role (80).

TABLE 3. *Relative binding affinity (RBA) for calf estrogen receptor and luciferase expression potency and efficacy in MCF-7-luc cells*

		MCF-7-luc	
Compound	RBA IC_{50} (nM)	Potency (ED_{50}, nM)	Efficacy (fold induction relative to solvent controls)
17β-Estradiol	6.9	0.006	4.5
Coumestrol	24	2500	2.8
17β-Ethynyl-estradiol	37.3	0.055	4.5
Nonyl-phenol	2,300	480	1.9
Octyl-phenol	12,300	320	2.3
Bisphenol A	24,300	3,640	4.3
Indole-3-carbinol	2.3×10^{10}	21,100	2.3
Atrazine	$>2.3 \times 10^{10}$	260,000	2.8
o,p′-DDE	$>2.3 \times 10^{10}$	NE	—
p,p′-DDE	$>2.3 \times 10^{10}$	NE	—
2,3,7,8-TCDD	$>2.3 \times 10^{10}$	NE	—

IC_{50}-concentration of competitor to reduce total binding control counts by 50%; ED_{50}-dose to elicit 50% of maximal effect. NE, potency estimate was not possible based on the dose-response obtained.

There are also ligand-independent pathways for modulating the transcriptional activity of the ER through activation of protein kinases (see Fig. 1, pathway #2). For example, on treatment with growth factors, such as EGF, insulin-like growth factor 1, and platelet-derived growth factor, or agents that increase cAMP levels, the ER can be phosphorylated, bind to EREs, and activate transcription in the absence of ligand (81–83). The fact that the pure antiestrogen ICI 164,384 blocks the effects of these agents supports an ER-mediated mechanism (83). Likewise, protein-kinase inhibitors block the effects of these agents and E_2 (69). ICI 164,384 also stimulates phosphorylation of the ER without a similar increase in transcriptional activation. This indicates that overall phosphorylation does not necessarily result in increased transcriptional activity (69), although this could be due to impaired receptor dimerization by ICI 164,384 (55). Given the fact that some xenobiotics can modulate protein phosphorylation (26, 84, 85), this ligand-independent mechanism for modulating ER function has a potentially important impact on screening for chemicals with estrogenic activity, which is discussed later, especially if the emphasis is placed on receptor binding alone as a first-tier screen. For example, a compound like 2,3,7,8-TCDD, which is a potent antiestrogen (86), would be missed in an initial screen because it does not bind to the ER (87). Similarly, the triazine herbicide atrazine, which does not bind to the ER, can cause estrogen-like responses in in-vitro expression assays (Table 3). Neither of these in-vivo or in-vitro effects would be predicted from receptor binding assays.

Rapid, Nongenomic Effects of Estrogens

In addition to the mechanisms described previously, estrogens produce rapid (within seconds to minutes), nontranscriptional responses that are similar to those evoked by growth factors. Recent evidence suggests the existence of a membrane ER (88) that may play a role in these rapid, nongenomic effects of estrogens, which include prolactin release in GH3/B6 rat pituitary tumor cells (88), intracellular calcium release in chicken granuloma cells (89), stimulation of protein tyrosine phosphorylation in MCF-7 mammary carcinoma cells (90), activation of the $p21^{ras}$/mitogen-activated protein (MAP) kinase pathway in MCF-7 cells (91), and stimulation of adenylate cyclase and cAMP-regulated gene transcription (75).

Interactions and Cross-Talk from Other Signaling Pathways

Cross-talk between the estrogen receptor and other signaling pathways provides important mechanisms for modulating biologic responses. Interactions and cross-talk with the ER have been described for the progesterone receptor (69), aryl hydrocarbon receptor (92), EGF receptor (81), insulin-like growth factor I (69), and, as discussed above, pathways involving agents affecting protein-kinase activities, particularly those affecting protein kinase C, cAMP-dependent protein kinase, and tyrosine kinases (81, 83).

As interest in monitoring the environment for environmental estrogens increases, it is likely that more than one mechanism will need to be assessed. Some environmental mixtures contain compounds that modulate the responsiveness of multiple receptor-mediated pathways, such as the AhR, estrogen receptor, androgen receptor, and EGF receptor. Thus, the complexities of environmental mixtures require innovative methods and approaches to assess the potential for adverse effects.

SCREENING AND MONITORING

Concern over xenoestrogens has created a need to both screen and monitor for compounds which can modulate endocrine effects. This need is underscored by recent legislation mandating that chemicals and formulations be screened a priori for potential estrogenic activity before they are manufactured or used in certain processes (Safe Drinking Water Act Amendments of 1995—Bill Number S.1316; Food Quality Protection Act of 1996—Bill Number P.L. 104-170). Monitoring a posteriori is needed in order to assess concentrations of estrogenic compounds in both abiotic matrices like soil, sediments, and water, and biotic matrices such as human and wildlife tissues and food stuffs (93, 94).

Both biologic and instrumental methods can be applied to monitor or screen for estrogenic compounds. Instrumental methods are useful for measuring the uptake, disposition, and concentrations of specific compounds (monitoring). They also can be applied to help discern metabolic pathways. They are generally not useful, however, for discerning biological efficacy (screening). A variety of bioassays, both in vitro and in vivo, can be used to screen individual compounds, formulations, or environmental samples for potential estrogenic or antiestrogenic activity. Models may also be used to predict the potential estrogenicity based on the physicochemical properties of the compound of interest.

A number of techniques for screening and monitoring for the effects of xenoestrogens are discussed here, with most attention being focused on bioassay methods. The advantages and disadvantages of each technique are discussed. Although this discussion is focused on assays used to determine the estrogenicity or antiestrogenicity of a given compound or sample, the same types of techniques can be applied to other receptor-mediated processes, as long as the mechanism of action is known.

Predictive Quantitative Structure–Activity Relationships

The ideal screening tool is one that is rapid, inexpensive, objective, applicable for a wide range of compounds acting through a given mechanism of action, and capable of accurately predicting a compound's potential to elicit adverse effects in vivo. One of the most attractive screening tools is computer modeling, based on quantitative structure–activity relationships (QSARs). Once a suitable computer model is constructed, compounds can potentially be screened for possible activity in less time than it takes to conduct even the most simple in-vitro bioassays, and at a fraction of the cost. For this reason, development of an accurate QSAR model is certain to be a goal of any large-scale screening program. Some of the disadvantages of such models are the difficulty in collecting necessary input parameters that are representative of all potential ER ligands and the reliance on receptor-binding assays in many of these models rather than looking at ER function (to be discussed in more detail in following sections).

Structure–activity relationships for estrogenic compounds have been studied extensively (50, 51, 95, 96). Correlations have been made between certain structural features and both affinity for the ER, and expression of estrogen-modulated genes (50, 95, 96). Evidence of stereochemical recognition and stereospecific modulation of gene expression has been reported (50). Although estrogenic substances vary widely in structure, some common characteristics of most estrogens include a sterically unhindered phenol group and a hydrophobic substituent of greater than three carbons bonded *para* to the phenolic hydroxyl (56, 97). The aromatic A ring with its free hydroxyl group has been cited as a key for affinity of endogenous steroid hormones to the estrogen receptor (95, 96). Structure–activity relationships identified have been based on both simple structural features and more advanced computer-based models such as the electrostatic models used by VanderKuur et al. (96), comparative molecular field analysis (CoMFA) used by Waller et al. (98, 99), and computer-graphic and energy-based models for fit into DNA used by Hendry et al. (100, 101).

QSAR-based computer models for predicting estrogenic or androgenic potential have been developed. Two of the most promising models are discussed here. Models based on CoMFA and three-dimensional (3D) QSAR for predicting ER or AR binding affinity have been proposed (98, 99). These models consider the overall steric and electrostatic properties of the compound of interest. Empirically derived ER-binding affinity for seven classes of potentially estrogenic compounds, both natural and synthetic, were used for the construction and validation of the model for ER-binding affinity (98). Average errors of less than 2 log units for predicting empirically derived ER or androgen-receptor affinity based on the CoMFA/3D-QSAR model have been reported (98, 99). Such a range of error may or may not be within the range of certainty required by a tier I screening tool.

Another QSAR approach has been proposed by Hendry et al. (100, 101). This approach is based on the hypothesis that ligand fit into DNA (or DNA complementarity) is an important property affecting hormone-like activity in vivo (100, 101). The model developed uses both 3D computer graphics and energy (electrostatic and van der Waals) calculations to estimate fit into DNA (100, 101). Degree of fit into DNA has been found to correlate with in-vivo responses to estrogens such as uterotropic activity (101) and in-vitro responses like proliferation of MCF-7 human breast-cancer cells (100). Hendry et al. found that, without exception, compounds that fit into DNA better than E_2 were more active in vivo than E_2; those that fit more poorly exhibited less uterotropic activity in vivo (101).

Both of the models discussed appear to hold promise as potential tools for screening compounds for potential estrogenic or endocrine-modulating activity. QSAR approaches could be used in a tiered screening system for direct-acting estrogenic compounds acting as agonists or antagonists. QSAR models based on receptor-binding affinity alone have little or no utility in identifying estrogenic compounds acting through non–ER mediated mechanisms. Furthermore, additional validation is necessary before QSAR techniques can be applied with certainty. Both models need to be validated with a larger set of compounds. Additional in-vitro and in-vivo studies are needed to establish relationships between predicted receptor or DNA fit and relevant in-vivo endpoints in multiple species. Finally, if they are to be applied, such models should be used as a tier I tool to narrow the range of compounds to be examined more thoroughly using in-vitro and in-vivo bioassays. Regardless of their utility as screening tools, QSAR approaches are useful for an understanding of mechanisms and scaling doses in experiments.

In-vitro Assays

There are a number of types of in-vitro assays available for measuring the estrogenic activity of single compounds or complex mixtures (102). The range of responses includes everything from simple receptor binding to expression of endogenous or exogenous genes to cell proliferation and differentiation (52). In-vitro systems are attractive as screening tools because they are rapid, inexpensive, and fairly reproducible. For these reasons precise estimates of the relative potency of a great many samples or compounds can be obtained in a short period of time. One of the greatest utilities of in-vitro assays is for studying mechanisms of action of compounds or assessing the potential for synergisms among direct-acting estrogen agonists. Using simple in-vitro cell systems it often is possible to determine the mechanism of action of a compound. Knowledge of a compound's mechanism of action can reveal ways to monitor using in-vivo biomarkers.

In-vitro assays are limited by the fact that they do not completely represent the in vivo situation. Pharmacokinetics, biotransformation, and binding to carrier proteins may not be accurately represented by in-vitro systems. In addition, some xenoestrogens may be activated or deactivated by enzymatic conversion during metabolism, conjugation, or excretion. Although there are methods to adjust for and minimize some of these limitations, they must be considered in applying the results of in-vitro screening tests.

Receptor-Binding Assays

For direct-acting estrogen agonists or antagonists to exert an effect, it is necessary for them to bind to the ER (103). The affinity with which such compounds bind to the estrogen receptor may be related to the potency of the compound relative to endogenous estrogen. For this reason, receptor-binding assays have been used as a method to screen for potential estrogenic compounds (16, 44). It is important to note, however, that although binding of the ligand to the ER is necessary, it alone may not be sufficient for eliciting estrogenic responses in tissues.

Receptor-binding assays are conducted as competitive binding assays in which a compound with unknown binding affinity is allowed to compete for ER-binding sites with a labeled standard of known affinity (Fig. 2). Operationally, a known amount of the compound of interest is added at varying concentrations to a fixed amount of receptor and competitor. The compound of interest is allowed to compete with radiolabeled competitor for a fixed number of binding sites (see Fig. 2a). The receptor–ligand

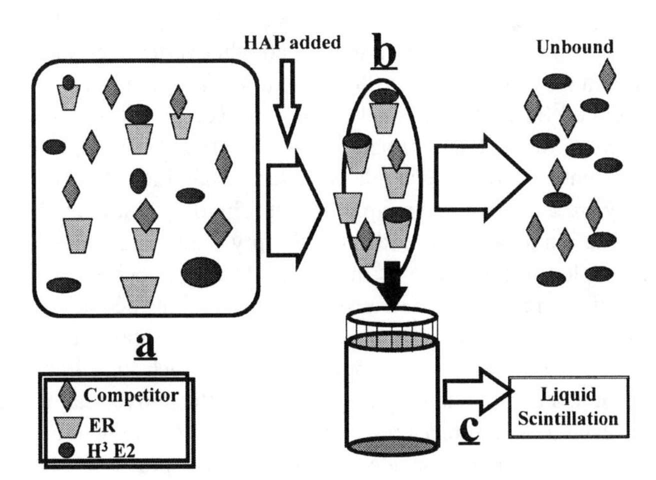

FIG. 2. Competitive receptor binding assay. *a*) Competitor (compound being tested) competes with tritiated 17-β-estradiol ($^3HE_2$) for binding to calf uterine estrogen receptor (ER) in a test tube. *b*) Hydroxyapatite (HAP) is added to complex with proteins (and bound ligands). Proteins with their bound ligands are trapped on filter paper while unbound ligand passes through. *c*) Filter paper transferred to scintillation vial and counted. Successful competitors cause reduction in counts relative to a control using $^3HE_2$ only.

complexes are separated from suspension by filtration (see Fig. 2b) or centrifugation, and the amount of activity is determined via liquid scintillation (see Fig. 2c) (104). Compounds with affinity for the ER will "displace" or compete for more of the available binding sites, thereby decreasing the radioactivity of the receptor–ligand complex fraction accordingly. Scatchard (105) or Woolf (106) plots can be used to determine the concentration of compound required to displace 50% of the endogenous compound (hormone) or its synthetic analogue. Relative binding affinities can be derived for compounds of interest. These values, generally reported as the effective concentration to reduce the binding of the labeled competitor by 50% (IC_{50}), are measures of the relative potency of the compound. For instance, if the concentration of a compound required to compete for 50% of the binding sites is 1000 times (on a molar basis) greater than the endogenous or synthetic reference compound, it has a relative potency 1000-fold less than the reference compound. In theory, this information can be used to predict concentrations of competing compounds in organisms that would be required to cause a given level of effect. Several competitors have been used in receptor-binding assays. These include endogenous ligands such as E_2 (14, 63), in the case of the ER. Alternatively, synthetic competitors such as ethinylestradiol (EE_2) or diethystilbestrol (DES) can be used. Synthetic competitors are useful because they often are more stable and are not as likely to bind nonspecifically to endogenous nonreceptor proteins.

Receptor-binding assays are attractive because they are simple and inexpensive to conduct. The only materials necessary are a preparation containing the receptor of interest and an agonist of known binding affinity. In the case of the ER the receptor is similar among species (107). Thus, results obtained with a preparation from one vertebrate species can generally be extrapolated to other species. This may not be true for all receptor-mediated responses, however, ER preparations from estrogen-responsive tissues of rodents, calves, humans, and other vertebrates such as fish have been used in binding assays (44, 47, 108). In addition, receptor does not necessarily need to be harvested from animals or their tissues. Receptor also can be harvested from in-vitro cell cultures. Transformed yeast cells (109, 110) and transfected cell lines (14) have been used to produce relatively large quantities of uniform receptor.

Although they are attractive and simple, receptor-binding assays are limited in the amount of information they can provide. The affinity of binding to a receptor is only one step in a complex series of events that occur during endocrine modulation (see Fig. 1). In addition, there needs to be transformation and translocation of the receptor followed by binding to accessory ligands or proteins and subsequent transcription (see Fig. 1). Receptor-binding assays cannot account for such steps in the expression process. In addition, a compound may be present in sufficiently great concentrations to cause effects even though it has a relatively weak affinity for the receptor. In-vitro receptor-binding assays do not consider pharmacokinetic processes that are important determinants of exposure in vivo. Enzymatic modification, differential turnover rates, and binding of endogenous and exogenous ligands are often different in the in-vitro and in-vivo situation. Finally, from binding affinity one can not infer whether a compound will be an agonist or antagonist. One can only identify the potential for modulation of the receptor-mediated process.

Relationships Between In-vitro Receptor Binding and In-vivo Assays

For simple in-vitro bioassays such as receptor binding to be useful for screening, it is necessary for the responses to be correlated to physiologically relevant responses in vivo. Risk assessment based on receptor-binding affinity would be meaningless without such correlation. Thus, the use of receptor binding affinity as a predictive tool is predicated on the assumption that endocrine modulation proceeds through a receptor-mediated process (111). There may be, however, alternative signal transduction pathways. Here, the relationship between receptor-binding affinity and in vitro gene expression is discussed.

In an attempt to elucidate the potential for receptor-binding assays to predict effects in vivo, affinity for binding to calf uterine ER was compared with responses in a simple in-vitro gene expression assay (discussed in more detail later in this chapter). The expression assay used MCF-7-luc cells, which are MCF-7 human breast tumor cells (ATCC #HTB-22) stably transfected with a DNA construct that includes an exogenous reporter gene, luciferase, under control of EREs and a mouse mammary tumor virus (mmtv) promoter (74). The comparison was made for a range of potentially estrogenic compounds including endogenous estrogen, synthetic estrogen, xenoestrogens, and a natural product (see Table 3). The hypothesis that receptor-binding affinity can be used to predict ERE-mediated gene expression in vitro was tested, to gain insight regarding the potential predictive power of receptor binding assays. Results indicated that relative binding affinity (RBA) for calf uterine ER was correlated with the potency for expression in the MVLN bioassay ($r^2 = 0.711$; see Table 3). RBA was not very predective, however, of efficacy or magnitude of expression, ($r^2 = -0.356$). Furthermore, there was no relationship between potency and efficacy among the compounds studied using the expression assay ($r^2 = -0.165$). Because the response of an organism is a complex interaction between both potency and efficacy, these results suggest that screening of compounds or mixtures with a simple receptor binding assay is not very predictive of other in-vitro or in-vivo responses.

Results from this in-vitro study correspond with reports in the literature that suggest a lack of correlation between RBA and ligand activity in vivo (112, 113). It has been stated that an interaction with the ER may not necessarily be an absolute, or sole requirement for the expression of estrogenic activity (114). It also has been reported that there is little relationship between receptor binding of E_2 analogues and either the character or extent of response (95). Similarly, a poor relationship between receptor binding and biologic activity of nonsteroidal DES analogues has been found. Researchers have concluded that ligand structure may be important in processes other than binding to and activating the ER (50). Alkaline degradation products of some steroids are potent estrogens, even though they do not bind with great affinity to the ER (115). One of the most active estrogen analogues developed has a receptor-binding affinity of less than 1% of that of E_2 (116). The triazine herbicide atrazine was found to have no measurable affinity for the ER, yet caused a significant response in the MCF-7-luc assay (see Table 3). Ligand-independent activation of the ER by growth-factor signaling and protein-kinase activation can occur (69). Some potent antiestrogens bind the ER poorly or not at all (117, 118). In some cases there is a need for metabolic activation of compounds, and thus receptor-binding affinity of the parent compound may not relate to in-vivo activity (112, 119).

In-vitro and in-vivo estrogenicity of *o,p′*-DDT, *o,p′*-DDE, and chlordecone, based on vitellogenesis in fish, has been compared (33). Although both in-vitro (ER-binding affinity) and in-vivo (plasma vitellogenin) assays indicated that these compounds were weakly estrogenic, the potencies of the substances, based on the two assays, were not correlated. ER-binding studies indicated that *o,p′*-DDT and *o,p′*-DDE were less potent competitors for moxestrol (synthetic estrogen) than was chlordecone. In-vivo studies, however, based on plasma vitellogenin normalized for tissue residue levels, indicated that *o,p′*-DDT and *o,p′*-DDE were more potent than chlordecone.

Together, the results of our studies and reports in the literature suggest that the assumption on which application of receptor-binding assays is based may not be valid. The information available suggests that ER binding may be a poor predictor of more complex in-vitro and in-vivo responses. One reason for the discrepancy between receptor binding and hormonal activity in vitro could be the multiple steps involved in expression beyond receptor binding. These include both transcriptional and post-transcriptional events. Lack of concordance between receptor-binding assays and in-vivo responses may also be due to nongenomic mechanisms or influences through other pathways, which were discussed earlier in this chapter. Slight differences in receptors or receptor expression among different species and tissues (79, 120, 121) also must be considered a potential source of the discrepancy. Although the structure of the ER is well conserved among species there are species-specific, tissue-specific, and even temporal differences in the affinity to estrogen agonists (69, 121).

One alternative working hypothesis that could explain the lack of correlation between binding affinity and response is the "ligand insertion hypothesis" (122). This hypothesis states that on binding to both the ligand and the DNA (at the ERE), the receptor shifts conformation in a way that facilitates insertion and release of the ligand into the DNA, where it acts as an additional transcription factor (122). Thus, the characteristic structure of the compound and its fit into DNA influence transcriptional activity in a manner not necessarily related to its receptor-binding affinity (122). The results reported here do not provide a rigorous test of this hypothesis, but they are consistent with it. Binding affinity could determine the concentration of ligand needed to produce a response (potency), while the structure of the ligand itself would determine its effectiveness as a transcription factor (which would have bearing on the magnitude of response generated: efficacy). This hypothesis is being developed only now and tested with computer modeling and simple expression assays. If it is validated it would further support the contention that measures of receptor binding are poor predictors of the effects of endocrine disrupters. In light of the information reviewed here, the use of empirically determined receptor-binding affinity is unlikely to be very accurate for screening and risk-assessment purposes. Furthermore, QSAR relationships based solely on ER binding may not be sufficiently accurate to be used in a tiered screening approach.

Cell Proliferation and Differentiation

One of the in-vitro assays most frequently used to determine the relative potency of EDCs acting through the ER is cell proliferation of estrogen-responsive tissues (123). The MCF-7 and ZR-75 cell lines are human breast-cancer cells that have been used extensively for screening of estrogenic effects (34, 44, 124). The primary cell line used to assess estrogenic effects is the MCF-7 line, which originally was derived from hormone-dependent metastatic breast cancer (125). These cells were demonstrated to contain ER (126) and express a number of responses that are under control of the ER. MCF-7 cells have been used in the *E-screen* to screen for estrogenicity of a number of compounds (34, 51, 127, 128). This assay is based on E_2-dependent proliferation of MCF-7 cells.

In the *E-screen* assay, MCF-7 cells are cultured in a medium stripped with dextran-coated charcoal so that it is E_2-deficient. A test compound is interpreted as an estrogen agonist if it significantly increases cell proliferation relative to a control on its addition to the stripped medium. In the original *E-screen* assay the response was determined by counting cell nuclei in trypsinized cells (129). Other responses of cell proliferation such as metabolic reduction of dimethylthiazolyldiphenyltriazolium bromide, reduction of thiamine blue, and incorporation of [^{3}H]thymidine have also been used.

The *E-screen* assay has been extensively used (15, 34, 42, 124), but it has some limitations. Because it measures cell proliferation, the *E-screen* assay is technically a mitogenicity assay. Thus, a positive response cannot be attributed strictly to estrogen agonists. Variation in media preparation and relative concentrations of E_2 and other growth factors can greatly affect the responses obtained by different laboratories. Also, because it depends on a proliferation response, ER antagonists are not easily detected using this assay. Furthermore, secondary types of endocrine modulators that could result in responses that are similar to those involving E_2 at the ER (such as antiandrogens) are not measured by this assay. For these reasons, if used alone, the *E-screen* assay could result in a significant number of false-negative determinations of EDCs. A positive response in the *E-screen* assay should be confirmed by in-vivo studies before a conclusion about environmental activity of a compound is made.

Expression Assays

Because it has been demonstrated that, in some situations, binding of a ligand to the ER is necessary but not sufficient for endocrine modulation, while in other situations modulation can occur without significant ER binding, more complex in-vitro assays have been developed. Knowledge of ER-binding characteristics is only part of the information necessary to interpret the potential endocrine modulatory effects of compounds. A compound that binds with high affinity may be either an agonist or an antagonist. Furthermore, proliferation is a potentially nonspecific response. For this reason a series of in-vitro receptor-dependent transcriptional expression assays have been developed. Those designed to measure potency of estrogenic, antiestrogenic, androgenic, and antiandrogenic compounds are the best developed. A number of these types of systems have been developed for ER-dependent responses.

MCF-7-luc Bioassay Method

Expression assays examine induction or suppression of proteins encoded by genes whose transcription is thought to be modulated through an ER-mediated mechanism. Increases or decreases in the activity of the protein of interest on exposure to a single compound or complex mixture, such as an environmental extract, suggest the presence of one or more ligands with the potential to modulate a broad range of genomically controlled estrogenic responses. The MCF-7-luc bioassay method (see the discussion of receptor binding in this chapter) is detailed here as a prototypic expression assay (see Fig. 1). MCF-7-luc cells are seeded into 96-well microplates. Compounds or environmental samples then are delivered to the plates. Active compounds enter the cell, where they affect the transcription of ERE-regulated genes, presumably through mechanisms discussed earlier in this chapter. Estrogen agonists cause upregulation of transcription of the luciferase reporter gene. Because there is sufficient E_2 in the medium to allow constitutive luciferase expression, the assay also can be used to detect antagonists (103). Luciferase mRNA is produced and translated at the ribosomes into the luciferase enzyme. The activity of this enzyme then is measured by adding the exogenous substrate luciferin, which interacts with luciferase in a light-producing reaction. The magnitude of light production, measured by a luminometer, serves as a gauge of reporter-gene expression and thus a gauge of the estrogenic potential of the sample. Potency of the sample can be in terms of the concentration of test substance needed to yield a significant amount of light or concentration needed to yield 50% of maximal light production. The maximum amount of light produced provides a measure of the compound's efficacy. This means that relative potencies of both agonists and antagonists can be determined singly or as a net potency for a complex mixture of both agonists and antagonists. Although there are other expression assays (6), the MCF-7-luc method illustrates many of the features common to expression assays (i.e., examines changes in expression of a gene whose activity is modulated by a known mechanism of action on exposure to a ligand). Most of the differences among expression assays are the result of differences in the particular cells or reporter gene or protein, used in the assay system.

Applications of the MCF-7-luc Bioassay (A Model Expression Assay)

The MCF-7-luc bioassay has been successfully utilized to screen for both estrogen agonists and antagonists, assess interactions within mixtures, conduct mechanistic studies, and perform environmental monitoring of a wide variety of biotic and abiotic matrices. Recently, there has been controversy over the potential for mixtures of weakly estrogenic pesticides to act in a synergistic (130) or additive manner (131). When tested in the MCF-7-luc bioassay, (Fig. 3) dieldrin, endosulfan I, and a 1:1

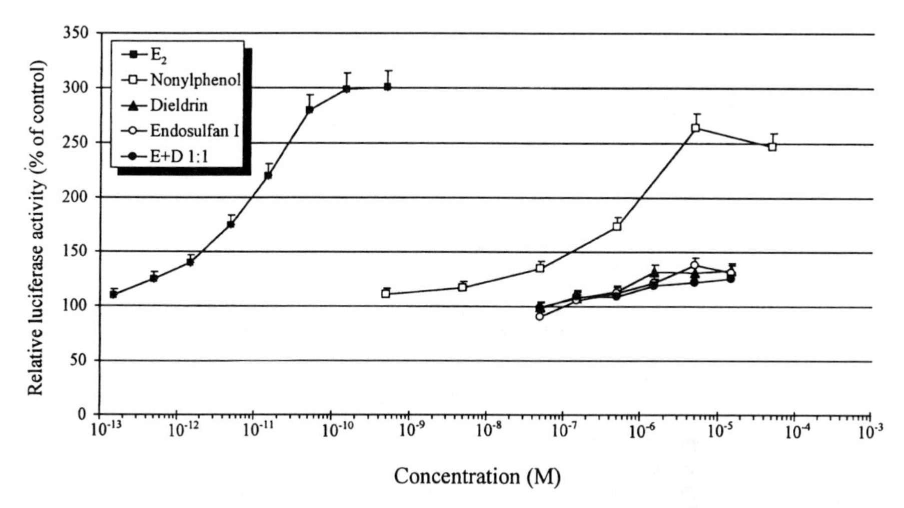

FIG. 3. Estrogen-responsive element–mediated induction of luciferase activity in MCF-7-luc cells. Cells were treated with E_2, nonylphenol, dieldrin, endosulfan I, dieldrin plus endosulfan I (E+D 1:1 mixture), or solvent only for 3 days. Relative luciferase activity is expressed as a percentage of control, with each point representing the mean of at least three replicates (standard deviations are represented by error bars).

mixture of endosulfan I and dieldrin all displayed similar dose-response curves. Potencies for individual compounds and the binary mixture were not significantly different and were approximately a million-fold less potent than E_2. Thus, synergism was not observed between endosulfan I and dieldrin in MCF-7-luc cells. In addition, it is important to point out that the magnitude of response (efficacy) elicited by these chlorinated pesticides is considerably less than that caused by E_2. Another environmental estrogen, nonylphenol, was found to possess an efficacy approximately 90% of E_2 but with a relative potency approximately 10,000-fold less than E_2. Therefore, in comparing estrogenic activities of potential xenoestrogens it is important to consider both potency and efficacy.

Because estrogen-antagonist potency can be determined using the MCF-7-luc bioassay, the dose dependence for the antiestrogenicity of TCDD was evaluated for the full dose–response curve of E_2 (Fig. 4). In general, parallel dose–response curves were obtained and there was a dose-dependent increase in the EC_{50} for E_2 (i.e., a decrease in potency) with increasing concentrations of TCDD.

Other Expression Assays

Cell lines used for expression assays include chicken fibroblasts, both primary and transformed rainbow trout hepatocytes (6, 44, 132, 133), MCF-7 (42, 131), HeLa (6, 14), yeast (6, 130, 131), and others. There is considerable evidence that in-vitro expression of ER-modulated genes is influenced by cell and promoter context (69,

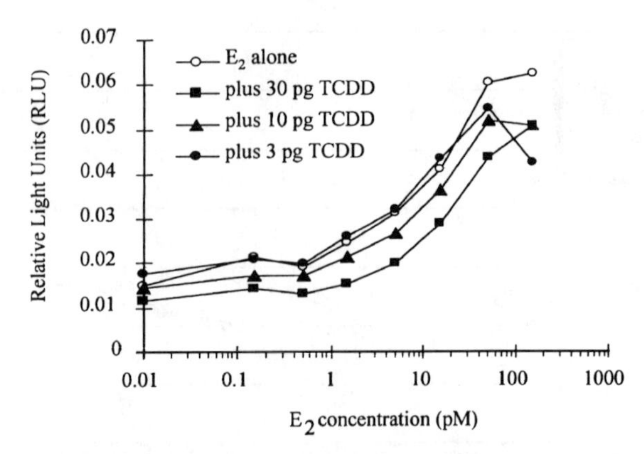

FIG. 4. Dose-dependent inhibition of estrogen-responsive element–mediated induction of luciferase activity by 2,3,7,8-tetrachlorodibenzo-*p*-dioxin (TCDD) in MCF-7-luc cells. Cells were treated with E_2 either alone or in combination with TCDD for 3 days. Relative luciferase activity is expressed as relative light units, with each point representing the mean of at least three replicates (standard deviations are represented by error bars).

121, 134, 135). The type of cell used in the expression assay system largely determines the number of confounding factors that must be considered in interpreting results and determining the relevance of the assay for predicting effects in vivo. This is due to differences in the intracellular "machinery" found in different types of cells. Some cells, such as yeast, contain none of the receptors, DNA response elements, or steroid-responsive genes involved in the gene expression assay. These cells simply act as housing for exogenous expression systems (6). Others contain parts of the machinery but lack one or more components, such as receptors or regulatory sequences of DNA. Finally, some cells contain all the machinery necessary for expression.

Yeasts, such as *Sacchromyces cerevisiae*, represent the stripped-down type of cell. Although yeasts contain no steroid receptors of their own, steroid receptors expressed in yeast have been shown to function normally (136, 137). Thus, they must be transformed with plasmids coding for ER and a reporter gene to be used for expression assays. Yeast-based expression assays have been used by a number of researchers (6, 130, 136, 138–140). They have a number of advantages. First, they tend to be relatively simple and rapid assays because yeast is more easily cultured than more complex cells. Because they have no endogenous hormone responsive machinery, they are specific for compounds that act directly through ER-mediated control of gene expression. Responses are not modulated by other regulatory factors. This also means that yeast cells are isolated from many of the confounding factors that affect more complex cells (140). The tradeoff, however, is the inability of yeast-based expression systems to account for many of the factors at the cellular level that may affect a compound's activity in-vivo. Furthermore, yeast cells have some morphologic (e.g., cell walls) and physiologic properties not found in most animal cells (6). These factors may affect the accumulation and response dynamics of the xenobiotics. Indeed, significant differences in response between yeast cell-based assays and mammalian-cell assays have been reported (6).

Another level of complexity is expression assays that use complex cells such as mammalian hepatoma cells that do not express steroid receptors of their own. Such cells transfected with plasmids coding for ER and an appropriate reporter gene retain the specificity of yeast-based assays but incorporate some of the more complex pharmacokinetics, metabolic potential, and so forth of more complicated cells. Such cells are considered particularly useful for studying receptor function (140).

Cells containing all the necessary machinery such as MCF-7 cells or rainbow trout hepatocytes are the most realistic in terms of cellular level factors that can influence expression. This, however, increases the complexity of interpretation by expanding the number of possible avenues via which compounds in a sample may influence the endpoint measured. The choice of whether to use an exogenous or endogenous reporter gene or protein as an endpoint provides another level of complexity and realism within cells containing a complete set of endogenous estrogen-modulated gene-expression machinery.

The different reporter proteins used as endpoints for expression assays are nearly as numerous as the cell types used. Common reporters include endogenous proteins such as pS2 (42, 141), vitellogenin (44, 132, 133), and sex hormone–binding globulins (6) and exogenous reporters like luciferase (74), β-galactosidase (130, 140), and chloramphenicol acetyltransferase (14). Like cell type, the type of reporter protein used as an endpoint can affect the number of confounding factors to consider. Endogenous proteins may have a direct affinity for the ligands being studied, may be linked to metabolic processes or feedback mechanisms not directly controlled by the estrogen expression mechanism, may be affected by other steroid hormones, and so forth. Exogenous proteins like luciferase are less likely to be affected by such factors.

The type of reporter protein or endpoint used also has a major impact on the sensitivity and responsiveness of the assay, because it affects the type of instruments or analytic methods that can be used to detect it. Because of the availability of sensitive detectors for light and the high quantum efficiency of the luciferase reaction, the light-producing endpoint for a luciferase-based expression assay can be very sensitively detected using a luminometer (142). The sensitivity of MCF-7 assays using a luminescent endpoint are in the low picomolar ranges (6). Colorimetric endpoints, such as the β-galactosidase endpoint used in many yeast-based assays tend to be nearly 100-fold less sensitive (140).

Expression assays have several advantages over other potential in-vitro screening assays. First, they tend to be more sensitive than receptor-binding assays (6) (Fig. 5), although their degree of sensitivity varies considerably with both cell type and reporter protein used. Unlike receptor binding or cell proliferation assays, expression assays can be used to detect both estrogen agonists and antagonists (14, 74, 143). Because protein expression requires the transcription of a gene to mRNA followed by translation at the ribosomes, expression assays incorporate transcriptional and post-transcriptional processes that may be important determinants of ligand-influenced response. This means that expression assays can detect ligands that influence any

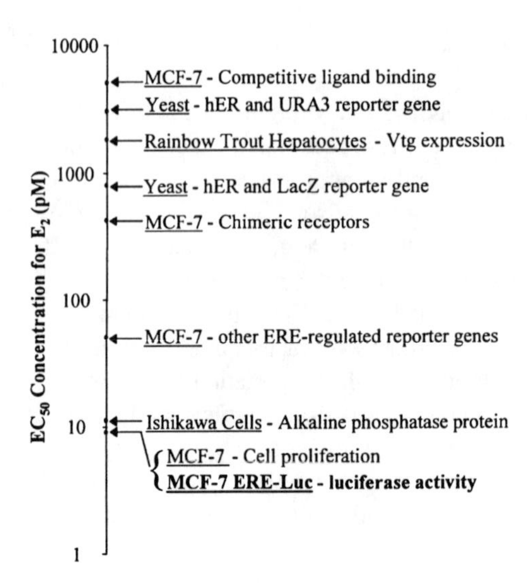

FIG. 5. A comparison of several in-vitro bioassays for estrogenic activities. Adapted from Zacharewski (6).

part of the gene-expression pathway, not just those that bind to the estrogen receptor. This makes such assays more robust in light of uncertainties that remain regarding mechanisms of action. Expression assays also should be more specific than proliferation assays, because many factors that can affect proliferation (e.g., temp, pH, serum quality, other mitogens) should have little or no bearing on expression of estrogen-modulated genes.

The primary disadvantage of expression assays are those common to all in-vitro bioassays. Like all in-vitro systems, expression assays do not incorporate the entire range of metabolic and pharmacokinetic factors that can affect the disposition of target compounds in a whole organism. The inability to account for such complexities makes the in vivo relevance of such tests unclear. More work is needed to examine and develop correlations between expression-assay results and effects in vivo if indeed they exist. Relative to other in-vitro bioassays, potential disadvantages include the need to develop and characterize recombinant cell lines, in some cases longer assay duration (6), and greater system complexity. Expression assays vary considerably in sensitivity, specificity, and responsiveness (in terms of fold induction), because the wide variety of constructs used (6). This can make it difficult to compare results from different assay systems and laboratories.

ENVIRONMENTAL MONITORING

In addition to the need to screen chemicals being produced and used, there is a need to monitor levels of estrogenic compounds in the environment. Monitoring is an essential part of the risk-assessment process. It also is necessary to evaluate the need for and effectiveness of remediation and regulation efforts. Tools such as instrumental and analytical chemistry analyses and in-vivo biomarkers, although not particularly suitable for screening applications, are useful for monitoring. Many of the in-vitro methods discussed previously can be applied to monitor the environment for estrogenic compounds as well.

Instrumental analysis can be used to monitor the environment for known estrogen agonists or antagonists. Extracts can be prepared from water, sediments, soil, and biological matrices. The extracts can then be analyzed using gas chromatograph (GC), high pressure liquid chromatography (HPLC), GC/mass spectroscopy (GC/MS), and so forth. Such analysis can be very sensitive for common estrogen agonists, particularly owing to the ability to concentrate samples during the extraction and clean-up process. The key advantage to this type of analysis is the ability to precisely quantify concentrations of compounds of interest in the environmental matrix being studied.

Instrumental analyses alone, however, are rather limited as monitoring tools. Although quantitative, they provide no information regarding the biologic activity of the sample. Only compounds that have been identified previously as estrogen agonists or antagonists can be monitored in this fashion. This problem is further complicated by the fact that complex interactions may occur among various agonists, antagonists, and other compounds found in a complex environmental mixture. Analytic methods provide no mechanism by which to evaluate the overall activity of a mixture, unless strict additivity of effects is assumed. Nontarget compounds in a complex environmental mixture also may interfere with proper resolution and quantification of target compounds. Thus, although instrumental analyses are powerful and sensitive tools, they alone cannot provide the information needed for comprehensive monitoring and risk assessment.

Some of the in-vitro bioassays used for screening are also useful for monitoring. Extracts from a variety of biotic and abiotic matrices can be tested. Unfortunately, as for instrumental analysis, the extraction and clean-up process often is the most tedious and time-consuming step in monitoring using in-vitro bioassays. In some cases, however, less extensive sample preparation is necessary because the need to resolve individual compounds is not an issue for bioassays. The primary advantage to using in-vitro bioassays for monitoring is that they should respond to any and all active compounds in the sample, not just a limited set of known or target compounds. Because of this, they can integrate the activity of the entire mixture, accounting for all potential interactions, whether they are additive or not.

One distinct advantage of monitoring with in-vitro bioassays is the potential to derive and compare relative potencies. The concentration or volume of environmental extract needed to elicit a given level of effect can be compared with that of a standard, such as E_2, and expressed relative to it (i.e., E_2 equivalents = EC_{50} sample/EC_{50} E_2). This type of approach has been used extensively for dioxin-like compounds (144–147). It provides a simple method for comparing estrogenic potency from sample to sample, even if sample composition cannot be determined. Comparison of relative potencies of different fractions from the same sample can also be used to describe complex interactions that may be occurring in mixtures.

Unfortunately, however, it is not always easy or straightforward to calculate comparable potencies for environmental samples based on bioassay results. For comparison of potencies to be accurate and useful, the same response must be achieved with each sample. That is, the concentration associated with a particular magnitude of response needs to be estimated, and that magnitude must be the same for all samples being compared. This generally requires obtaining a complete dose–response curve for all samples being compared, or at least assuming that given a sufficient dose of sample, all will reach the same maximum level of response. In practice, however, these conditions are often difficult to meet and the assumptions usually do not hold. The inability to control the concentration of active compounds in the extract beyond the limits of sample concentration and/or dilution often precludes the ability to generate a complete dose–response curve such that maximal activity is achieved. Variation in the efficacy of different estrogen agonists within their ranges of solubility generally precludes the ability to assume that all samples will reach the same maxima (see Fig. 3). This generally means that in order to estimate or compare potencies, the concentration needed to elicit a minimum statistically significant response or threshold is about the only comparable point to use for all samples. This is not ideal, however, because there is minimal statistical confidence at this point.

This discussion illustrates the advantages and deficiencies of both instrumental and in-vitro bioassay methods for monitoring estrogenic compounds in the environment. From the discussion, it also should be apparent that each type of tool complements the other. Analytical and in-vitro tools can be used effectively together to provide information needed for monitoring, as well as risk assessment, via bioassay-directed fractionation and identification. Such complementary application of instrumental and bioassay techniques often is referred to as *toxicant identification and evaluation* (TIE) (Fig. 6).

In the first step of such a scheme, samples from an abiotic or biotic matrix of interest are collected and chemicals are selectively extracted and fractionated, chromatographically, to separate the complex mixture from the matrix and into groups of compounds with nominal characteristics. In monitoring for the presence of estrogen agonists in water, samples of water could be passed through a solid-phase sampling device, such as Sep-pac or Empore™ disks. Alternatively, passive samplers such as semi-permeable membrane devices (SPMDs) can be placed in the environment to collect a time-integrated sample of the materials of interest (148). Once a sample has been separated from its matrix it can be analyzed in an in-vitro assay. This generally is a functional assay such as the MCF-7-luc assay used to test for estrogenicity. A battery of assays based on different functional endpoints can be used (52). If there is a positive response in the assay, the sample can be further fractionated based on molecular size, polarity, or a combination of the two. Each fraction then is subjected to bioassay. In this way the types of compounds contributing to a positive response can be narrowed to members of operationally defined functional classes. Instrumental analyses then are applied to the fractions that elicited activity in vitro. Typically, active fractions are analyzed by several methods, such as HPLC with both fluorescence and UV-visible detection, gas chromatography with flame-ionization detection, electron-capture detection, or mass-selective detection. The particular methods used depend on

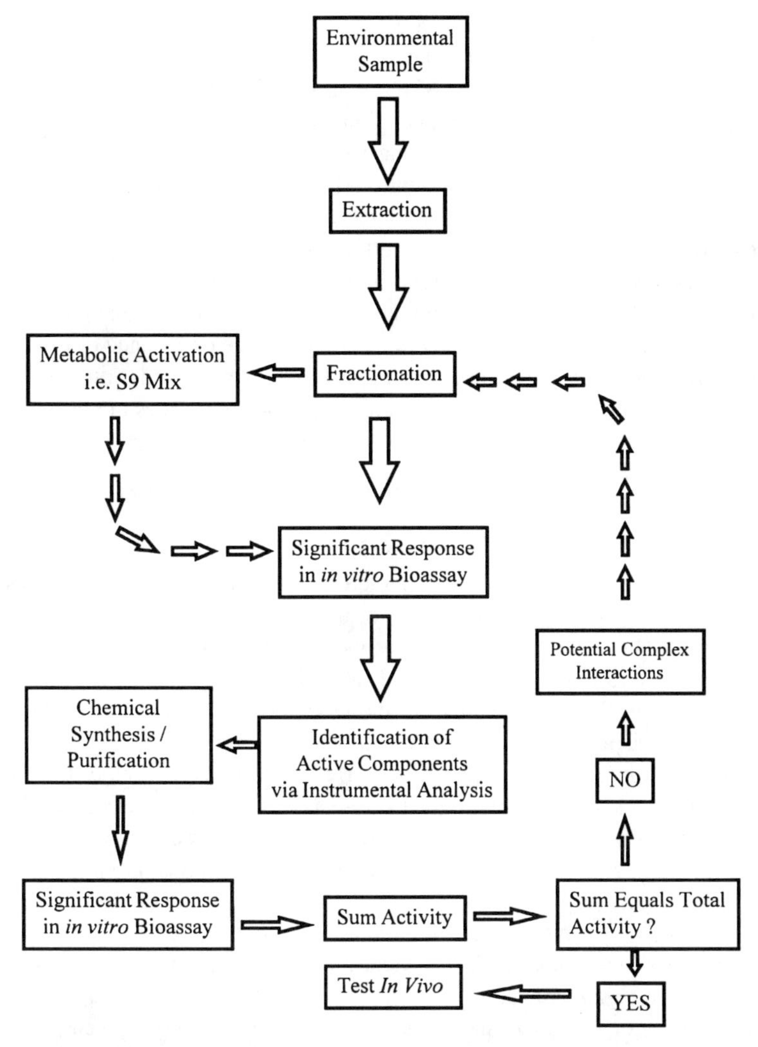

FIG. 6. Generalized toxicant identification and evaluation scheme for environmental samples using in-vitro bioassay-directed fractionation and instrumental analysis.

the properties of the fraction of interest. For instance, if the activity is in the nonpolar fraction, it can be analyzed by gas chromatography directly. If a more polar fraction contains the activity, however, it might be necessary to use HPLC-based analysis or derivatization techniques. Through additional iterations of fractionation, bioassay, and instrumental analysis, it should be possible to identify the specific bioactive agents present in the original extract. Once a tentative identification is achieved, standards are used to quantify the mass of material in the sample. If authentic standards are not available commercially they can be synthesized. Pure compounds can then be analyzed by in-vitro bioassay to determine if they are, indeed, active. If so, relative potency factors are derived and total equivalents determined for the sample by multiplying these factors by the molar concentrations of the compound in the sample. Predicted and measured activity in the assay are compared in a mass balance of potency. If they are equal, it indicates that all of the activity has been accounted for by the assay and that there are no interactions among compounds. Alternatively there could be more activity or less activity in the measured or predicted values. More activity in the assay than predicted could indicate that there were additional unidentified compounds or super additive interactions in the mixture. Less activity in the bioassay might indicate an infra additive interaction among compounds. Use of selective isobolar additions and selective removal of compounds can confirm the presence or absence of interactions or unidentified compounds. A similar technique has been proposed for use to separate the relative contributions of endogenous and exogenous hormones in plasma (124).

In-vivo Biomarkers

In-vivo biomarkers are also viable monitoring tools. Reports of in-vivo vitellogenin induction (149), altered secondary sex characteristics (150, 151, 152), lowered sperm counts (44), and so forth are responsible for much of the attention that EDCs have drawn recently. These and other in-vivo responses can be monitored as indicators of potential exposure to estrogenic and other EDCs. Such indicators have the advantage of incorporating all the complex pharmacokinetic and metabolic factors that can affect uptake and disposition of compounds in a whole organism, giving them more biologic significance. In some cases, however, it is not known whether the changes observed are adverse or not. This limits the risk assessment utility of some biomarkers. Because in-vivo endpoints are influenced by so many additional variables, both physiologic and environmental, it is much harder to establish clear cause–effect relationships between responses and specific compounds in the environment. Thus, in-vivo biomarkers are most useful for monitoring and risk assessment as part of a comprehensive, tiered approach that incorporates both instrumental analyses and rapid in-vitro screening assays.

For each receptor-mediated mechanism, an agonist/antagonist screen potentially can be developed. This can serve as the first tier in an integrated monitoring approach. In-vivo models can provide a second tier for evaluating the toxicologic significance of active samples. In conjunction with chemical analysis, this tiered approach can be used to monitor for and identify specific bioactive compounds in the environment and evaluate their potential toxicologic relevance.

ACKNOWLEDGMENTS

Portions of the research presented herein and preparation of the chapter were supported by the Chlorine Chemistry Council of the Chemical Manufacturers Association, cooperative agreement NO. CR 822983-01-0 between Michigan State University and the US EPA Office of Water Quality, the NIEHS–Superfund Basic Research Program (NIH-ES-04911), a Michigan State University Distinguished Fellowship, and an NIEHS postdoctoral fellowship for financial support. We thank Dr. M.D. Pons for providing MCF-7-luc cells. We acknowledge Rebecca Zmyslo and Emily Nitsch and the members of MSU's Aquatic Toxicology Laboratory Estrogen Workgroup for their assistance and comments on the manuscript. Finally, we acknowledge the editors for their patience.

REFERENCES

1. Wolff MS, Toniolo PG, Lee EW, Rivera M, Dubin N. Blood levels of organochlorine residues and risk of breast cancer. *J Natl Cancer Inst* 1993;85:648–652.
2. Dibb S. Swimming in a sea of estrogens, chemical hormone disrupters. *Ecologist* 1995;25:27–31.
3. McLachlan JA, Arnold SF. Environmental estrogens. *Am Sci* 1996;84:452–461.
4. Colborn T, Vom Saal FS, Soto AM. Developmental effects of endocrine-disrupting chemicals in wildlife and humans. *Environ Health Perspect* 1993;101:378–384.
5. Adlercreutz H. Phytoestrogens: Epidemiology and a possible role in cancer protection. *Environ Health Perspect* 1995;103(suppl 7):102–112.
6. Zacharewski T. *In-vitro* bioassays for assessing estrogenic substances. *Environ Sci Technol* 1997;31:613–623.
7. Mazeaud MM, Mazeaud F. Adrenergic responses to stress in fish. In: Pickering AD, ed. *Stress in Fish*. New York: Academic Press, 1982;49–75.
8. Selye H. *Stress in Health and Disease: I. History and General Outline of the Stress Concept*. Boston: Butterworths, 1976;3–34
9. Ness DK, Schantz SL, Moshtaghian J, Hansen LG. Effects of perinatal exposure to specific PCB congeners on thyroid hormone concentrations and thyroid histology in the rat. *Toxicol Lett* 1993;68:311–323.
10. Madhukar BV, Brewster DW, Matsumura F. Effects of *in-vivo* administration of 2,3,7,8-TCDD on receptor binding of epidermal growth factor in the hepatic plasma membrane of rat, guinea pig, mouse, and hamster. *Proc Natl Acad Sci USA* 1984;81:7407–7411.
11. Makela S, Santti R, Salo L, McLachlan JA. Phytoestrogens are partial estrogen agonists in the adult male mouse. *Environ Health Perspect* 1995;103(suppl 7):123–127.
12. Mellanen P, Petanen T, Lehtimaki J, Makela S, Bylund G, Holmbom B, Mannila E, Oikari A, Santti R. Wood-derived estrogens: studies *in-vitro* with breast cancer cell lines and *in-vivo* in trout. *Toxicol Appl Pharmacol* 1996;136:381–388.
13. Barrett J. Phytoestrogens: Friends or foes? *Environ Health Perspect* 1996;104:478–482.
14. Miksicek RJ. Commonly occurring plant flavonoids have estrogenic activity. *Mol Pharmacol* 1993;44:37–43.
15. Kelce WR, Monosson E, Gamcsik MP, Laws SC, Gray LE. Environmental hormone disruptors: evidence that vinclozolin developmental toxicity is mediated by antiandrogenic metabolites. *Toxicol Appl Pharmacol* 1994;126:276–285.
16. Kelce WR, Stone CR, Laws SC, Gray LE, Kemppainen JA, Wilson EM. Persistent DDT metabolite p,p'-DDE is a potent androgen receptor agonist. *Nature* 1995;375:581–585.
17. Astroff B, Romkes M, Safe S. Mechanism of action of 2,3,7,8-TCDD and 6-methyl-1,3,8-trichlorodibenzofuran (MCDF) as antiestrogens in the female rat. *Chemosphere* 1989;19:785–788.
18. Rozman K, Rozman T, Greim H. Effect of thyroidectomy and thyroxine on 2,3,7,8-tetrachlorodibenzo-ρ-dioxin (TCDD) induced toxicity. *Toxicol Appl Pharmacol* 1981;72:372–376.
19. Janz DM, Bellward GD. In ovo 2,3,7,8-tetrachlorodibenzo-ρ-dioxin exposure in three avian species: effects on thyroid hormones and growth during the perinatal period. *Toxicol Appl Pharmacol* 1996;139:281–291.
20. Liu H, Biegel L, Narasimhan TR, Rowlands C, Safe S. Inhibition of insulin-like growth factor-I responses in MCF-7 cells by 2,3,7,8-tetrachlorodibenzo-ρ-dioxin and related compounds. *Mol Cell Endocrinol* 1992;87:19–28.
21. Choi E, Toscano D, Ryan J, Riedel N, Toscano W. Dioxin induces transforming growth factor-β in human keratinocytes. *J Biol Chem* 1991;266:9591–9597.

22. Carrier F, Owens RA, Nebert DW, Puga A. Dioxin-dependent activation of murine Cyp1a-1 gene transcription requires protein kinase C-dependent phosphorylation. *Mol Cell Biol* 1992;12:1856–1863.
23. Reiners JJ, Cantu AR, Schöller A. Phorbol ester-mediated supression of cytochrome P450 CYP1a1 induction in murine skin: involvement of protein kinase C. *Biochem Biophys Res Commun* 1992;186:970–976.
24. Aoki Y, Matsumoto M, Suzuki K. Expression of glutathione S-transferase P-form in primary cultured rat liver parencymal cells by coplanar biphenyl congeners is supported by protein kinase inhibitors and dexamethasone. *Fed Eur Biochem Soc* 1993;333:114–118.
25. DeVito MJ, Ma X, Babish JG, Menache M, Birnbaum LS. Dose-response relationships in mice following subchronic exposure to 2,3,7,8-tetrachlorodibenzo-ρ-dioxin: CYP1A1, CYP1A2, estrogen receptor and protein tyrosine phosphorylation. *Toxicol Appl Pharmacol* 1994;124:82–90.
26. Blankenship A, Matsumura F. 2,3,7,8-Tetrachlorodibenzo-p-dioxin (TCDD) causes an Ah receptor-dependent and ARNT-independent increase in membrane levels and activity of p60Src. *Environ Toxicol Pharmacol* 1997;3:211–220.
27. Jobling S, Sumpter JP. Detergent compounds in sewage effluents are weakly oestrogenic to fish: an *in-vitro* study using rainbow trout (Onchorhyncus mykiss) hepatocytes. *Aquatic Toxicol* 1993; 27:361–372.
28. Hoare SA, Jobling S, Parker MG, Sumpter JP, White R. Environmental persistent alkylphenolic compounds are estrogenic. *Endocrinology* 1994;135:175–182.
29. Guillette LJ, Crain DA, Rooney AA, Pickford DB. Organization versus activation: the role of endocrine disrupting contaminants (EDCs) during embryonic development in wildlife. *Environ Health Perspect* 1995;103(suppl 7):157–164.
30. Colborn T. Environmental estrogens: Health implications for humans and wildlife. *Environ Health Perspect* 1995;103(suppl 7):135–136.
31. Feldman D, Krishnan A. Estrogens in unexpected places: Possible implications for researchers and consumers. *Environ Health Perspect* 1995;103(suppl 7):129–133
32. Duby RT, Travis HF, Terrill CE. Uterotropic activity of DDT in rats and mink and its influence on reproduction in the rat. *Toxicol Appl Pharmacol* 1971;18:348–355.
33. Donohoe RM, Curtis LR. Estrogenic activity of chlordecone, o,p′-DDT and o,p′-DDE in juvenile rainbow trout: induction of vitellogenesis and interaction with hepatic estrogen binding sites. *Aquatic Toxicol* 1996;36:31–52.
34. Soto AM, Chung KL, Sonnenschein C. The pesticides endosulfan, toxaphene and dieldrin have estrogenic effects in human estrogen-sensitive cells. *Environ Health Perspect* 1994;102:380–383.
35. Eroschenko VP. Estrogenic activity of the insecticide Chlordecone in the reproductive tract of birds and mammals. *J Toxicol Environ Health* 1981;8:731–742.
36. Korach KS, Sarver P, Chae K, McLachlan JA, McKinney JD. Estrogen receptor-binding activity of polychlorinated hyroxybiphenyls: conformationally restricted structural probes. *Mol Pharmacol* 1988;33:120–126.
37. Spink DC, Johnson JA, Connor SP, Aldous KM, Gierthy JF. Stimulation of 17 β-estradiol metabolism on MCF-7 cells by bromochloro- and chloromethyl-substituted dibenzo-p-dioxins and dibenzofurans: correlations with antiestrogenic activity. *J Toxicol Environ Health* 1994;41:451–466.
38. Jansen HT, Cooke PS, Porcelli J, Liu TC, Hansen LG. Estrogenic and antiestrogenic actions of PCBs in the female rat: *in vitro* and *in-vivo* studies. *Reproduct Toxicol* 1993;7:237–248.
39. Santodonato J. Review of the estrogenic and antiestrogenic activity of polycyclic aromatic hydrocarbons: relationship to carcinogenicity. *Chemosphere* 1997;34:835–848.
40. Birnbaum LS. Developmental effects of dioxins. *Environ Health Perspect* 1995;103 (suppl 7):89–94.
41. Mobly T.A, Moore RW, Peterson RE. In utero and lactational exposure of male rats to 2,3,7,8-tetrachloro-dibenzo-p-dioxin 1. Effects on Androgenic status. *Toxicology* 1992;114:97–107.
42. Olea N, Pulgar R, Perez P, Olea-Serrano F, Rivas A, Novillo-Fertrell A, Pedraza V, Soto AM, Sonnenschein C. Estrogenicity of resin-based composites and sealants used in dentistry. *Environ Health Perspect* 1996;104:298–305.
43. Sharpe RM, Fisher JS, Millar MM, Jobling S, Sumpter JP. Gestational and lactational exposure of rats to xenoestrogens results in reduced testicular size and sperm production. *Environ Health Perspect* 1995;103:1136–1143.
44. White R, Jobling S, Hoare SA, Sumpter JP, Parker MG. Environmentally persistent alkylphenolic compounds are estrogenic. *Endocrinology* 1994;135:175–182.

45. Sumpter JP, Jobling S, Tyler CR. Toxicology of aquatic pollution: physiological, cellular and molecular approaches, Taylor EW, ed., Cambridge Press, NY. *Physiology and Molecular Approaches* 1996;205–224.
46. Aherne GW, Briggs R. The relevance of the presence of certain synthetic steroids in the aquatic environment. *J Pharm Pharmacol* 1989;41:735–736.
47. Jobling S, Sheahan D, Osborne JA, Matthiessen P, Sumpter JP. Inhibition of testicular growth in rainbow trout (Oncorhyncus mykiss) exposed to estrogenic alkylphenolic chemicals. *Environ Toxicol Chem* 1996;15:194–202.
48. Sheehan DM, Young M. Diethylstilbestrol and estradiol binding to serum albumin and pregnancy plasma of rat and human. *Endocrinology* 1979;104:1442–1446.
49. Solmssen UV. Synthetic estrogens and the relation between their structure and their activity. *Chem Rev* 1945;37:481–598.
50. Korach KS, Levy LA, Sarver PJ. Estrogen receptor stereochemistry: receptor binding and hormonal responses. *J Steroid Biochem* 1987;27:281–290.
51. Soto AM, Lin TM, Justicia H, Silvia RM, Sonnennschein C. An "in culture" bioassay to assess the estrogenicity of xenobiotics (E-screen). In: Colburn T, Clement C, eds. *Chemically Induced Alterations in Sexual and Functional Development: The Wildlife/Human Connection*. Princeton, NJ: Princeton Scientific Publishing, 1992;295–309.
52. McLachlan JA. Functional toxicology: A new approach to detect biologically active xenobiotics. *Environ Health Perspect* 1993;101:386–387.
53. Kumar V, Chambon P. The estrogen receptor binds tightly to its responsive element as a ligand induced homodimer. *Cell* 1988;55:145–156.
54. Bondy KL, Zacharewski TR. ICI 164,384: A control for investigating estrogen responsive genes. *Nucleic Acids Res* 1993;21:5277–5278.
55. Fawell SE, White R, Hoare S, Sydenham M, Page M, Parker MG. Inhibition of estrogen receptor DNA binding by the "pure" anti-estrogen ICI 164,384 appears to be mediated by impaired receptor dimerization. *Proc Natl Acad Sci USA* 1990;87:6883–6887.
56. Jordan VC, Kock R, Lieberman ME. Structure-activity relationships of nonsteroidal estrogens and anti-estrogens. In: Jordan VC, ed. *Estrogen/Anti-estrogen Action and Breast Cancer Therapy*. Madison, WI: University of Wisconsin Press, 1986;19–41.
57. Farooqi ZH, Aboul-Enein HY. Conformational flexibility of cyclohexylaminoglutethimide: a potent aromatase inhibitor. Comparison of the three configurations of the cyclohexyl moiety. *Cancer Lett* 1994;87:121.
58. Giorgi EP, Stein WD. The transport of steroids in animal cells in culture. *Endocrinology* 1981;108:688–697.
59. Price TM, Blauer KL, Hansen M, Stanczyk F, Lobo R, Bates GW. Single-dose pharamacokinetics of sublingual versus oral administration of micronized 17 β-estradiol. *Obstet Gynecol* 1997;89:340–345.
60. Dunn JF, Nisula BC, Rodbard D. Transport of steroid hormones: binding of 21 endogenous steroids to both testosterone-binding globulin and corticosteroid-binding globulin in human plasma. *J Clin Endocrinol Metab* 1981;53:58–68.
61. Abramson FP, Miller HC. Bioavailability, distribution and pharmacokinetics of diethylstibestrol produced from stilphostrol. *J Urol* 1982;128:1336–1339.
62. Tew B, Xu X, Wang H, Murphy PA, Hendrich S. A diet high in wheat fiber decreases the bioavailability of soybean isoflavones in a single meal fed to women. *J Nutr* 1996;126:871–877.
63. Martin PM, Horwitz KB, Ryan DS, McGuire WL. Phytoestrogen interaction with estrogen receptors in human breast cancer cells. *Endocrinology* 1978;103:1860–1867.
64. World Health Organization. *Environmental Health Criteria for DDT and its Derivatives: Environmental Aspects*. Geneva: WHO, 1989.
65. Skalsky HL, Guthrie FE. Binding of insecticides to human serum proteins. *Toxicol Appl Pharmacol* 1978;43:229–235.
66. Arnold SF, Klotz DM, Collins BM, Vonier PM, Guillette LJ, Jr, McLachlan JA. Synergistic activation of estrogen receptor with combinations of environmental chemicals. *Science* 1996;272:1489–1492.
67. Nagel SC, vom Saal FS, Thayer KA, Dhar MG, Boechler M, Welshons WV. Relative binding affinity-serum modified access (RBA-SMA) assay predicts the relative *in-vivo* bioactivity of the xenoestrogens bisphenol A and octylphenol. *Environ Health Perspect* 1997;105:70–76.
68. Mangelsdorf DJ, Thummel C, Beato M, Herrlich P, Schutz G, Umesono K, Blumberg B, Kastner P, Mark M, Chambon P, Evang RM. The nuclear receptor superfamily: the second decade. *Cell* 1995;83:835–839.

69. Katzenellenbogen BS. Estrogen receptors: bioactivities and interactions with cell signaling pathways. *Biol Reprod* 1996;54:287–293.
70. Kuiper GG, Carlsson B, Grandien K, Enmark E, Haggblad J, Nilsson S, GustaFsson J-A. Comparison of the ligand binding specificity and transcript tissue distribution of estrogen receptors α and β. *Endocrinology* 1997;138:863–870.
71. Gorski J, Furlow JD, Murdoch FE, Fritsch M, Kanenko K, Ying C, Malayer JR. Perturbations in the model of estrogen receptor regulation of gene expression. *Biol Reprod* 1993;48:8–14.
72. Smith DF, Toft DO. Steroid receptors and their associated proteins. *Mol Endocrinol* 1993;7:4–11.
73. Landel CC, Kushner PJ, Greene GL. Estrogen receptor accessory proteins: effects on receptor-DNA interactions. *Environ Health Perspect* 1995;103(suppl 7):23–28.
74. Pons M, Gagne D, Nicolas JC, Mehtali M. A new cellular model of response to estrogens: A bioluminescent test to characterize (anti) estrogen molecules. *Biotechniques* 1990;9:450–456.
75. Aronica SM, Katzenellenbogen BS. Stimulation of estrogen receptor-mediated transcription and alteration in the phosphorylation state of the rat uterine estrogen receptor by estrogen, cyclic adenosine monophosphate, and insulin-like growth factor. *Mol Endocrinol* 1993;7:743–752.
76. Le Goff P, Montano MM, Schodin DJ, Katzenellenbogen BS. Phosphorylation of the human estrogen receptor. *J Biol Chem* 1994;269:4458–4466.
77. Arnold SF, Obourn JD, Jaffe H, Notides AC. Serine 167 is the major estradiol-induced phosphorylation site on the human estrogen receptor. *Mol Endocrinol* 1994;8:1208–1214.
78. Arnold S, Vorojeikina, Notides AC. Phosphorylation of Tyrosine 537 on the human estrogen receptor is required for binding to an estrogen response element. *J Biol Chem* 1995;270(50):30205–30212.
79. Kuiper GG, Brinkman AO. Steroid hormone receptor phosphorylation: is there a physiological role? *Mol Cell Endocrinol* 1994;100:103–107.
80. Meek DW, Street AJ. Nuclear protein phosphorylation and growth control. *Biochem J* 1992;287:1–15.
81. Ignar-Trowbridge DM, Pimentel M, Parker MG, McLachlan JA, Korach KS. Peptide growth factor cross-talk with the estrogen receptor requires the A/B domain and occurs independently of protein kinase C or estradiol. *Endocrinology* 1996;137:1735–1744.
82. Ingnar-Trowbridge DM, Nelson KG, Bidwell MC, Curtis SW, Washborn TF, McLachlan JA, Korach KS. Coupling of dual signaling pathways: Epidermal growth factor action involves the estrogen receptor. *Proc Natl Acad Sci USA* 1992;89:4658–4662.
83. Cho H, Katzenellenbogen BS. Synergistic activation of estrogen receptor-mediated transcription by estradiol and protein kinase activators. *Mol Endocrinol* 1993;7:441–452.
84. Archuleta MM, Schieven GL, Ledbetter JA, Deanin GG, Burchiel SW. 7,12-Dimethylbenz[*a*]anthracene activates protein-tyrosine kinases Fyn and Lck in the HPB-ALL human T-cell line and increases tyrosine phosphorylation of phospholipase C-γ1, formation of inositol 1,4,5-trisphosphate, and mobilization of intracellular calcium. *Proc Natl Acad Sci USA* 1993;90:6105–6109.
85. Enan E, Matsumura F. Activation of phosphoinositide/protein kinase C pathway in rat brain tissue by pyrethroids. *Biochem Pharmacol* 1993;45:703–710.
86. Safe S, Astroff B, Harris M, Zacharewski T, Dickerson R, Romkes M, Bieqel L. 2,3,7,8-Tetrachlorodibenzo-ρ-dioxin (TCDD) and related compounds as antiestrogens: characterization and mechanism of action. *Pharmacol Toxicol* 1991;69:400–409.
87. Romkes M, Piskorska-Pliszczynska J, Safe S. Effects of 2,3,7,8-tetrachlorodibenzo-ρ-dioxin on hepatic and uterine estrogen receptor levels in rats. *Toxicol Appl Pharmacol* 1987;87:306–314.
88. Pappas, TC, Gametchu B, Watson CS. Membrane estrogen receptors identified by multiple antibody labeling and impeded-ligand binding. *FASEB J* 1995;9:404–410.
89. Morley P, Whitfield JF, Vanderhyden BC, Tsang BK, Scwartz JL. A new, nongenomic estrogen action: the rapid release of intracellular calcium. *Endocrinology* 1992;131:1305–1313.
90. Migliaccio A, Pagano M, Auricchio F. Immediate and transient stimulation of protein tyrosine phosphorylation by estradiol in MCF-7 cells. *Oncogene* 1993;8:2183–2191.
91. Migliaccio A, Di Domenico M, Castoria G, de Falco A, Bontempo P, Nola E, Auricchio F. Tyrosine kinase/p21ras/MAP kinase pathway activation by estradiol-receptor complex in MCF-7 cells. *EMBO J* 1996;15:1292–1300.
92. Kharat I, Saatcioglu F. Antiestrogenic effects of 2,3,7,8-tetrachlorodibenzo-ρ-dioxin are mediated by direct transcriptional interference with the liganded estrogen receptor. *J Biol Chem* 1996;271:10533–10537.
93. Kavlock RT, Daston GP, DeRosa C, Fenner-Crisp P, Gray LE, Kaattari S, Lucier G, Luster M, Mac MJ, Maczka. Research needs for the risk assessment of health and environmental effects of endocrine disruptors: a report of the US EPA sponsored workshop. *Environ Health Perspect* 1996;104:715–740.

94. Kavlock RJ, Ankley GT. A perspective on the risk assessment process for endocrine-disruptive effects on wildlife and human health. *Risk Analysis* 1996;16:731–739.
95. Brooks SC, Wappler NL, Corombos JD, Doherty LM, Horwitz JP. Estrogen structure-receptor function relationships. In: Moudgil VK, ed. *Recent Advances in Steroid Hormone Action*. New York: W. De Gruyer, 1987;443–366.
96. VanderKuur JA, Wiese T, Brooks SC. Influence of estrogen structure on nuclear binding and progesterone receptor induction by the receptor complex. *Biochemistry* 1993;32:7002–7008.
97. Dodds EC, Lawson W. Molecular structure in relation to oestrogenic activity: compounds without a phenanthrene nucleus. *Proc Royal Soc London B Biology* 1937;125:222–232.
98. Waller CL, Oprea TI, Chae K, Park HK, Korach KS, Laws SC, Wiese TE, Kelce WR, Gray LE. Ligand based identification of environmental estrogens. *Chem Res Toxicol* 1996;9:1240–1248.
99. Waller CL, Booker WJ, Gray LE, Kelce WR. Three-dimensional quantitative structure-activity relationships for androgen receptor ligands. *Toxicol Appl Pharmacol* 1996;137:219–227.
100. Hendry LB, Chu CK, Copland JA, Mahesh VB. Antiestrogenic piperidinediones designed prospectively using computer graphics and energy calculations of DNA-ligand comlexes. *J Steroid Biochem Molec Biol* 1994;48:495–505.
101. Hendry LB, Chu CK, Rosser ML, Copland JA, Wood JA, Mahesh VB. Design of novel antiestrogens. *J Steroid Biochem Molec Biol* 1994;49:269–280.
102. Endocrine Screening Methods Workshop: Meeting Report. July 1997, Nicholas School of the Environment, Duke University, Durham, NC, 1–84.
103. Kramer VK, Helferich WG, Bergman A, Klasson-Wehler E, Giesy JP. Hydroxylated polychlorinated biphenyl metabolites are anti-estrogenic in a stably transfected human breast adenocarcinoma (MCF7) cell line. *Toxicol Appl Pharmacol* 1997;144:363–376.
104. Ireland JS, Mukku VR, Robinson AK, Stancel GM. Stimulation of uterine deoxyribonucleic acid synthesis by 1,1,1-trichloro-2-(ρ-chlorophenyl)-2-(O-chlorophenyl)ethane(o,p′-DDT). *Biochem Pharmacol* 1980;24:1469–1474.
105. Scatchard G. The attraction of proteins for small molecules and ions. *Ann N Y Acad Sci* 1949;51:660–672.
106. Cressie NAC, Keightly DD. Analyzing data from hormone receptor assays. *Biometrics* 1981;37:235–249.
107. Pakdel F, Le-Gac F, Le-Goff P, Valotaire Y. Full-length sequence and *in-vitro* expression of rainbow trout estrogen receptor cDNA. *Mol Cell Endocrinol* 1990;71:195–204.
108. Mani SK, Allen JMC, Clark JH, Blaustein JD, O'Malley BW. Convergent pathways for steroid hormone and neurotransmitter induced rat sexual behavior. *Science* 1994;265:1246–1249.
109. Zysk JR, Johnson B, Ozenberger BA, Bingham B, Gorski J. Selective uptake of estrogenic compounds by Saccharomyces cerevesiae: a mechanism for antiestrogen resistance in yeast expressing the mamallian estrogen receptor. *Endocrinology* 1995;136:1323–1326.
110. Hwang KJ, Carlson KE, Anstead GM, Katzenellenbogen JA. Donor-acceptor tetrahydrochrysenes, inherently fluorescent, high affinity, ligands for the estrogen receptor: binding and fluorescence characteristics and fluorometric assay of receptor. *Biochemistry* 1992;31:11536–11545.
111. McDonnell DP, Vegeto E, Gleeson MA. Nuclear hormone receptors as targets for new drug discovery. *Biotechnology* 1993;11:1256–1261.
112. Bulger WH, Kupfer D. Estrogenic activity of pesticides and other xenobiotics on the uterus and male reproductive tract. In: Thomas JA, Korach KS, McLachlan JA, eds. *Endocrine Toxicology*. New York: Raven Press, 1985;1–33.
113. Wakeling AE. Anti-hormones and other steroid analogues. In: Green B, Leake RE, eds. *Steroid Hormones, A Practical Approach*. Washington: IRL Press, 1987;219–236.
114. Raynaud JP, Ojasoo T, Bouton MM, Bingnon E, Pons M, Craaastes de Paulet A. Structure-activity relationships of steroid hormones. In: McLachlan JA, ed. *Estrogens in the Environment*. Amsterdam: Elsevier, 1985;24–42.
115. Meyers CY, Kolb VM, Dandliker WB. Doisynolic acids: potent estrogens with very low affinity for the estrogen receptor. *Res Commun Chem Pathol Pharmacol* 1982;35:165–168.
116. Rosser ML, Muldoon TG, Hendry LB. Computer modeling of the fit of 11-beta-acetoxy-estradiol into DNA correlates with potent uterotropic activity but not with binding to the estrogen receptor. 73rd Annual Meeting of the Endocrine Society, Washington, 1991, Abstract No. 368:122.
117. Polossek T, Ambros R, Von Angerer S, Brandl G, Mannschreck A, Von Angerer E. 6-Alkyl-12-formylindolo[2,1a]isoquinolines: synthesis, estrogen receptor binding affinities and stereospecific cytostatic activity. *J Med Chem* 1992;35:3537–3547.

118. Thompson EW, Katz D, Shima TB, Wakeling WE, Lippman ME, Dickson RB. ICI 164,384: a pure antagonist of estrogen-stimulated MCF-7 cell proliferation and invasiveness. *Cancer Res* 1989;49:6929–6934.
119. Bitman J, Cecil HC, Harris SJ, Feil VJ. Estrogenic activity of o,p'-DDT metabolites and related compounds. *J Agric Food Chem* 1978;26:149–151.
120. Martin MB, Saceda M, Garcia-Morales P, Gottardis MM. Regulation of estrogen receptor expression. *Breast Cancer Res Treat* 1994;31:183–189.
121. McDonnell DP, Clevenger B, Dana S, Santiso-Mere D, Tzukerman MT, Gleeson MAG. The mechanism of action of steroid hormones: a new twist to an old tale. *J Clin Pharmacol* 1993;33:1165–1172.
122. Hendry LB, Mahesh VB. A putatitve step in steroid hormone action involves insertion of steroid ligands into DNA facilitated by receptor proteins. *J Steroid Biochem Mol Biol* 1995;55:173–183.
123. Soto AM, Sonnenschein C. Mechanism of estrogen action on cellular proliferation: Evidence for indirect and negative control on cloned breast tumor cells. *Biochem Biophys Res Commun* 1984;122:1097–1103.
124. Soto AM, Sonnenschein C, Chung KL, Fernandez MF, Olea N, Serrano FO. The E-screen assay as a tool to identify estrogens: An update on estrogenic environmental pollutants. *Environ Health Perspect* 1995;103(suppl 7):113–122.
125. Soule HD, Vazquez, Long A, Alberts S, Brennan MJ. A human cell line from a pleural effusion derived from a breast carcinoma. *J Natl Cancer Inst* 1973;51:1409–1413.
126. Brooks SC, Locke ER, Soule HD. Estrogen receptor in a human cell line (MCF-7) from breast carcinoma. *J Biol Chem* 1973;248:6251–6253.
127. Soto AM, Justicia H, Wray JW, Sonnenschein C. p-Nonyl phenol: An estrogenic xenobiotic released from "modified" polystyrene. *Environ Health Perspect* 1991;92:167–173.
128. Welshons WV, Rottinghaus GE, Nonneman DJ, Dolan-Timpe M, Ross PF. A sensitive bioassay for detection of dietary estrogens in animal feeds. *J Vet Diagn Invest* 1990;2:268–273.
129. Soto AM, Sonnenshein C. The role of estrogens on the proliferation of human breast tumor cells (MCF-7). *J Steroid Biochem* 1985;23:87–94.
130. Arnold SF, Robinson MK, Notides AC, Guillette LJ, McLachlan JA. A yeast estrogen screen for examining the relative exposure of cells to natural and xenoestrogens. *Environ Health Perspect* 1996;104:544–548.
131. Ramamoorthy K, Wang F, Chen I, Norris JD, McDonnell DP, Leonard LS, Gaido KS, Bocchinfuso WP, Korach KS, Safe S. Estrogenic activity of a dieldrin/toxaphene mixture in the mouse uterus, MCF-7 human breast cancer cells, and yeast-based estrogen receptor assays: no apparent synergism. *Endocrinology* 1997;138:1520–1527.
132. Flouriot G, Vaillant C, Salbert G, Pelissero C, Guiraud JM, Valotaaire Y. Monolayer and aggregate cultures of rainbow trout hepatocytes: long-term and stable liver-specific expression of aggregates. *J Cell Sci* 1993;105:407–416.
133. Anderson MJ, Miller MR, Hinton DE. *In vitro* modulation of 17-β-estradiol-induced vitellogenin synthesis: effects of cytochrome P4501A1 inducing compounds on rainbow trout (*Oncorhynchus mykiss*) liver cells. *Aquatic Toxicol* 1996;34:327–350.
134. Tzukerman MT, Esty A, Santiso-Mere D, Danielen P, Parker MG, Stein RB, Pike JW, McDonnell DP. Human estrogen receptor transactivational capacity is determined by both cellular and promoter context and mediated by two functionally distinct intrmolecular regions. *Mol Endocrinol* 1994;9:21–30.
135. Berry M, Metzger D, Chambon P. Role of the two activating domains of the oestrogen receptor in the cell-type and promoter-context dependent agonistic activity of the anti-oestrogen 4-hydroxytomoxifen. *EMBO J* 1990;9:2811–2818.
136. Metzger D, White JH, Chambon P. The human oestrogen receptor functions in yeast. *Nature* 1988;334:31–36.
137. McDonnell DP, Nawaz Z, Densmore C, Weigel NL, Pham TA, Clark JH, O'Malley BW. High level expression of biologically active estrogen receptor in Sacharomyces cerevisiae. *J Steroid Biochem Mol Biol* 1991;39:291–297.
138. Connor K, Howell J, Chen I, Liu H, Berhane K, Sciarretta C, Safe S, Zacharewski T. Failure of chloro-S-triazine-derived compounds to induce estrogen receptor-mediated responses *in-vivo* and *in-vitro*. *Fundam Appl Toxicol* 1995;30:93–101.
139. Routledge EJ, Sumpter JP. Estrogenic activity of surfactants and some of their degradation products assessed using a recombinant yeast screen. *Environ Toxicol Chem* 1996;15:241–248.

140. Gaido KW, Leonard LS, Lovell S, Gould JC, Babai D, Portier CJ, McDonnell DP. Evaluation of chemicals with endocrine modulating activity in a yeast-based steroid hormone receptor gene transcription assay. *Toxicol Appl Pharmacol* 1997;143:205–212.
141. Sathyamoorthy N, Wang TTY, Phang JM. Stimulation of pS2 expression by diet-derived compounds. *Cancer Res* 1994;54:957–961.
142. Roelant CH, Burns DA, Scheirer W. Accelerating the pace of luciferase reporter gene assays. *Biotechniques* 1996;20:914–917.
143. Demirpence E, Pons M, Balaguer P, Gagne D. Study of an antiestrogenic effect of retinoic acid in MCF-7 cells. *Biochem Biophys Res Commun* 1992;183:100–106.
144. Villeneuve DL, Crunkilton RL, DeVita WM. Aryl hydrocarbon receptor-mediated toxic potency of dissolved lipophilic organic contaminants collected from Lincoln Creek, Milwaukee, Wisconsin, USA, to PLHC-1 (*Poeciliopsis lucida*) fish hepatoma cells. *Environ Toxicol Chem* 1997;16:977–984.
145. Tillitt DE, Ankley GT, Verbrugge D, Giesy JP, Ludwig JP, Kubiak TJ. H4IIE rat hepatoma cell bioassay derived 2,3,7,8-tetrachlorodibenzo-ρ-dioxin equivalents in colonial fish-eating waterbird eggs from the Great Lakes. *Arch Environ Contam Toxicol* 1991;21:91–101.
146. Jones PD, Giesy JP, Newsted JL, Verbrugge DA, Beaver DL, Ankley GT, Tillitt DE, Lodge KB, Niemi GJ. Determination of 2,3,7,8-tetrachlorodibenzo-ρ-dioxin equivalents in tissues of birds at Green Bay, Wisconsin, USA. *Arch Environ Contam Toxicol* 1993;24:345–354.
147. Sanderson JT, Aarts JMMJG, Brouwer A, Froese KL, Denison MS, Giesy JP. Comparison of Ah receptor-mediated luciferase and ethoxyresorufin *O*-deethylase induction in H4IIE cells: implications for their use as bioanalytical tools for the detection of polyhalogenated aromatic hydrocarbons. *Toxicol Appl Pharmacol* 1996;137:316–325.
148. Huckins JN, Tubergen MW, Manuweera GK. Semipermeable membrane devices containing model lipid: a new approach to monitoring the bioavailability of lipophilic contaminants and estimating their bioconcentration potential. *Chemosphere* 1990;20:533–552.
149. Sumpter JP, Jobling S. Vitellogenesis as a biomarker for estrogenic contamination of the aquatic environment. *Environ Health Perspect* 1995;103(suppl 7):173–178.
150. Miles-Richardson SR, Fitzgerald SD, Render J, Barbee S, Giesy JP, Kramer VJ. Effects of waterborne exposure of 17-β-estradiol on secondary sex characteristics and gonads of fathead minnow (*Pimephales promelas*). 1997, submitted.
151. Crews D, Bergeron JM, McLachlan JA. The role of estrogen in turtle sex determination and the effect of PCBs. *Environ Health Perspect* 1995;103(suppl 7):73–77.
152. Gimeno S, Gerritsen A, Bowmer T, Komen H. Feminization of male carp. *Nature* 1996;384:221–222.

PART II

Xenobiotics and Transcription Factors

Toxicant–Receptor Interactions
Edited by Michael S. Denison and William G. Helferich

5

Barbiturate-Inducible Gene Expression

Armand J. Fulco

Department of Biological Chemistry, UCLA School of Medicine, Los Angeles, California, USA

BIOLOGIC EFFECTS OF BARBITURATES IN MAMMALS

Phenobarbital (PB) and many other barbiturates have a variety of biologic effects in mammals, some of which apparently are unrelated. The sedative and anticonvulsant effects of barbiturates are well known, and barbiturates (especially PB) have been used medically to relieve insomnia and to prevent seizures, particularly in epileptics. Other medical uses of specific barbiturates include the treatment of a variety of anxiety–tension states and, for exceptionally short-acting derivatives, utilization as general anesthetics. Mechanistically, barbiturates can be classified as lipophilic gamma aminobutyric acid (GABA)–receptor agonists that function to inhibit the excitable cells of the central nervous system and other tissues. Most of the neurologic effects of barbiturates can be attributed, in part at least, to their interaction with GABA receptors or, in some cases, to their physical interaction with and modification of biologic membrane lipids (1). Other biologic alterations mediated by barbiturates, however, cannot be explained easily by GABA-receptor or membrane-interaction mechanisms. These include tumor promotion, especially in the liver in some mammalian species, and the induced expression of specific P450 cytochromes and a number of other proteins in humans and other higher animals.

PB and Other Barbiturates as Mammalian Tumor Promoters

Tumor promoters are agents that do not directly react with genetic material (i.e., DNA) but rather affect its expression by a variety of mechanisms to enhance the effectiveness of tumor formation caused by carcinogens (i.e., initiating agents or their metabolic products that directly react with DNA to cause neoplasms). PB long has been recognized as a hepatic-tumor promoter in rodents and in some other animal species but, at therapeutic levels, apparently not in humans (2–5). The exact mechanism by which PB and other barbiturates function as tumor promoters still is unknown, although a number of correlations have been established. In general, PB and other barbiturates elicit a "pleiotropic response" in the livers of treated animals, that is, stimulation of cellular growth, proliferation of smooth endoplasmic reticulum (ER), hepatocytomegaly, and liver hypertrophy with accompanying increases in DNA and protein synthesis (including induction of specific enzymes). The relationship between the potency of a given barbiturate to elicit a pleiotropic response and its effectiveness as a tumor promoter is a strong one, and this correlation seems to hold as well for barbiturate analogs and nonbarbiturate tumor promoters (3, 6–10).

Induction of Specific Proteins in Mammals by Barbiturates

As noted, in addition to a variety of more general biologic effects mediated by barbiturates they also induce specific proteins in mammals. These proteins may be divided conveniently into four distinct categories. These are specific cytochrome P450 monooxygenases, cytochrome P450 reductases, the so-called "phase II" enzymes that normally function in concert with P450s, and other proteins not obviously related to the metabolism of xenobiotics.

Cytochrome P450s constitute ancient, multigene families of heme proteins (11–13) that, in mammals, are found in greatest variety and concentration in the ER of the liver. In the presence of O_2, NAD(P)H, and one or more electron-transfer proteins, they can catalyze the oxygenation or oxidative transformation of a wide variety of hydrophobic xenobiotics that, in turn, usually act as inducers of the P450s that act on them. In rat liver, the most exhaustively studied source of xenobiotically inducible P450s, the primary P450 genes inducible by barbiturates are CYP2B1 and CYP2B2 (P450b and P450e in the older literature), although the expression of several other P450 genes in rat liver also is enhanced by barbiturates but usually to lesser degree (14–16). Analogous barbiturate-inducible P450s are found in other rodents and in most other mammals tested including nonhuman primates and man (14, 15, 17, 18).

Cytochrome P450 reductases are membrane-associated proteins, containing both FMN and FAD, that serve to transfer electrons from NADPH to the heme moieties of the various P450s of the ER. Although there may be scores of distinct P450 cytochromes in the livers of most mammalian species, there usually is only one major reductase in any given species that interacts catalytically with all of the various P450s of the ER (19). Interestingly, this contrasts with the situation in higher plants, in which multiple forms of P450 reductase have been demonstrated to function in one plant species (20).

Phase II enzymes of mammalian liver include species of epoxide hydrolases, glutathione S-transferases, aldehyde dehydrogenases, glucuronosyltransferases, N-acetyltransferases, and sulfotransferases among others. In addition to directly transforming a variety of primary xenobiotics, they also may act on either the immediate products of P450 oxygenation or the products produced by other phase II enzymes. In general, phase II enzymes are induced by the same xenobiotics that induce specific P450s, but their degree of inducibility by a given xenobiotic is almost always significantly lower than the inducibility of the specific P450s induced by the same xenobiotic. This is true for the major PB-inducible mammalian P450s (14, 15).

Other proteins not obviously related to the metabolism of xenobiotics but nevertheless inducible by barbiturates have been the subject of numerous reports, but by far the best-characterized is a major acute-phase protein in rats and humans, α-1-acid glycoprotein (21). The mechanism of induction of this protein by PB is discussed later in this chapter. The investigation of other proteins apparently not involved in xenobiotic metabolism but induced or repressed by PB is just beginning. A hint of what the future may hold in this area is contained in the abstract of a presentation [Extent and character of phenobarbital-mediated gene expression] by F. Früh, J. Ourlin, and U. Meyer at the XIth International Symposium on Microsomes and Drug Oxidations, Los Angeles, 1996. These authors reported that, in a nonmammalian system (chick embryo), they could detect more than fifty cDNA fragments reflecting PB-mediated increases or decreases in the levels of mRNAs after in-vivo treatment with the barbiturate. About 30 of these showed increases, the remainder decreases. Among the genes shown for the first time to be barbiturate-regulated were fibrinogen β- and γ-chains, complement factor H, retinal glutamine synthetase, elongation factor 1δ, and apolipoprotein B; several previously shown to be barbiturate-induced (P4502H1, glutathione S-transferase, and UDP-glucuronosyl-transferase) also were detected, and more than 30 others remained to be identified.

Species and Strain Differences in Response to Barbiturates and Barbiturate Analogs

One of the more puzzling aspects of the induction of P450 cytochromes and other proteins in mammals by barbiturates and "barbiturate-like" compounds is the great structural diversity among these compounds and especially the often striking species- and strain-dependent differences observed in their potency as inducers (14). For example, 1,4-bis[2-(3,5-dichloropyridyloxy)]benzene (TCPOPOP), a PB-like inducer (22) is a potent P450 inducer in mice but not in rats or guinea pigs (23, 24), although PB can induce P450s in all of these rodents (25, 26). Nevertheless, marked strain-dependent differences (within the same species) in the effectiveness of PB and other barbiturates and barbiturate analogs as inducers in liver also is a common observation (27–29). It should also be noted that, in common with most other inducible proteins, barbiturate-inducible proteins generally are tissue-specific, with the liver usually the major site of induction (15, 30).

INDUCTION OF SPECIFIC PROTEINS BY BARBITURATES IN NONMAMMALIAN ORGANISMS

For many years, the barbiturate-mediated induction of proteins (specifically P450 cytochromes and phase II enzymes) was assumed tacitly to be the province of higher animals. It was with some surprise, therefore, that Reichhart and coworkers reported, in 1979 (31), that the cytochrome P450 content in Jerusalem artichoke tuber cells could be induced by PB. Although barbiturate-mediated induction of P450s has not been intensively studied in plants, Borlakoglu and John (32) demonstrated that in several plant tissues other than Jerusalem artichoke, microsomal P450 content could be induced by PB. Barbiturate-mediated induction of P450 cytochromes and in some cases phase II enzymes also has been demonstrated in a variety of other nonmammalian organisms including fowl (33–35), fish (36), invertebrates and eukaryotic microorganisms (37–46), and bacteria (47-49). Nevertheless, with the exception of the chicken, barbiturate-inducible enzymes identified in nonmammalian organisms are few in number relative to those characterized from mammals. In part, this reflects the focus on higher animals of those researchers who are interested in barbiturate-mediated induction of enzymes. It also should be noted, however, that P450s, because of their unique spectral properties, are relatively easy to detect and often are the only proteins that are analyzed with respect to barbiturate inducibility. Thus the perception that most nonmammalian species lack barbiturate-inducible proteins may be a consequence of the relative scarcity of multiple forms of P450 cytochromes in most lower organisms.

Induction by Barbiturates of P450 Cytochromes in *Bacillus megaterium*

Background

Barbiturate-Inducible P450 Cytochromes of B. Megaterium

To date, the two most intensively studied barbiturate-inducible systems are those from the gram-positive bacterium, *Bacillus megaterium* (induction of cytochromes $P450_{BM-3}$ [CYP102] and $P450_{BM-1}$ [CYP106]), and the rat (induction of CYP2B1 and CYP2B2 and α-1-acid glycoprotein). Because the *B. megaterium* system, studied chiefly in the author's laboratory, is probably the simpler and better characterized of the two and has provided an induction-mechanism model for comparison with higher animal systems, it is considered first. Cytochromes $P450_{BM-3}$ and $P450_{BM-1}$ were discovered, characterized and later cloned (including their regulatory regions) and expressed in *Escherichia coli*, primarily in the author's laboratory. Structure–function studies on these monooxygenases have been reviewed in detail (50) and are summarized only briefly here.

$P450_{BM-3}$, a soluble protein, has an apoprotein molecular weight of 117,641 Da (1048 amino acid residues) and functions as a catalytically self-sufficient long-chain fatty-acid monooxygenase. In contrast to the microsomal P450 systems of mammalian liver, $P450_{BM-3}$ combines both the substrate oxygenation function (in a heme domain) and electron-transport functions (in a reductase domain containing both FAD and FMN) in one large polypeptide with the domains separated by a linker that can be cleaved by trypsin without loss of the individual domain functions (51–59). As a consequence, perhaps, $P450_{BM-3}$ is the most catalytically active P450 characterized to date. The $P450_{BM-3}$ heme and reductase domains show significant structural homology to the two-component microsomal P450-reductase systems of eukaryotes but not to known bacterial systems. In *B. megaterium*, $P450_{BM-3}$ can be induced several hundred–fold by barbiturates. $P450_{BM-1}$ (molecular weight 47,479 Da) shows sequence similarity to $P450_{CAM}$ and to several other bacterial P450s and can be moderately induced by barbiturates (approximately 20-fold) in *B. megaterium*. To date, no substrates or reductase components have been identified for $P450_{BM-1}$ (60, 61). A third barbiturate-inducible fatty-acid monooxygenase in *B. megaterium*, cytochrome $P450_{BM-2}$ (48), has been partially characterized (52, 60) but not yet sequenced or cloned.

Early Studies

Experiments showing that the P450s of *B. megaterium* were inducible by PB were first described in 1982 (47). Since then, it has been demonstrated that a large number of other barbiturates as well as disubstituted acyl ureas, many peroxisome proliferators, and at least three nonsteroidal anti-inflammatory drugs are also inducers (62–66). Most of these are more effective than PB itself on a molar basis and, within the first two groups (barbiturates, disubstituted acyl ureas), there are good correlations between lipophilicity and inducer potency (63, 64).

Although the *B. megaterium* P450s (including $P450_{BM-2}$) are all barbiturate-inducible, the relative levels of induction of the three different P450s in response to inducers vary with inducer structure (64, 67, 68). This observation suggested that the regulation of expression of the three genes differed from each other to some degree, a hypothesis later shown to be correct. Observable stimulation of P450 synthesis in *B. megaterium* after barbiturate inducers were added to log-phase cultures was rapid (less than 5 minutes), with maximum rates often achieved in less than 30 minutes. Inhibitors of protein or RNA synthesis strongly inhibited P450 induction, but DNA synthesis inhibitors did not (53, 68, 69); specific mRNAs were induced by barbiturates, but RNA stability was not affected (68; Ashby M, He JS, Fulco AJ, unpublished experiments). It therefore seemed evident that barbiturate-mediated induction of P450s in *B. megaterium* was regulated, in part at least, at the level of transcription.

Barbiturate-Mediated Regulation of P450 Expression in B. megaterium at the Transcriptional Level

Barbiturate-Responsive cis-Acting Regulatory Elements: Barbie Boxes and Related Sequences in the 5′-Flanking Regions of the B. megaterium P450 Genes

Once the 5′-flanking regulatory regions of the $P450_{BM\text{-}3}$ and $P450_{BM\text{-}1}$ genes had been sequenced (56, 61), it became possible to compare the two for homologies that might reveal putative cis-acting sequences involved in barbiturate-mediated regulation. The only homologies apparent were almost identical 17–base pair DNA sequences, one in the 5′-flanking regulatory region of $P450_{BM\text{-}3}$ and the other in the analogous region of the $P450_{BM\text{-}1}$ gene. These two elements, now called "Barbie boxes" and shortened to 15 base pairs (70), were identified by a computer-assisted search of the 5′-flanking regions of the $P450_{BM\text{-}3}$ and $P450_{BM\text{-}1}$ genes and also in the analogous 5′-flanking regions of the rat CYP2B1/2 genes. We later were able to demonstrate experimentally that all four Barbie boxes could bind barbiturate-responsive proteins from *B. megaterium* and the rat (71). Once the potential importance of Barbie boxes in the *B. megaterium* system was recognized, we asked two questions. First, were Barbie box elements found in the regulatory regions of other genes that encoded barbiturate-inducible proteins, and second, if present, did they play significant roles in the barbiturate-mediated induction of these genes? The unequivocal answer to the first question is yes; Barbie box elements have been found in the regulatory regions of essentially all barbiturate-inducible genes whose 5′-flanking sequences have been reported (Table 1). The second question remains to be comprehensively answered for most barbiturate-inducible genes but, as discussed in the next section, Barbie boxes are involved in the barbiturate-mediated induction of cytochromes $P450_{BM\text{-}3}$ and $P450_{BM\text{-}1}$ (72, 73) and rat α-1-acid glycoprotein (74).

In addition to the Barbie-box elements, three other cis-acting sequences (operators) that show similarity (especially in a conserved poly A motif) to the Barbie box consensus sequence have been identified experimentally by gel retardation and nuclease protection (footprinting) assays carried out in the presence of various trans-acting regulatory factors. These operator sites include O_{III}, a 20–base pair (bp) perfect palindromic sequence just 3′ of the transcription start site of in the 5′-flanking region of the $P450_{BM\text{-}3}$ gene that serves as a repressor (Bm3R1 protein) binding site (75), and two imperfect palindromic sequences, O_I and O_{II} (73). O_I, located just 5′ to the transcription start site of the $P450_{BM\text{-}1}$ gene, and O_{II}, which straddles this site, can bind to both Bm3R1 repressor and several positive regulator proteins (73). The relative location of all known cis-acting sequences in the regulatory regions of the $P450_{BM\text{-}3}$ and $P450_{BM\text{-}1}$ genes is shown schematically in Fig. 1. Although

Table 1. *Comparison of 5′-flanking sequence from genes encoding barbiturate-inducible enzymes with a 15 bp consensus sequence (Barbie box) that binds barbiturate-responsive proteins from rat or Bacillus megaterium*

Gene	5′ Location[a]	15-Base sequence															Identity	Notes, references
		*	*			*	*	*	*	*		*	*		*	*		
Percentage of sequences containing consensus base		63	56	50	38	100	100	100	100	63	50	69	56	31	56	69		
		|	|	|	|	|	|	|	|	|	|	|	|	|	|	|		
Consensus Sequence		A	T	C	A	**A**	**A**	**A**	**G**	C	T	G	G	A	G	G	**15**	
P450BM-3 (*B. megaterium*)	−227	A	T	C	A	**A**	**A**	**A**	**G**	C	T	G	G	t	G	G	**14**	Involved in transcription[b], 72
P450BM-1 (*B. megaterium*)	−302	A	T	a	A	**A**	**A**	**A**	**G**	C	T	G	G	t	G	c	**12**	Involved in transcription[b], 72
P450b [CYP2B1] (rat)	−73	A	g	C	t	**A**	**A**	**A**	**G**	C	a	G	G	A	G	G	**12**	Involved in transcription, 100
P450e [CYP2B2] (rat)	−73	A	g	C	c	**A**	**A**	**A**	**G**	C	a	G	G	A	G	G	**12**	Involved in transcription, 100
P450 CYP2C1 (rabbit)	−228	t	T	C	A	**A**	**A**	**A**	**G**	a	g	G	G	c	t	t	**9**	*Not tested*
Epoxide hydrolase (rat)	−84	g	T	C	t	**A**	**A**	**A**	**G**	t	c	c	a	g	G	G	**8**	*Not tested*
Glutathione-S-transferase (mouse)	−166	A	g	g	g	**A**	**A**	**A**	**G**	g	T	G	G	t	G	G	**10**	*Not tested*
Aldehyde dehydrogenase (rat)	−168	A	T	t	t	**A**	**A**	**A**	**G**	g	c	a	a	A	G	G	**9**	*Not tested*
P450 CYP6A1 (house fly)	−188	A	a	a	A	**A**	**A**	**A**	**G**	C	T	G	a	A	t	G	**11**	*Not tested*
Aldehyde dehydrogenase (mouse)	−574	A	T	C	A	**A**	**A**	**A**	**G**	C	T	t	G	g	G	a	**13**	*Not tested*
	−58	A	T	a	A	**A**	**A**	**A**	**G**	g	a	G	c	A	a	G	**10**	*Not tested*
P450 CYP3A2 (rat)	−1166	A	T	a	g	**A**	**A**	**A**	**G**	C	a	t	t	c	t	G	**8**	*Not tested*
	−1007	c	a	t	t	**A**	**A**	**A**	**G**	C	c	t	G	t	G	G	**8**	*Not tested*
	−48	t	c	C	c	**A**	**A**	**A**	**G**	C	T	G	t	g	t	G	**9**	*Not tested*
α_1-acid glycoprotein (rat)	−127	g	c	C	c	**A**	**A**	**A**	**G**	C	T	G	G	c	t	t	**9**	Involved in transcription[b], 74
SU1 [128bp] (*S. griseolus*)	−21	c	T	t	t	**A**	**A**	**A**	**G**	g	T	G	a	g	a	a	**7**	Protected in footprinting, 49

Note. *More than 50% of the sequences contain the consensus base at this position.
[a]Counting from the translation start site except for α_1-acid glycoprotein and the CYP2B1/B2 genes (transcription start site).
[b]Determined by CAT assays using mutated and wild-type Barbie boxes.

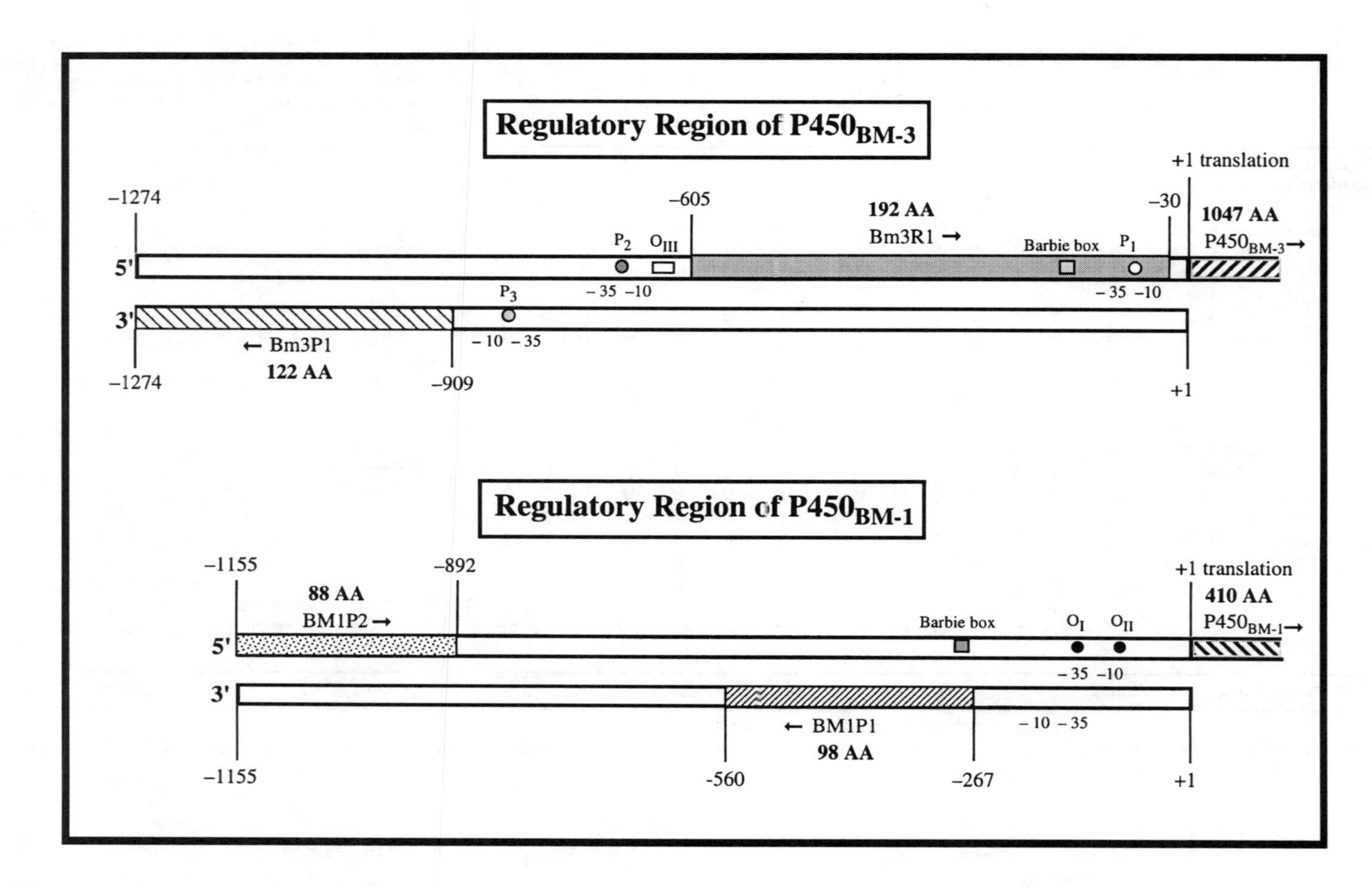

Regulatory Region of P450BM-3
+1 translation
-1274
-605
192 AA
Bm3R1 →
-30
1047 AA
P450BM-3 →
P2
OIII
Barbie box
P1
5'
-35 -10
P3
-35 -10
3'
- 10 - 35
← Bm3P1
122 AA
-1274
-909
+1
Regulatory Region of P450BM-1
-1155
-892
+1 translation
88 AA
BM1P2 →
410 AA
Barbie box
OI
OII
P450BM-1 →
5'
-35 -10
3'
← BM1P1
98 AA
- 10 - 35
-1155
-560
-267
+1

site-specific mutagenesis studies on O_I and O_{II} have not yet been carried out, circumstantial evidence indicates that their integrity is important in the regulation of transcription by barbiturates.

Significance of Barbie Box Sequences in the Barbiturate-Mediated Induction of the $P450_{BM\text{-}1}$ and $P450_{BM\text{-}3}$ Genes in B. megaterium

To evaluate the significance of Barbie box sequences in the regulation of the cytochrome P450 genes in *B. megaterium*, we carried out site-specific mutagenesis studies on these elements (72). A four-bp sequence (AAAG) is invariant and found in the same relative position in all Barbie box elements (see Table 1); we reasoned that if the Barbie box was functionally important to barbiturate-mediated induction, then mutations in this conserved subsequence should affect P450 expression. One constraint was imposed by the location of the *B. megaterium* Barbie boxes in critical open reading frames as illustrated in Fig. 1. In the $P450_{BM\text{-}3}$ Barbie box, the two base changes selected did not result in amino-acid changes in the sequence of the repressor, Bm3R1; the two base changes in the $P450_{BM\text{-}1}$ Barbie box resulted in two conservative changes in the amino-acid sequence of a positive regulatory factor, BM1P1, but with no detectable change in its binding properties (72). Mutation of the Barbie box located in the $P450_{BM\text{-}1}$ gene led to constitutive synthesis of cytochrome $P450_{BM\text{-}1}$ and a 10-fold increase of expression of *BM1P1*; mutation of the $P450_{BM\text{-}3}$ Barbie box significantly increased the expression of both $P450_{BM\text{-}3}$ and *Bm3P1* in response to pentobarbital induction but left the basal levels unaffected. The mutated Barbie boxes showed a decreased binding affinity for Bm3R1 compared with their wild-type (unmutated) counterparts. In these same series of experiments, proteins induced by pentobarbital (but different from BM1P1, BM1P2, and Bm3P1) in extracts of *B. megaterium* cells also specifically interacted with Barbie boxes. These putative positive factors also were present at high levels during late stationary phase of

←

FIG. 1. Regulatory features of the 5′-flanking regions of the $P450_{BM\text{-}3}$ and $P450_{BM\text{-}1}$ genes of *B. megaterium.* Two proteins are encoded in the 5′-flanking region of the cytochrome $P450_{BM\text{-}3}$ structural gene; these are Bm3R1, a transcriptional repressor for $P450_{BM\text{-}3}$ encoded in the same strand as $P450_{BM\text{-}3}$, and Bm3P1, a positive regulator of $P450_{BM\text{-}3}$ transcription encoded in the complementary strand. Bm3R1 and $P450_{BM\text{-}3}$ form a cotranscriptional unit regulated by the bicistronic operator, O_{III} (−646 CGGAATGAACGTTCATTCCG −627). The strong, barbiturate-activated promoter for this unit is labeled P_2; a second, much weaker, promoter, just upstream from the translation start site for cytochrome $P450_{BM\text{-}3}$ is labeled P_1. The promoter (P_3) for Bm3P1 in the complementary strand is ajacent to but does not overlap P_2. The $P450_{BM\text{-}3}$ Barbie box (−241 ATCAAAAGCTGGTGG −227) also is shown. Two positive regulator proteins, BM1P1 and BM1P2, are encoded in the 5′-flanking region of the structural gene encoding $P450_{BM\text{-}1}$. The promoters for the $P450_{BM\text{-}1}$ and *BM1P1* genes overlap in the −35 regions in opposite strands as indicated; the promoter for BM1P2 has not yet been located. The $P450_{BM\text{-}1}$ Barbie box (−316 ATAAAAAGCTGGTGC −302) and two operator sites, O_I (−188 TTTCCTTGATAACCAAG-TAA −168) and O_{II} (−154 TATTAGTACATTTTTATACTAATTGTATAAAAATGTACTAAT −113) are also shown.

B. megaterium cell cultures grown in the absence of barbiturates, presumably as the result of an endogenous inducer analogous to xenobiotic inducers such as barbiturates. Surprisingly, the mutated Barbie box sequences had greater binding affinity for these positive factors than did wild-type Barbie boxes. DNase I footprinting experiments showed that these positive factors protected a segment of DNA in the $P450_{BM-1}$ Barbie box, yet they did not protect sequences in the $P450_{BM-3}$ Barbie box region. It thus appears that these uncharacterized positive factors compete with Bm3R1 for binding to the Barbie box region, especially in the 5′-flanking region of the $P450_{BM-1}$ gene. Further support for this conclusion is based on nested-deletion analyses in the Barbie box region of $P450_{BM-1}$ (71, 72) that show that repressor and positive-factor binding sites overlap at the Barbie box sequence. These and other experimental results indicate that, in *B. megaterium*, Barbie boxes are important cis-acting elements involved in the coordinate regulation of the barbiturate-mediated expression of the $P450_{BM-1}$ and $P450_{BM-3}$ genes and the genes encoding their positive regulatory factors.

Barbiturate-Responsive Trans-Acting Regulatory Factors

In addition to the five known cis-acting sequences involved in barbiturate-mediated induction, there are at least six barbiturate-responsive protein factors that interact with one or more of these sequences. These trans-acting factors include three positive transcription factors (BM1P1, BM1P2 and Bm3P1) (73, 76), a critical repressor, Bm3R1, involved in the regulation of both $P450_{BM-3}$ and $P450_{BM-1}$ (70, 75), and two as yet uncharacterized Barbie box binding proteins that are putative positive regulatory factors (72). The first four factors as well as a point mutant of Bm3R1 have been cloned, sequenced, expressed in *E. coli*, and characterized with respect to their apparent function. The general properties and functions of these cloned proteins are summarized in Table 2; the location of their coding regions in the 5′-regulatory portions of the $P450_{BM-3}$ gene (encoding Bm3R1 and Bm3P1) and in the $P450_{BM-1}$ gene (encoding BM1P1 and BM1P2) are shown in Fig. 1.

Significance of Bm3R1 Repressor and Positive Regulatory Proteins in the Barbiturate-Mediated Induction of the $P450_{BM-1}$ and $P450_{BM-3}$ Genes in B. megaterium

The central role that Bm3R1 plays in the barbiturate-mediated regulation of the $P450_{BM-3}$ gene was clearly demonstrated by the results of experiments we published several years ago (70, 75). We recognized, prior to conducting these experiments, that about one kilobase of 5′ flanking sequence of the $P450_{BM-3}$ gene was required for barbiturate-inducible expression of the cytochrome $P450_{BM-3}$ gene in *B. megaterium* but the reason for this was unknown (69). By analysis of deletion and frame shift derivatives of this region (75), however, it became apparent that an open reading frame immediately upstream of the *B. megaterium* cytochrome $P450_{BM-3}$ structural gene (see Fig. 1) coded for a protein, Bm3R1, that contained a helix-turn-helix DNA binding motif (see Table 2) and negatively controlled the expression of

TABLE 2. *Barbiturate-responsive[a] proteins involved in the regulation of $P450_{BM-1}$ and $P450_{BM-3}$ gene expression at the level of transcription in Bacillus megaterium*

Protein[a]	Residues (molecular weight)	Protein type	Function
Bm3R1	192 (21,884)	Helix-turn-helix motif and dimerization domain; binds to *$P450_{BM-1}$* and *$P450_{BM-3}$* operators and Barbie boxes	Repressor that inhibits the expression of itself, *$P450_{BM-1}$*, *$P450_{BM-3}$* and three positive regulator proteins (BM1P1, BM1P2, Bm3P1)
Mutant Bm3R1 (39G → E)	192 (21,924)	The turn of the HTH motif is lost with the G → E change at residue 39; Does not bind to Barbie box or operator sites	No longer functions effectively as a repressor; *B. megaterium* containing this mutation is constitutive for $P450_{BM-1}$ and $P450_{BM-3}$
BM1P1	98 (11,506)	HTH motif; alone, binds to O_I and O_{II} in the 5′ flanking regions of *$P450_{BM-1}$* but not to Barbie boxes; interacts with Bm3R1 and positive regulator proteins.	A positive regulator for expression of *$P450_{BM-1}$* and *$P450_{BM-3}$*, inhibits binding of Bm3R1 to operators and Barbie boxes
BM1P2	88 (10,297)	DNA binding protein; alone, binds weakly to O_I and O_{II} in the 5′ flanking regions of *$P450_{BM-1}$* but not to Barbie boxes; enhances the positive effects of BM1P1	In concert with Bm1P1, stimulates the expression of $P450_{BM-1}$ and $P450_{BM-3}$ by inhibiting binding of Bm3R1 to *cis*-acting sequences
BM3P1	122 (13,657)	May interact with BM1P1, BM1P2 and/or Bm3R1; does not seem to bind to DNA	A positive regulator for *$P450_{BM-1}$* and *$P450_{BM-3}$* expression

[a]All of these proteins, including Bm3R1 are induced by barbiturates. In each case, primary regulation is at the transcriptional level as determined by direct measurements of specific mRNAs but regulation at the level of translation may also be implicated, especially for Bm3R1.

the *$P450_{BM-3}$* gene at the transcriptional level. Confirmation of the regulatory role of Bm3R1 was obtained by characterizing a *B. megaterium* mutant that constitutively produced cytochrome $P450_{BM-3}$ (75) and showed a dramatic amplification of $P450_{BM-1}$ synthesis (73, 77).[1] Complementation of this mutant by a DNA fragment containing the wild-type *Bm3R1* gene indicated that the mutation in this locus was trans-dominant. Sequence analysis of the *Bm3R1* gene and its upstream region from

[1]Although the *B. megaterium* mutant containing the G39E mutation of Bm3R1 expresses a much higher level of $P450_{BM-1}$ than wild-type *B. megaterium* if both are grown in the absence of barbiturates, growth of the mutant in the presence of pentobarbital still results in significant induction of $P450_{BM-1}$. This suggests that G39E–Bm3R1, which has no affinity for the O_{III} operator site in *$P450_{BM-3}$* still can bind as a repressor to one or more sites on the regulatory region of $P450_{BM-1}$. It also is possible that other factors (either barbiturate-induced proteins that can stimulate transcription by a mechanism not dependent on antagonizing repression by Bm3R1 or another barbiturate-responsive repressor specific for *$P450_{BM-1}$*) are involved in regulation of *$P450_{BM-1}$* expression.

this mutant revealed a single base change that resulted in a G to E substitution in the β-turn region of the DNA-binding motif (Table 2). A 20-bp perfect palindromic operator site (O_{III} in Fig. 1), located between the promoter sequences and the *Bm3R1* structural gene, was defined by both in-vivo titration of Bm3R1 repressor and electrophoretic mobility gel-shift assays with wild-type or mutant Bm3R1 protein.

Subsequent experiments (70) delineated the effect of barbiturate inducers on the interaction of Bm3R1 with a portion of *B. megaterium* DNA containing the bicistronic operator (O_{III}) and the promoter sequences. We showed that Bm3R1 protein bound specifically to a segment of DNA containing the promoter–operator region of the *Bm3R1* gene, and that Bm3R1 protected from DNase I digestion a region of DNA that covered and flanked the palindromic operator sequence. The interaction between Bm3R1 repressor and the O_{III} operator sequence, in vitro, was strongly inhibited by the addition of 2-mmol pentobarbital or 2-mmol methohexital (strong in vivo inducers of $P450_{BM-3}$) but not by the same concentration of PB (a relatively weak inducer) or by mephobarbital (a noninducer). A detailed comparison of pentobarbital and methohexital at concentrations lower than 2 mmol indicated that methohexital was 5 to 10 times more effective as an inhibitor of Bm3R1 binding in vitro compared with its seven-fold greater inducer potency in vivo (70). The striking correlation between the in-vitro inhibition effects of barbiturates on the interaction of Bm3R1 with O_{III} and their in-vivo potency as inducers of cytochrome $P450_{BM-3}$ strongly suggested that the barbiturate inducers functioned by interfering with the binding of Bm3R1 repressor to its operator site. More recently, English et al. (65) reported that many peroxisome proliferators also are potent inducers of cytochrome $P450_{BM-3}$ in *B. megaterium* and that the mechanism of induction likewise appears to involve the direct binding of the peroxisome proliferator to Bm3R1 repressor to cause its dissociation from the operator site. Other observations from our laboratory, however, indicated that a mechanism involving only the direct interaction of inducers with Bm3R1 could not account for barbiturate-mediated induction of $P450_{BM-1}$ (also regulated by Bm3R1 binding) and, indeed, may not be the whole story with $P450_{BM-3}$ induction. In particular, this simple hypothesis cannot explain the putative role of the *$P450_{BM-3}$* and *$P450_{BM-1}$* Barbie box sequences or the other operator sequences (O_I and O_{II}) in the 5′-flanking region of the *$P450_{BM-1}$* gene (see Fig. 1). Bm3R1 in vitro bound to all of these sequences, but in no case was binding directly inhibited by barbiturates (73). It also was apparent that other trans-acting factors were positive regulators of barbiturate-mediated induction and were themselves induced by barbiturates (73, 76). The roles of some of these positive regulators have now been clarified. Thus, analysis of a 1.3-kilobase segment of DNA immediately upstream from the cytochrome $P450_{BM-1}$ structural gene revealed two open reading frames (see Fig. 1). One, *BM1P1*, is located 267 base pairs upstream from the sequence encoding cytochrome $P450_{BM-1}$ but in the opposite orientation. The second, *BM1P2*, is 892 base pairs upstream from the $P450_{BM-1}$ coding sequence and in the same coding strand.

As Fig. 1 shows, the promoter sequences for *$P450_{BM-1}$* and *BM1P1* overlap with each other and with Barbie box and operator (O_I and O_{II}) sequences, a circumstance facilitating the coordinate regulation of the two genes. The expression of *BM1P1* and

BM1P2 as well as *P450*$_{BM-1}$ is strongly stimulated in *B. megaterium* cells grown in the presence of pentobarbital and the *BM1P1* gene product exerts positive control on expression of *P450*$_{BM-1}$. If a 177-bp fragment encompassing the overlapping promoter regions of the *P450*$_{BM-1}$ and *BM1P1* genes is used as a probe in DNA binding assays, the *BM1P1* and *BM1P2* gene products and Bm3R1 repressor can bind individually, but the addition of BM1P1 or BM1P2 to a binding mixture containing Bm3R1 completely prevented the appearance of a Bm3R1 binding band. If a 208-bp fragment containing a Barbie box sequence (but not O_I and O_{II}) and located upstream of the 177-bp fragment is used as a probe, only a Bm3R1 binding band is detected. Although neither BM1P1 and BM1P2 appear to bind to this 208-bp fragment, their presence strongly inhibits the binding of Bm3R1 to the same probe. Thus BM1P1 and BM1P2 probably act as positive regulatory proteins involved in the expression of the *P450*$_{BM-1}$ gene by interacting with and interfering with the binding of the repressor protein, Bm3R1, to the regulatory regions of *P450*$_{BM-1}$. At the same time, it should be remembered that at least two as-yet-uncharacterized barbiturate-inducible proteins can bind to Barbie box elements and directly compete with Bm3R1 for binding at these sites.

Mechanisms of Barbiturate-Mediated Regulation of Expression of the P450 genes of B. megaterium

The author's working hypothesis concerning the mechanism of barbiturate-mediated induction of cytochromes P450$_{BM-3}$ and P450$_{BM-1}$ can be summarized as follows (with the caveats that at least two trans-acting protein factors and the gene-encoding cytochrome P450$_{BM-2}$ have yet to be cloned and characterized and that certainly more cis-acting sequences involved in transcriptional regulation of the *B. megaterium* P450s remain to be identified).

Repression. In the absence of barbiturates or analogous inducers (or the endogenous inducers that barbiturates may mimic), Bm3R1 repressor binds strongly to the P450$_{BM-3}$ operator site, O_{III}, a perfect 20-bp palindromic sequence, and to a segment of DNA that includes the *P450*$_{BM-3}$ Barbie box sequence; in *P450*$_{BM-1}$ it also binds to O_I, O_{II} and the Barbie box (as already noted, Barbie box sequences show significant identity, particularly in the highly conserved poly-A motifs, to the three operator sequences). As a consequence of Bm3R1 binding to these cis-acting sites, the transcription of both P450 genes is repressed. Bm3R1 also regulates its own expression and suppresses the expression of the positive factors. How are these multiple repressions brought about by Bm3R1 binding? Although the explanations of the detailed mechanism remain at present unproved hypotheses, a careful analysis of the Bm3R1 binding sites does suggest how this might be accomplished. The binding of Bm3R1 to O_{III} and also to the Barbie box region of *P450*$_{BM-3}$ may cause a "looping out" of the intervening region that contains most of the DNA encoding the *Bm3R1* portion of the bicistronic message including the Shine–Dalgarno sequence (Fig. 2). Furthermore, the DNA encoding the mRNA start site, three bases upstream from O_{III}, is masked by contact with Bm3R1 (75). The weak promoter located between the Barbie box and the translation

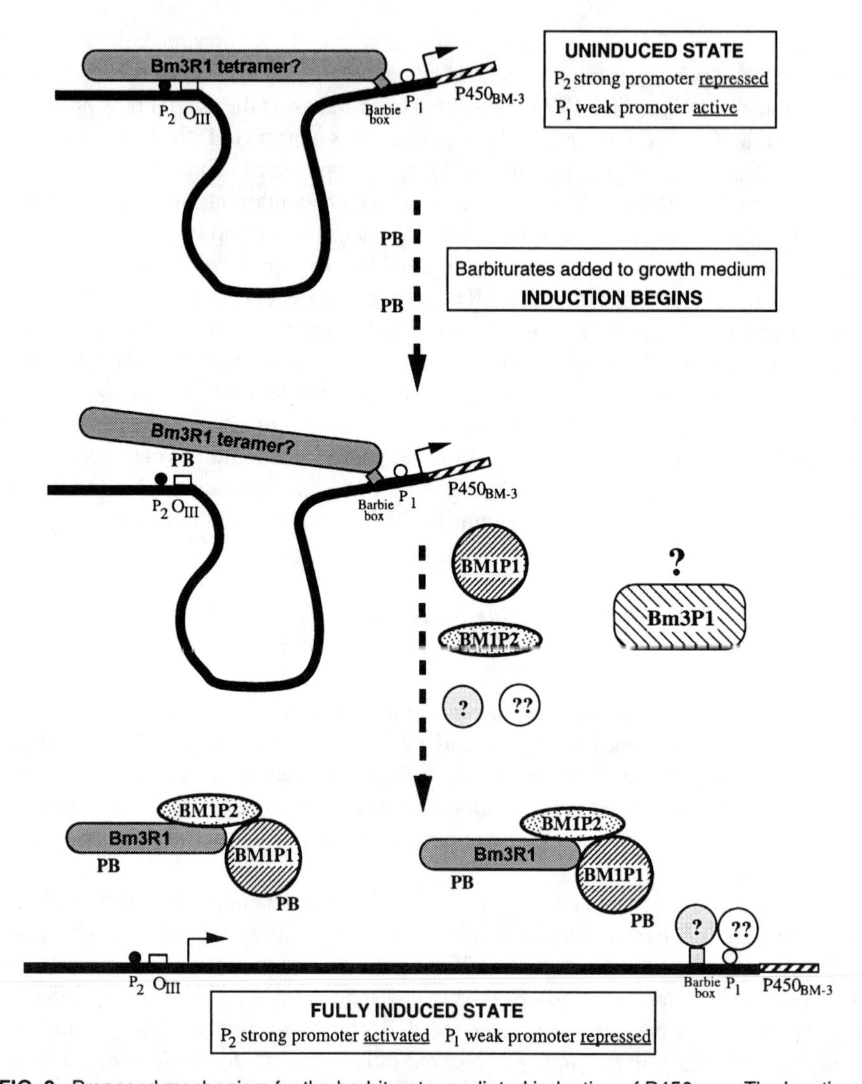

FIG. 2. Proposed mechanism for the barbiturate-mediated induction of $P450_{BM-3}$. The locations and properties of the various *cis*- and *trans*-acting factors in the scheme are shown in Fig. 1 and Table 2. The two proteins labeled *?* and *??* are positive regulator proteins that have not yet been cloned or characterized but have both been shown to bind to Barbie box sequences.

start site in $P450_{BM-3}$ is inhibited by barbiturates but may well be responsible for the extremely low basal level of cytochrome $P450_{BM-3}$ seen in the absence of inducers. For the $P450_{BM-1}$ gene a similar mechanism may be operating. Binding of Bm3R1 to the cis-acting sequences would loop out or mask the overlapping promoter regions of both $P450_{BM-1}$ and *BM1P1* to inhibit the transcription of both genes. Although two other barbiturate-inducible genes (*Bm3P1* and *BM1P2*), both encoding positive regulatory proteins, are located respectively in the 5′-flanking regions of $P450_{BM-3}$ and $P450_{BM-1}$, it is not clear from data presently available how Bm3R1 binding inhibits their transcription. It is possible that additional binding sites for Bm3R1 may exist in the regions of these genes, that different repressor proteins may bind in these regions or even that activation by barbiturate-inducible positive regulatory factors may be involved. The promoter regions for Bm3P1 and for Bm3R1/P450$_{BM-3}$ do not overlap but are close enough (see Fig. 1) that they could easily be coordinately regulated by protein binding in that region. Regardless, if pentobarbital (or other inducer barbiturates) are added to growing cultures of *B. megaterium*, all of the regulatory proteins examined to date (Bm3R1, Bm3P1, BM1P1, BM1P2, and two as-yet-uncharacterized Barbie box binding proteins) are induced as are cytochromes P450$_{BM-3}$ and P450$_{BM-1}$.

Induction. In the presence of inducer barbiturates, Bm3R1 repressor no longer binds to its bicistronic operator (O_{III}) and the $P450_{BM-3}$ Barbie box. Inducer barbiturates appear to interact with Bm3R1 to directly inhibit its ability to bind to O_{III}. Bm3R1 binding to its own Barbie box and also to the $P450_{BM-1}$ Barbie box and operator sites (O_I and O_{II}), however, is not directly inhibited by barbiturates; instead, complete release of repression seems to involve competitive binding either to cis-acting elements or to Bm3R1 itself by positive regulatory proteins. Although these proteins are themselves induced by barbiturates, they also may be activated by barbiturates without protein synthesis (71). In any event, they act in a concerted manner with the direct effect of barbiturates on Bm3R1 to cooperatively inhibit its binding to Barbie box and operator sites. Under these conditions, the rate of transcription of the P450 mRNAs is greatly accelerated, and barbiturate-mediated induction of $P450_{BM-1}$ and $P450_{BM-3}$ follows. As already noted, present evidence suggests that Bm3R1 binding to the Barbie box and operator sites causes a looping out of critical portions of the regulatory regions of both P450 genes. A putative mechanism for the barbiturate-mediated induction of cytochrome P450$_{BM-3}$, which is based on both published and unpublished evidence available to date and which incorporates a looping out of critical elements of the regulatory region, is presented in Fig. 2.

BARBITURATE-MEDIATED INDUCTION OF SPECIFIC PROTEINS IN RODENTS

Induction of Rat α-1-Acid Glycoprotein (Orosomucoid) by PB

α-1-Acid glycoprotein (AGP) is a major positive acute-phase protein in rodents, humans, and other mammals (78) whose concentration in plasma is increased in response to a variety of physical or physiologic insults including trauma, inflammation, infection, rheumatoid arthritis, certain neoplasms, and other malignancies (79–81). AGP gene expression is induced by a variety of cytokines and other stimulatory factors including interleukins (IL-1, IL-6, and IL-11), tumor necrosis factor α, leukemia inhibitory factor, and glucocorticoids (74, 82–84). In 1982 it was shown that there was increased binding of desmethylimipramine by AGP in plasma of PB-treated rats (85), and later it was clearly established that the administration of PB to rats causes an absolute increase in serum AGP content (86, 87). Porquet and coworkers (21, 88) demonstrated that the effect of PB was at the transcriptional level, as indicated by a greater than three-fold increase in AGP mRNA after PB treatment of rats.

When, in 1991 (71), homologous DNA sequences (Barbie box elements) in the 5′-flanking regions of genes encoding cytochromes $P450_{BM-1}$ and $P450_{BM-3}$ of *B. megaterium* and P4502B1 and P4502B2 of the rat were shown to bind barbiturate-responsive proteins, Porquet's laboratory began experiments to determine whether a similar sequence might play a role in the barbiturate-mediated induction of AGP. In 1994 they reported that a 17-bp Barbie box element was located at positions −140 to −124, counting from the transcription start site, in the 5′-flanking region of AGP (74). They showed, using electrophoretic mobility gel-shift assays, that a 17-bp oligonucleotide probe specific for the AGP Barbie box showed slight binding to liver nuclear protein from untreated animals. This binding, however, was strongly and specifically increased with protein extracts from PB-treated rats.

Transfection of rat primary hepatocytes with a construct (pAGPcat) containing a portion of the 5′-flanking sequence of the AGP gene (−763 to +20) that included the promoter and Barbie box inserted into a chloramphenicol acetyltransferase (CAT) reporter vector resulted in basal expression of CAT activity. Treatment of the cells with PB or dexamethasone caused a significant induction of CAT activity over this basal level. Induction of CAT activity by PB was abolished, however, when hepatocytes were transfected by constructs with a point mutation or deletion of the Barbie box sequence. Interestingly, basal activity and the response to dexamethasone also were decreased in these mutants. This could reflect the proximity of the glucocorticoid-responsive element (−120 to −107) (89) or perhaps indicate that both required the integrity of the glucocorticoid-responsive unit that partially overlaps the Barbie box (90). Nonetheless, these experiments established a significant role for the Barbie box sequence in AGP gene regulation by PB.

One could argue that the results described here also suggest a possible mechanistic relationship between induction of proteins by PB and by glucocorticoids. This hypothesis received support in a series of experiments reported by Shaw et al. (91), who examined the effect of the antiprogestin–antiglucocorticoid drug, RU486, on the induction of CYP2C6 (a moderately PB-inducible P450 in normal rat liver) in rat hepatoma cells (Fao and its derivatives). In general, liver-derived cell lines show no induction of proteins in response to barbiturate treatment. Although Fao cells show no PB-mediated induction of CYP2B1 (the major barbiturate-inducible P450 of normal rat liver or derived primary hepatocytes), they do induce CYP2C6 in response to either PB or dexamethasone (92). Shaw and coworkers (91) showed that the response to dexamethasone (as measured by CYP2C6 mRNA) was rapid, but induction by PB (a three to four-fold increase in CYP2C6 gene transcription) occurred only after an 8 to 10-hour lag period. Induction by PB of both accumulation of CYP2C6 mRNA and transcription of the gene was blocked by RU486, suggesting that a steroid receptor could be involved in the induction process. Transfection of Fao cells by promoter constructs containing a reporter gene whose expression was driven by a 1.4-kilobase 5′-flanking segment of the CYP2B1 or CYP2B2 genes led to about a three-fold increase in reporter gene activity in the presence of PB, a process that also was inhibited by RU486. On the basis of these findings, the authors proposed a general scheme in which PB acts indirectly to cause the accumulation of an endogenous steroid, and this molecule then acts via its receptor to directly induce the P450 cytochromes.

Results from Porquet's laboratory (93) involving the use of adrenalectomized or turpentine-treated rats clearly indicate, however, that PB-mediated induction of AGP and, by inference, of CYP2B1/2, are independent of both glucocorticoid and cytokine pathways. Most recently, in the abstract of a presentation "The phenobarbital-regulated binding of nuclear proteins to *cis* Barbie box sequences of the rat CYP2B1-2 and α-1-acid glycoprotein genes: a common molecular mechanism for the multiple gene-inducing effects of phenobarbital" by D. Porquet, N. Mejdoubi, E. Bui, and G. Durand at the XIth International Symposium on Microsomes and Drug Oxidations, Los Angeles, July 1996, it was reported that, in a strain of Zucker rat (fa/fa) resistant to PB for the induction of the CYP2B1/2 genes, the α-1-acid glycoprotein gene also failed to show induction. Porquet et al. concluded from their data that in Sprague-Dawley and Zucker rats, at least, the Barbie box is important in the PB-mediated regulation of both the CYP2B1/2 genes and the α-1-acid glycoprotein gene, and that the binding of PB-responsive proteins to Barbie box sequences is a critical step of a common molecular transduction pathway involved in PB-mediated induction. Porquet and coworkers also indicated that protein synthesis following administration of PB was not necessary to observe PB inductive effects, suggesting that a post-translational modification of the transactivating factor is the basis of the increase of binding to the Barbie box sequences (93). This hypothesis is supported by their results that demonstrated that binding of transactivating proteins to Barbie box sequences was abolished after a phosphatase treatment of nuclear protein extracts, and increased after incubation of the same extracts with protein kinase A, and also that PB treatment enhanced phosphorylation of the transactivating factor that bound to Barbie box sequences. The research findings from Porquet's laboratory, primarily with AGP, are generally consistent with the results obtained by Padmanaban and his colleagues with the PB-mediated regulation of the CYP2B1/2 genes, as described subsequently.

Induction of Rat CYP2B1 and CYP2B2 (Cytochromes P450b/P450e) by Phenobarbital

The regulation of the major PB-inducible proteins of rat liver microsomes, cytochromes P450B and P450E (CYP2B1/2), has been a subject of intense study for the past 15 years (15, 30). Nevertheless, as pointed out previously, the lack of immortalized cell-culture systems that retain barbiturate inducibility of proteins and the rapid loss of such inducibility in primary hepatocytes hindered progress for many years. More recently, however, the use of special basement-membrane matrices coupled with improved serum-free media (94, 95) has facilitated transfection experiments in primary liver-cell cultures, and refinements in the use of in-vitro transcription systems (15) have accelerated studies designed to elucidate the roles of specific cis- and trans-acting factors in the barbiturate-mediated induction of the CYP2B1/2 genes. Although many groups have made seminal contributions to our understanding of the regulation of these genes, a comprehensive coverage of their work in this field is not possible within the space limitations of this chapter. Instead, the reader is referred to a number of excellent reviews (14–18, 30, 96) that, together, cover most of this work through 1992. This chapter concentrates on the research published during the past few years as it relates to the regulation by barbiturates at the transcriptional level of the rat CYP2B1/2 genes.

Studies by Padmanaban and coworkers since 1989 have focused on the near 5′ up-stream region of the CYP2B1/2 genes with emphasis on identifying the barbiturate-responsive binding sites in this region, the proteins that bind to these cis-acting elements and the nature of their interactions as affected by PB. Rangarajan and Padmanaban (97) initially detected one or more PB-modulated nuclear protein transcription factors from rat liver that could bind to a portion of the 5′-flanking region of the CYP2B1/2 gene (nt −179 to nt +1, the transcription start site) in a manner that correlated well with barbiturate-mediated transcription of these genes. More intriguing, from our point of view, was their demonstration that nuclear extracts from the livers of rats treated with PB generated a DNase protection footprint that spanned base pairs −54 to −89. This protected region contained a complete Barbie box sequence that we later showed could bind both a putative positive regulator protein found in nuclear extracts from liver cells of PB-treated rats and barbiturate-responsive proteins from *B. megaterium* (71). On the other hand, Rangarajan and Padmanaban estimated that the binding protein in their system was about 85 kDa, a size significantly greater than that of the barbiturate-responsive Barbie box binding protein we detected in nuclear extracts from liver cells of PB-injected rats, and three times the size of a 26 to 28-kDa protein that Padmanaban and coworkers later identified as binding to the same region (see subsequent text). Furthermore, in the Rangarajan-Padmanaban experiments, cycloheximide, a protein-synthesis inhibitor, blocked the PB effect if it was administered to rats before the drug was injected. In our system, apparently the same binding protein found in rats treated with PB, as judged by the results of gel-retardation assays, could also be generated in nuclear extracts from untreated rats by prolonged preincubation of the extracts with the barbiturate.

In 1992, Upadhya et al. (98) in Padmanaban's laboratory described a somewhat different positive cis-acting DNA element in the near 5′-upstream regions of the CYP2B1/B2 genes in rat liver that they implicated in the barbiturate-mediated regulation of these genes. This element spanned bases −69 to −98 and thus overlapped the footprinted region described previously (97) and included the complete Barbie box (−73 to −87 nt). An oligonucleotide spanning the −69 to −98 nt region mimicked the electrophoretic mobility gel-shift pattern given by a larger fragment (−179 to +29 nt) that conferred PB-mediated regulatory features on these genes. Two major complexes were observed; the slower became intense under uninduced conditions while the faster-moving complex was strongly enhanced if nuclear extracts from the livers of PB-treated rats were used. A series of in-vitro transcription experiments using whole nuclei and cell-free transcription extracts clearly established that the −69 to −98 nt region was acting as a positive *cis*-acting element in the transcription of the CYP2B1/B2 genes that was involved in mediating the inductive effects of PB.

More recent studies from Padmanaban's laboratory have further clarified the roles of various *cis*- and *trans*-acting factors involved in the PB-mediated regulation of the CYP2B1/2 genes. Ram et al. (99) showed that the region −160 to −127 nt of the upstream of CYP2B1/B2 gene appears to function as a negative *cis*-acting element as deduced from DNase I footprint, gel shift, and cell-free transcription assays. They observed a reciprocal relationship in the interaction of this negative element and the previously described (97, 98) positive elements with their respective protein factors under repressed and PB-induced conditions of the CYP2B1/B2 gene. Thus, in gel-retardation assays, the −160 to −127 negative element interacted strongly with proteins in nuclear extracts from untreated controls but not with nuclear extracts from PB-injected animals. At the same time, PB treatment promoted complex formation with the −98 to −69 positive *cis*-acting element (containing the Barbie box sequence) described previously (98). The negative element also contains the core glucocorticoid responsive sequence (TGTCCT) and appeared to not only mediate the repressed state of the CYP2B1/B2 gene in the absence of PB but also the repressive effect of dexamethasone if given along with PB. Under the conditions used in all experiments (Wistar rats were routinely sacrificed 6 hours after drug injection), dexamethasone antagonized the effects of PB at a concentration of 100 μg/kg body weight.

Most recently, Prabhu et al. (100) comprehensively analyzed the PB response of a sequence (nt −179 to +1 in the 5′-flanking region of CYP2B1/2 gene) that, as described previously, was previously shown to contain subsequences that could bind barbiturate-responsive *trans*-acting factors. They first linked this sequence to the human growth-hormone gene as a reporter and targeted this construct to liver in vivo by injecting it into rats as an asialoglycoprotein–DNA complex. Their results, from rats either injected with PB or sham-injected and then sacrificed 12 hours later, clearly established that this 179-nt sequence 5′ upstream of the CYP2B1/2 gene represents a minimal PB-responsive promoter under in-vivo conditions.

They also showed, by competition analyses of the three nuclear protein–DNA complexes formed in gel shift assays with the positive (nt −69 to −98) and negative (nt −126 to −160) *cis*-acting subsequence elements, that the same protein is probably binding to both. Furthermore, nuclear protein preparations that were separately subjected to purification on affinity columns prepared from either the positive and negative *cis*-acting elements each yielded predominately a 26- to 28-kDa component that was indistinguishable from the other by SDS/polyacrylamide gel electrophoresis. A trace of a 94- to 100-kDa species also was detected in each purified preparation. The functionality of oligo affinity-purified protein from either PB-treated or uninduced rat liver nuclei was probed in a "mix-and-match" series of experiments using cell-free transcription with a construct containing the minimal PB-responsive promoter element of the CYP2B1/B2 gene linked to the first exon of P450E in a cell-free system from liver nuclei also from PB-treated or untreated rats. To summarize the results of these transcription experiments, the PB-transcription extract was about five-fold more efficient than the control extract; affinity-purified protein from PB-treated cells was almost twice as effective in stimulating transcription in control extracts (11-fold) as protein from uninduced cells (six-fold).

As in the studies on the α-1-acid glycoprotein gene reported by Porquet and coworkers, phosphorylation–dephosphorylation also appears to play a role in the PB-mediated regulation of transcription of the CYP2B1/2 genes. In gel-shift experiments, the affinity-purified protein mimicked the crude nuclear extract, under appropriate conditions, by giving rise to three discrete complexes with the positive regulatory element (nt −69 to −98). In-vitro phosphorylation–dephosphorylation experiments indicated that both phosphorylated and dephosphorylated forms of the protein could bind to the positive element while the dephosphorylated form had significantly higher affinity for the negative element (nt −126 to −160). PB treatment of the rats resulted in significant increases in the phosphorylation of both the 26- to 28-kDa and 94-kDa proteins.

Based on these results, Padmanaban and his colleagues hypothesize that the 26- to 28-kDa protein, binding predominantly to the negative element in the dephosphorylated state, mediates the basal level of transcription of the CYP2B1/B2 gene. On the other hand, PB treatment leads to the induced state by significantly increasing phosphorylation of this protein and thus shifting the equilibrium toward binding to the positive element containing the Barbie box subsequence. They speculate that such binding to the positive element promotes interaction with an upstream enhancer through other proteins such as the 94-kDa species, with a consequent significant increase in the rate of transcription of the CYP2B1/2 gene.

The research team of Sheppard and Phillips, and their coworkers, using many of the same techniques as those employed in Padmanaban's laboratory, also have studied the effects of phenobarbital on the regulation of CYP2B1/2 gene transcription. Their recent experiments aimed at identifying the barbiturate-responsive *cis*- and *trans*-acting factors involved in PB-mediated induction of the CYP2B2 gene of the rat produced results that differed significantly from those published in 1995 by Padmanaban's group (100). Shephard et al. (101, 102) used in-vitro transcription, gel retardation, and DNase I footprinting assays to characterize regulatory protein binding sites within a CYP2B2 gene promoter. The showed that a region between −368 and −4 nt contains sequences that are involved in the phenobarbital-mediated regulation of transcription of the CYP2B2 gene. Within this region, two DNA sequences (−183 to −199 and −31 to −72) were shown to bind to rat-liver nuclear proteins that were either enriched or activated in vivo by phenobarbital. Evidence that these two sequences bound different proteins was obtained by gel-retardation competition experiments; in-vitro transcription competition experiments indicated that the sequences and the proteins that interact with them are involved in regulating CYP2B2 gene transcription. There was no indication that the Barbie box sequence (−73 to −87 nt) that lies between these two sequences (and is immediately upstream of one) could bind to a PB-responsive protein. These researchers concluded that the two DNA sequences and their cognate binding proteins may play a role in the PB-mediated induction of CYP2B2 gene expression. Thus, Padmanaban's group and that of Shephard and Phillips differ in their identification of both the *cis*- and *trans*-acting factors involved in barbiturate-mediated regulation, although they agree that such elements lie in approximately the same region in the 5′-flanking region of the CYP2B1/2 gene.

Ramsden and colleagues have taken a somewhat different approach by using transgenic mice (103) as well as primary hepatocytes in their studies of barbiturate-mediated induction of CYP2B1/2. They developed transgenic mouse strains incorporating the rat CYP2B2 gene and analyzed its expression in mouse tissues using two series of CYP2B2 gene constructs (19- and 39-kbp in total length). Each contained the entire coding region, introns, and 3′-flanking sequences of CYP2B2, but differed in the lengths of 5′-flanking sequences. Mice whose transgene included the complete CYP2B2 gene but only 800 bp of 5′-flanking sequence were not PB-inducible in any mouse tissue but expressed CYP2B2 at high levels constitutively in kidney and liver. Mice with the transgene containing an additional 19 kbp of 5′-flanking sequence expressed CYP2B2 at high levels only after PB treatment and then exclusively in the liver. They concluded that the presence of sequences between −800 and −20 kbp of the 5′-flanking region of the rat CYP2B2 gene confer critical regulatory information necessary for PB-mediated induction in vivo. They also inferred from their results that the −800 bp 5′-flanking region containing a 17-bp Barbie box sequence (71) and the encompassing larger sequence identified by Rangarajan and Padmanaban (99) was not sufficient to mediate the PB response.

Most recently, Sidhu and Omiecinski (106) used primary rat hepatocytes, cultured with an extracellular matrix overlay to maintain the PB-response, to study cAMP-associated effects on PB-mediated induction of P450 (104). Their results in this area agree with those of Prabhu et al. (100) on the critical role that phosphorylation/dephosphorylation plays in mediating the PB response although superficially they may appear in conflict with respect to the specific effects of phosphorylation and dephosphorylation. Sidhu and Omiecinski (104) showed that cAMP analogs or activators of intracellular cAMP-dependent pathways inhibited PB-mediated induction of CYP2B1/2 genes and that phosphodiesterase inhibitors or a potent protein kinase A activator potentiated these effects; Prabhu et al. (100) clearly established that PB-stimulated phosphorylation of a transcription factor decreased its binding to a negative PB-response element and increased its affinity for binding to a positive PB-response element and stimulated the rate of CYP2B1/2 gene transcription. Because the results from Omiecinski's laboratory apply to the first steps in a putative transduction cascade, however, while those from Padmanaban's group presumably deal with the end result of this pathway, no conflict exists in this area at present. Other differences, however, among these and other groups studying the mechanism of barbiturate-mediated induction of the CYP2B1 and CYP2B2 genes (105, 106) in the rat are not so easily resolved.

Apparent Conflicts Among Various Laboratories Regarding the Mechanism of Induction of the Rat CYP2B1/2 Genes by Phenobarbital

As is readily apparent from the discussions with respect to research on the mechanism of PB-mediated induction of CYP2B1/2 genes in the rat, no two laboratories seem to obtain the same experimental results, and there seldom is agreement among them on the implications of their findings. Certainly the complexity of the barbiturate-mediated regulatory system of the CYP2B1/2 genes in higher animals makes exact duplication of experimental results between different laboratories difficult and permits a variety of possible explanations for the apparent conflicts that arise. For example, in whole-animal experiments using rodents, the nature of the results obtained may be significantly affected by the species used and, within a given species, the sex, age or strain; the dosage of PB (or other barbiturates) administered as an inducer; the mode of administration (i.e., multiple or single doses by injection, food, or drinking water); the time of sacrifice after administration of the drug (very important if PB-responsive transcription factors are the objects of study); the mode of preparation of tissues, cell cultures, or cell-free systems after the animal has been sacrificed; the concentrations and preparation (including purification procedures) of the proteins and oligonucleotides and the composition, ionic strength, and pH of the buffers, gels, and other reagents used in gel-retardation assays, footprinting studies, in-vitro transcription and enzymes assays, the nature of the vectors and other plasmids used in transfection experiments. Each laboratory, understandably, has standardized on certain procedures and animal subjects for internal consistency but not uncommonly at the expense of congruence with other laboratories. This caveat should be kept in mind in considering apparently conflicting experimental results from different laboratories. Although it is beyond the scope of this chapter (and well beyond the author's abilities and knowledge) to attempt to resolve all of the apparent conflicts mentioned (or implied) above, one is considered as an example.

The experimental results of Ramsden et al. (103) on the −800-bp CYP2B2 transgene (the complete rat CYP2B2 gene but only 800 base pairs of 5′-flanking sequence) in the transgenic mouse were interpreted by them as indicating that this sequence, containing the core promoter of the rat gene, was not sufficient to mediate a significant PB response. Nevertheless, an internal region of this same sequence, which contains the Barbie box element (71) and other barbiturate-responsive subsequences, has been shown by Padmanaban's group (97–100) and also by Shephard et al. (102) to do just that, that is, to significantly mediate the PB response. One explanation that we offered previously (72) for this apparent conflict was that this 800-bp sequence in the transgene may not be able to protect the transgene from the influence of its 5′-flanking genomic DNA sequences (such as enhancer or other promoter sequences). If the transgene integrates downstream of a promoter or an enhancer, the observed constitutive expression of the transgene may not reflect the function of the 800-bp CYP2B2 promoter in the rat genome, even assuming that it carried all of the PB-responsive *cis*-acting elements. A second possible explanation is that the mouse simply does not produce a protein that can substitute for the normal PB-responsive repressor of the

CYP2B2 gene in the rat that, in the absence of PB, binds *cis*-acting sequences in this 800-bp region. As a consequence of the absence of a functional repressor, the transgene was expressed constitutively at a very high level and was unresponsive to PB, which, in this region, probably mediates the inactivation of the repressor and the activation of positive transcription factors that displace it from its binding site (102).

Other apparent conflicts, such as those that exist between the published results from Padmanaban's laboratory (97–100) and those from Shephard and Phillips and their coworkers (101, 102) on the specific identities of the PB-responsive cis- and trans-acting factors in the near 5′-flanking region of the CYP2B1/2 genes are more difficult to explain. Still, one may note that many of the conditions used in obtaining the results differed between the two laboratories. These included differences in the rat strains used, in the doses of PB administered and times of sacrifice after PB administration, in the methods for preparing nuclear extracts, in the treatment of the protein fractions used in binding and footprinting experiments and in a variety of other perhaps less-critical procedures. Although there is no way at present of deciding how such differences in experimental approach may have affected the surprisingly divergent results, it would have been more surprising, perhaps, if the published results from the two laboratories had not shown significant incongruities.

CONCLUSIONS AND FUTURE PERSPECTIVES

Phenobarbital and many other barbiturates have numerous effects in mammals including a variety of neurologic activities related, at least in part, to their function as ABA-receptor agonists. Apparently unrelated to their neurologic effects are their roles as tumor promoters in some mammals, as elicitors of the pleiotropic response in the livers of treated animals, and as inducers of a variety of proteins, especially those related to xenobiotic catabolism such as cytochrome P450 monooxygenases, cytochrome P450 reductases, and phase II enzymes. Nevertheless, a growing number of other proteins not obviously related to the metabolic transformation of xenobiotics now are being recognized as inducible by barbiturates.

Although barbiturate-mediated induction of specific proteins (usually P450 cytochromes) has been studied primarily in the livers of higher animals, the phenomenon has been demonstrated in a variety of nonmammalian organisms including birds, insects, higher green plants, eukaryotic microorganisms, and bacteria. The most intensively studied of the nonmammalian systems relating to the mechanism of barbiturate-mediated induction is that of *B. megaterium*. Research from the author's laboratory on the regulation of the $P450_{BM\text{-}1}$ and $P450_{BM\text{-}3}$ genes in *B. megaterium* has centered on characterizing the *cis*- and *trans*-acting elements involved in their barbiturate-mediated induction. These include three positive transcription factors (BM1P1, BM1P2 and BM3P1), a critical repressor, Bm3R1, and various *cis*-acting barbiturate-responsive elements including two Barbie box elements and three other operator sites designated O_I, O_{II} and O_{III} (see Fig. 1 and Table 2). In the absence of barbiturates, Bm3R1 binds strongly to O_{III}, a bicistronic operator sequence, and to a segment of DNA that includes the Barbie box sequence, both residing in the 5′-flanking region of the *P450*$_{BM-3}$ structural gene; Bm3R1 also binds to O_I and O_{II} and to a Barbie box in the 5′-flanking region of the *P450*$_{BM-1}$ gene. The transcription of both P450 genes is repressed by Bm3R1, which also regulates its own expression and inhibits expression of the positive factors. Barbie box sequences (consensus sequence: ATCAAAAGCTGGAGG) show significant sequence similarity, particularly in the highly conserved poly-A motifs, to the three operator sequences. Barbie boxes have been found in the 5′-flanking regions of essentially all barbiturate-inducible genes whose regulatory regions have been reported, although their significance with respect to barbiturate-mediated induction has yet to be demonstrated in many of these systems.

In the presence of inducer barbiturates, Bm3R1 no longer binds to operator or Barbie box sequences. Barbiturates appear to interact with Bm3R1 and in so doing directly inhibit its ability to bind to O_{III}. Bm3R1 binding to Barbie box and other cis-acting elements, however, is not directly inhibited by barbiturates but involves either competitive binding to these elements or to Bm3R1 itself by the positive regulatory proteins. These proteins themselves are induced by barbiturates but also may be activated by them without protein synthesis. Present evidence suggests that, under noninducing conditions, Bm3R1 binds to the Barbie box and operator sites to cause a looping out of critical portions of the regulatory regions of both P450 genes and masking of the promoter regions (see Fig. 2); under inducing conditions mediated by barbiturates, these loops are opened with the consequence that the rates of transcription of the P450 genes and synthesis of cytochromes P450$_{BM-1}$ and P450$_{BM-3}$ are greatly accelerated.

Research on the barbiturate-mediated induction of proteins in the rat has centered on the regulation of two quite different genes, one encoding α-1-AGP, the major acute-phase protein of mammals and the other (actually two closely related genes) encoding cytochromes P450b and P450e (CYP2B1 and CYP2B2).

Research from Porquet's laboratory showed that the AGP gene, which contains a Barbie box sequence in its near 5′-flanking region, is induced by PB at the transcriptional level. Transfection assays in primary hepatocyte cultures showed that an intact Barbie box element was absolutely essential for PB induction of AGP. PB treatment of the hepatocytes also strongly enhanced the binding of a nuclear extract factor to the Barbie box sequence and, at the same time, led to an increase in the phosphorylation of this factor. This was congruent with the finding that the affinity of the transactivating factor for the Barbie box sequence of the AGP gene (and for the CYP2B1/2 gene Barbie box as well) was abolished after phosphatase treatment of nuclear extracts and enhanced by treatment of the extracts with protein kinase A. The demonstration that protein synthesis was not necessary for PB-mediated induction of the AGP gene further supported the concept that a post-translational modification of the *trans*-activating factor by phosphorylation was the basis for its increased binding to Barbie box sequences.

A PB-mediated molecular transduction pathway leading to the phosphorylation or dephosphorylation of transcription factors also is involved in the induction of the CYP2B1/2 genes by PB. In this case, there is evidence from Omiecinski's laboratory that the cAMP signal transduction pathway plays a negative regulatory role in this process. Several research groups have identified cis-acting elements in the near 5′-flanking (core promoter region) of the CYP2B1/2 genes, but they differ in assigning the locations of these sequences. The Shephard–Phillips group identified two DNA sequences (−183 to −199 and −31 to −72 nt) that interacted with several distinct proteins that were enriched or activated in response to PB. On the other hand, Padmanaban's laboratory group identified a negative element, −126 to −160 nt, and a positive element, −69 to −98 nt, containing within it a Barbie box (−73 to −87). They also isolated and purified a 26- to 28-kDa protein manifesting differential binding to these positive and negative elements depending on its phosphorylation state. In the dephosphorylated state it appears to mediate the basal level of transcription of the CYP2B1/2 genes by binding primarily to the negative element. PB treatment leads to the induced state by increasing the phosphorylation of this factor and thus shifting the equilibrium toward binding to the positive cis-acting element containing the Barbie box sequence. It has been hypothesized that binding of the 26- to 28-kDa protein and a second 94-kDa factor to this positive element promotes interaction with an upstream enhancer region (identified in Omiecinski's laboratory) to modulate transcription of CYP2B1/2.

ACKNOWLEDGMENT

The research reported in this chapter from the Laboratory of A. Fulco on the barbiturate-mediated regulation of the P450 cytochromes of *Bacillus megaterium* was supported by National Institutes of Health Research Grant GM23913 (United States Public Health Service).

REFERENCES

1. Ho IK, Harris RA. Mechanism of action of barbiturates. *Annu Rev Pharmacol Toxicol* 1981;21:83–111.
2. Peraino C, Fry RJM, Staffeldt E, Kisieleski WE. Effect of varying the exposure to phenobarbital on its enhancement of 2-acetylaminofluorene-induced hepatic tumorigenesis in the rat. *Cancer Res* 1973;33:2701–2705.
3. Pitot H. *Fundamentals of Oncology*, ed. 3. New York: Marcel Dekker, 1986.
4. McLean EM, Driver H, McDanell R. Nutrition and enzyme inducers in liver tumor promotion in human and rat. *Bull Cancer (Paris)* 1990;77:505–508.
5. McLean AE, Driver HE, Sutherland IA. Liver tumor promotion by phenobarbital: comparison of rat and human studies. *Prog Clin Biol Res* 1992;374:251–259.
6. Remmer H, Merker HJ. Drug-induced changes in the liver endoplasmic reticulum: association with drug-metabolizing enzymes. *Science* 1983;142:1657–1658.
7. Nims RW, Devor DE, Henneman JR, Lubet RA. Induction of alkoxyresorufin O-dealkylases, epoxide hydrolase, and liver weight gain: correlation with liver tumor-promoting potential in a series of barbiturates. *Carcinogenesis* 1987;8:67–71.
8. Busser MT, Lutz WK. Stimulation of DNA synthesis in rat and mouse liver by various tumor promoters. *Carcinogenesis* 1987;8:1433–14377.
9. Diwan BA, Rice JM, Nims RW, Lubet RA, Hu H, Ward JM. P-450 enzyme induction by 5-ethyl-5-phenylhydantoin and 5,5-diethylhydantoin, analogues of barbiturate tumor promoters phenobarbital

and barbital, and promotion of liver and thyroid carcinogenesis initiated by N-nitrosodiethylamine in rats. *Cancer Res* 1988;48:2492–2497.

10. Diwan BA, Nims RW, Ward JM, Hu H, Lubet RA, Rice JM. Tumor promoting activities of ethylphenylacetylurea and diethylacetylurea, the ring hydrolysis products of barbiturate tumor promoters phenobarbital and barbital, in rat liver and kidney initiated by N-nitrosodiethylamine. *Carcinogenesis* 1989;10:189–194.
11. Nelson DR, Strobel HW. Evolution of cytochrome P-450 proteins. *Mol Biol Evol* 1987;4:572–593.
12. Nebert DW, Nelson DR, Feyereisen R. Evolution of the cytochrome P450 genes. *Xenobiotica* 1989;19:1149–1160.
13. Nelson DR, Kamataki T, Waxman DJ, Guengerich FP, Estabrook RW, Feyereisen R, Gonzalez FJ, Coon MJ, Gunsalus IC, Gotoh O, Okuda K, Nebert DW. CYP450 superfamily: update on new sequences, gene mapping, accession numbers, and nomenclature. *Pharmacogenetics* 1996;6:1–42.
14. Okey AB. Enzyme induction in the cytochrome P-450 system. *Pharmacol Ther* 1990;45:241–298.
15. Waxman DJ, Azaroff L. Phenobarbital induction of cytochrome P450 gene expression. *Biochem J* 1992;281:577–592.
16. Nebert DW, Gonzalez FJ. P450 genes: structure, evolution and regulation. *Annu Rev Biochem* 1987;56:945–993.
17. Gonzalez FJ, Liu S-Y, Yano M. Regulation of cytochrome P450 genes: molecular mechanisms. *Pharmacogenetics* 1993;3:51–57.
18. Jones CR, Guengerich FP, Rice JM, Lubet RA. Induction of various cytochromes CYP2B, CYP2C and CYP3A by phenobarbitone in non-human primates. *Pharmacogenetics* 1992;2:160–172.
19. Strobel HW, Hodgson AV, Shen S. NADPH cytochrome P450 reductase and its structural and functional domains. In: Ortiz de Montellano PR, ed. *Cytochrome P450: Sructure, Mechanism and Biochemistry*, ed. 2. New York: Plenum Press, 1995;225–244.
20. Benveniste I, Lesot A, Hasenfratz M, Kochs G, Durst F. Multiple forms of NADPH-cytochrome P450 reductase in higher plants. *Biochem Biophys Res Commun* 1991;177:105–112.
21. Bertaux O, Fournier T, Chauvelot-Moachon L, Porquet D, Valencia R, Durand G. Modifications of hepatic alpha-1-acid glycoprotein and albumin gene expression in rats treated with phenobarbital. *Eur J Biochem* 1992;203:655–661.
22. Poland A, Mak I, Glover E, Boatman RJ, Ebetino FH, Kende AS. Bis[2-(3,5-dichloropyridyloxy)]benzene, a potent phenobarbital-like inducer of microsomal monooxygenase activity. *Mol Pharmacol* 1980;18:571–580.
23. Poland A, Mak I, Glover E. Species differences in responsiveness to 1,4-bis[2-(3,5-dichloropyridyloxy)]benzene, a potent phenobarbital-like inducer of microsomal monooxygenase activity. *Mol Pharmacol* 1981;20:442–450.
24. Raunio H, Kojo A, Juvonen R, Honkakoski P, Jarvinen P, Lang MA, Vahakangas K, Gelboin HV, Park SS, Pelkonen O. Mouse hepatic cytochrome P-450 isozyme induction by 1,4-bis[2-(3,5-dichloropyridyloxy)]benzene, pyrazole, and phenobarbital. *Biochem Pharmacol* 1988;37:4141–4147.
25. Yamada H, Kaneko H, Takeuchi K, Oguri K, Yoshimura H. Tissue-specific expression, induction, and inhibition through metabolic intermediate-complex formation of guinea pig cytochrome P450 belonging to the CYP2B subfamily. *Arch Biochem Biophys* 1992;299:248–254.
26. Oguri K, Kaneko H, Tanimoto Y, Yamada H, Yoshimura H. A constitutive form of guinea pig liver cytochrome P450 closely related to phenobarbital inducible P450b(e). *Arch Biochem Biophys* 1991;287:105–111.
27. Jones CR, Lubet RA. Induction of a pleiotropic response by phenobarbital and related compounds: response in various inbred strains of rats, response in various species and the induction of aldehyde dehydrogenase in Copenhagen rats. *Biochem Pharmacol* 1992;44:1651–1660.
28. Larsen MC, Brake PB, Parmar D, Jefcoate CR. The induction of five rat hepatic P450 cytochromes by phenobarbital and similarly acting compounds is regulated by a sexually dimorphic, dietary-dependent endocrine factor that is highly strain specific. *Arch Biochem Biophys* 1994;315:24–34.
29. Lubet RA, Nims RW, Dragnev KH, Jones CR, Diwan BA, Devor DE, Ward JM, Miller MS, Rice JM. A markedly diminished pleiotropic response to phenobarbital and structurally-related xenobiotics in Zucker rats in comparison with F344/NCr or DA rats. *Biochem Pharmacol* 1992;43:1079–1087.
30. Adesnik M, Atchison M. Genes for cytochrome P-450 and their regulation. *CRC Crit Rev Biochem* 1986:19:247–305.
31. Reichhart D, Salaun JP, Benveniste I, Durst F. Induction by manganese, ethanol, phenobarbital, and herbicides of microsomal cytochrome P-450 in higher plant tissues. *Arch Biochem Biophys* 1979;196:301–303.

32. Borlakoglu JT, John P. Cytochrome P-450-dependent metabolism of xenobiotics: a comparative study of rat hepatic and plant microsomal metabolism. *Comp Biochem Physiol* 1989;94:613–617.
33. Jondorf WR, MacIntyre DE, Powis G. Induction of liver microsomal drug metabolism in newly-hatched chicks. *Br J Pharmacol* 1973;47:624P–625P.
34. Lorr NA, Bloom SE, Park SS, Gelboin HV, Miller H, Friedman FK. Evidence for a PCN-P450 enzyme in chickens and comparison of its development with that of other phenobarbital-inducible forms. *Mol Pharmacol* 1989;35:610–616.
35. Hahn CN, Hansen AJ, May BK. Transcriptional regulation of the chicken CYP2H1 gene. Localization of a phenobarbital-responsive enhancer domain. *J Biol Chem* 1991;266:17031–17039.
36. Elskus AA, Stegeman JJ. Further consideration of phenobarbital effects on cytochrome P-450 activity in the killifish, *Fundulus heteroclitus. Comp Biochem Physiol C* 1989;92:223–230.
37. Fuchs SY, Spiegelman VS, Safaev RD, Belitsky GA. Xenobiotic-metabolizing enzymes and benzo[a]-pyrene metabolism in the benzo[a]pyrene-sensitive mutant strain of *Drosophila simulans. Mutat Res* 1992;269:185–191.
38. Julistiono H, Briand J. Microsomal ethanol-oxidizing system in *Euglena gracilis*: similarities between Euglena and mammalian cell systems. *Comp Biochem Physiol B* 1992;102:747–755.
39. Ndifor AM, Ward SA, Howells RE. Cytochrome P-450 activity in malarial parasites and its possible relationship to chloroquine resistance. *Mol Biochem Parasitol* 1990;41:251–257.
40. Clarke SE, Brealey CJ, Gibson GG. Cytochrome P-450 in the housefly: induction, substrate specificity and comparison to three rat hepatic isoenzymes. *Xenobiotica* 1989;19:1175–1180.
41. Agosin M, Cherry A, Pedemonte J, White R. Cytochrome P-450 in culture forms of *Trypanosoma cruzi. Comp Biochem Physiol C* 1984;78:127–32.
42. Hallstom I, Blanck A, Atuma S. Comparison of cytochrome P-450-dependent metabolism in different developmental stages of *Drosophila melanogaster. Chem Biol Interact* 1983;46:39–54.
43. Agosin M, Naquira C, Paulin J, Capdevila J. Cytochrome P-450 and drug metabolism in *Trypanosoma cruzi*: effects of phenobarbital. *Science* 1976;194:195–197.
44. Gil DL, Rose HA, Yang RS, Young RG, Wilkinson CF. Enzyme induction by phenobarbital in the Madagascar cockroach, *Gromphadorhina portentosa. Comp Biochem Physiol B* 1974;47:657–662.
45. Fuchs SY, Spiegelman VS, Belitsky GA. Inducibility of various cytochrome P450 isozymes by phenobarbital and some other xenobiotics in *Drosophila melanogaster. Biochem Pharmacol* 1994;47:1867–1873.
46. Scott JG, Lee SS. Tissue distribution of microsomal cytochrome P-450 monooxygenases and their inducibility by phenobarbital in the insecticide resistant LPR strain of house fly, *Musca domestica L. Insect Biochem Mol Biol* 1993;23:729–738.
47. Narhi LO, Fulco AJ. Phenobarbital induction of a soluble cytochrome P-450-dependent fatty acid monooxygenase in *B. megaterium. J Biol Chem* 1982;257:2147–2150.
48. Fulco AJ, Ruettinger RT. Occurrence of a barbiturate-inducible catalytically self-sufficient 119,000 dalton cytochrome P-450 monooxygenase in bacilli. *Life Sci* 1987;40:1769–1775.
49. Patel NV, Omer CA. Phenobarbital and sulfonylurea-inducible operons encoding herbicide metabolizing cytochromes P-450 in *Streptomyces griseolus. Gene* 1992;112:67–76.
50. Fulco AJ. $P450_{BM-3}$ and other inducible bacterial P450 cytochromes: biochemistry and regulation. *Annu Rev Pharm Toxicol* 1991;31:177–203.
51. Narhi LO, Kim BH, Stevenson PM, Fulco AJ. Partial characterization of a barbiturate-induced cytochrome P-450-dependent fatty acid monooxygenase from *B. megaterium. Biochem Biophys Res Commun* 1983;116:851–858.
52. Narhi LO, Fulco AJ. Characterization of a catalytically self-sufficient 119,000 dalton cytochrome p-450 monooxygenase induced by barbiturates in *Bacillus megaterium. J Biol Chem* 1986;261:7160–7169.
53. Wen LP, Fulco AJ. Cloning of the gene encoding a catalytically self-sufficient cytochrome P-450 fatty acid monooxygenase induced by barbiturates in *bacillus megaterium* and its functional expression and regulation in heterologous (*Escherichia coli*) and homologous (*Bacillus megaterium*) hosts. *J Biol Chem* 1987;262:6676–6682.
54. Narhi LO, Fulco AJ. Identification and characterization of two functional domains in cytochrome $P\text{-}450_{BM-3}$, a catalytically self-sufficient monooxygenase induced by barbiturates in *Bacillus megaterium. J Biol Chem* 1987;262:6683–6690.
55. Narhi LO, Wen L, Fulco AJ. Characterization of the protein expressed in *Escherichia coli* by a recombinant plasmid containing the cytochrome $P450_{BM-3}$ gene. *Mol Cell Biochem* 1988;79:63–71.

56. Ruettinger RT, Wen L, Fulco AJ. Coding nucleotide, 5′-regulatory, and deduced amino acid sequences of P450$_{BM-3}$, a single peptide cytochrome P450:NADPH-P450 reductase from *Bacillus megaterium. J Biol Chem* 1989;264:10987–10995.
57. Ravichandran KG, Boddupalli SS, Hasemann C A, Peterson JA, Deisenhofer J. Crystal structure of hemoprotein domain of P450$_{BM-3}$, a prototype for microsomal P450s. *Science* 1993;261:731–736.
58. Klein ML, Fulco AJ. Critical residues involved in FMN binding and catalytic activity in cytochrome P450$_{BM-3}$. *J Biol Chem* 1993;268:7553–7561.
59. Govindaraj S, Li H, Poulos TL. Flavin supported fatty acid oxidation by the heme domain of *Bacillus megaterium* cytochrome P450$_{BM-3}$. *Biochem Biophys Res Commun* 1994;203:1745–1749.
60. Schwalb H, Narhi LO, Fulco AJ. Purification and characterization of pentobarbital-induced cytochrome P-450$_{BM-1}$, from *Bacillus megaterium* ATCC 15481. *Biochim Biophys Acta* 1985;838:302–311.
61. He JS, Ruettinger RT, Liu HM, Fulco AJ. Molecular cloning, coding nucleotides and deduced amino acid sequence of P450$_{BM-1}$ from *Bacillus megaterium. Biochim Biophys Acta* 1989;1009:301–303.
62. Fulco AJ, Kim BH, Matson RS, Narhi LO, Ruettinger RT. Nonsubstrate induction of a soluble bacterial cytochrome P-450 monooxygenase by phenobarbital and its analogs. *Mol Cell Biochem* 1983;53/54:155–162.
63. Kim BH, Fulco AJ. Induction by barbiturates of a cytochrome P-450-dependent fatty acid monooxygenase in *Bacillus megaterium*: relationship between barbiturate structure and inducer activity. *Biochem Biophys Res Commun* 1983;116:843–850.
64. Ruettinger RT, Kim BH, Fulco AJ. Acylureas: a new class of barbiturate-like bacterial cytochrome P-450 inducers. *Biochim Biophys Acta* 1984;801:372–380.
65. English N, Hughes V, Wolf CR. Common pathways of cytochrome P450 gene regulation by peroxisome proliferators and barbiturates in *Bacillus megaterium* ATCC14581. *J Biol Chem* 1994;269:26836–26841.
66. English N, Hughes V, Wolf CR. Induction of cytochrome P-450$_{BM-3}$, (CYP 102) by nonsteroidal anti-inflammaory drugs in *Bacillus megaterium. Biochem J* 1996;316:279-283.
67. Wen LP, Fulco AJ. Induction of a cytochrome P-450-dependent monooxygenase in *Bacillus megaterium* by a barbiturate analog, 1-[2-phenylbutyryl]-3-methylurea. *Mol Cell Biochem* 1985;67:77–81.
68. Wen LP. Cloning, expression and regulation of a barbiturate-inducible cytochrome P-450 gene of *Bacillus megaterium*. Ph.D. diss., University of California, Los Angeles, 1988.
69. Wen LP, Ruettinger R, Fulco AJ. Requirements for a 1 kilobase 5′-flanking sequence for barbiturate-inducible expression of the cytochrome P450$_{BM-3}$ gene in *Bacillus megaterium. J Biol Chem* 1989;264:10996–11003.
70. Shaw GC, Fulco AJ. Inhibition by barbiturates of the binding of Bm3R1 repressor to its operator site on the barbiturate-inducible cytochrome P450$_{BM-3}$ gene of *Bacillus megaterium. J Biol Chem* 1993;268:2997–3004.
71. He JS, Fulco AJ. A barbiturate-regulated protein binding to a common sequence in the cytochrome P450 genes of rodents and bacteria. *J Biol Chem* 1991;266:7864–7869.
72. Liang Q, He JS, Fulco AJ. The role of Barbie box sequences as cis-acting elements involved in the barbiturate-mediated induction of cytochromes P450$_{BM-1}$ and P450$_{BM-3}$ in *Bacillus megaterium. J Biol Chem* 1995;270:4438–4450.
73. Liang Q, Fulco AJ. Transcriptional regulation of the genes encoding cytochromes P450$_{BM-1}$ and P450$_{BM-3}$ in *Bacillus megaterium* by the binding of Bm3R1 repressor to barbie box elements and operator sites. *J Biol Chem* 1995;270:18606–18614.
74. Fournier T, Mejdobui N, Lapoumeroulie C, Hamelin J, Elion J, Durand G, Porquet D. Transcriptional regulation of rat α 1-acid glycoprotein gene by phenobarbital. *J Biol Chem* 1994;269:27175–27178.
75. Shaw GC, Fulco AJ. Barbiturate-mediated regulation of expression of the cytochrome P450$_{BM-3}$ gene of *Bacillus megaterium* by Bm3R1 protein. *J Biol Chem* 1992;267:5515–5526.
76. He JS, Liang Q, Fulco AJ. The molecular cloning and characterization of BM1P1 and BM1P2 proteins, putative positive transcription factors involved in the barbiturate-mediated induction of the genes encoding cytochrome P450$_{BM-1}$ of *Bacillus megaterium. J Biol Chem* 1995;270:18615–18625.
77. Fulco A, He JS, Liang Q. The role of conserved 5′-flanking (barbie box) DNA sequences and barbiturate-responsive dna-binding proteins in the mechanism of induction by barbiturates of P450 cytochromes in *Bacillus megaterium* and other prokaryotic and eukaryotic organisms. In: Lechner MC,

ed. *Cytochrome P450: Biochemistry, Biophysics and Molecular Biology*. Paris: John Libby Eurotext, 1994;37–42.
78. Kushner I, Mackiewicz A. Acute phase proteins as disease markers. *Dis Markers* 1987;5:1–11.
79. Mackiewicz A, Mackiewicz K. Glycoforms of serum alpha 1-acid glycoprotein as markers of inflammation and cancer. *Glycocon J* 1995;12:241–247.
80. Dowton S, Colten HR. Acute phase reactants in inflammation and infection. *Semin Hematol* 1988;25:84–90.
81. Routledge PA. Clinical relevance of alpha 1 acid glycoprotein in health and disease. *Prog Clin Biol Res* 1989;300:185–198.
82. Schultz DR, Arnold PI. Properties of four acute phase proteins: C-reactive protein, serum amyloid A protein, alpha 1-acid glycoprotein, and fibrinogen. *Semin Arthritis Rheum* 1990;20:129–147.
83. Romette J, di Costanzo-Dufetel J, Charrel M. Inflammatory syndrome and changes in plasma proteins. *Pathol Biol (Paris)* 1986;34:1006–1012.
84. Baumann H, Jahreis GP, Morella K. Interaction of cytokine- and glucocorticoid-response elements of acute-phase plasma protein genes: importance of glucocorticoid receptor level and cell type for regulation of the elements from rat α 1-acid glycoprotein and β-fibrinogen genes. *J Biol Chem* 1990;265:22275–22281.
85. Brinkschulte M, Breyer-Pfaff U. Increased binding of desmethylimipramine in plasma of phenobarbital-treated rats. *Biochem Pharmacol* 1982;31:1749–1754.
86. Monnet D, Feger J, Biou D, Durand G, Cardon P, Leroy Y. Fournet effect of phenobarbital on the oligosaccharide structure of rat alpha-acid glycoprotein. *Biochim Biophys Acta* 1986;881:10–4.
87. Lin TH, Sugiyama Y, Sawada Y, Suzuki Y, Iga T, Hanano M. Effect of surgery on serum alpha 1-acid glycoprotein concentration and serum protein binding of DL-propranolol in phenobarbital-treated and untreated rats. *Drug Metab Dispos* 1987;15:138–140.
88. Fournier T, Mejdoubi N, Monnet D, Durand G, Porquet D. Phenobarbital induction of alpha 1-acid glycoprotein in primary rat hepatocyte cultures. *Hepatology* 1994;20:1584–1588.
89. Williams P, Ratajczak T, Lee SC, Ringold GM. AGP/EBP(LAP) expressed in rat hepatoma cells interacts with multiple promoter sites and is necessary for maximal glucocorticoid induction of the rat alpha-1 acid glycoprotein gene. *Mol Cell Biol* 1991;11:4959–4965.
90. Ratajczak T, Williams PM, DiLorenzo D, Ringold GM. Multiple elements within the glucocorticoid regulatory unit of the rat alpha 1-acid glycoprotein gene are recognition sites for C/EBP. *J Biol Chem* 1992;267:11111–11119.
91. Shaw PM, Adesnik M, Weiss MC, Corcos L. The phenobarbital-induced transcriptional activation of cytochrome P-450 genes is blocked by the glucocorticoid-progesterone antagonist RU486. *Mol Pharmacol* 1993;44:775–783.
92. Corcos L, Weiss MC. Phenobarbital, dexamethasone and benzanthracene induce several cytochrome P450 mRNAs in rat hepatoma cells. *FEBS Lett* 1988;233:37–40.
93. Fournier T, Vranckx R, Mejdoubi N, Durand G, Porquet D. Induction of rat alpha-1-acid glycoprotein by phenobarbital is independent of a general acute-phase response. *Biochem Pharmacol* 1994;48:1531–1535.
94. Schuetz EG, Li D, Omiecinski CJ, Muller-Eberhard U, Kleinman HK, Elswick B, Guzelian PS. Regulation of gene expression in adult rat hepatocytes cultured on a basement membrane matrix. *J Cell Physiol* 1988;134:309–323.
95. Waxman D, Morrissey JJ, Naik S, Jauregui HO. Phenobarbital induction of cytochromes P-450: high-level long-term responsiveness of primary rat hepatocyte cultures to drug induction, and glucocorticoid dependence of the phenobarbital response. *Biochem J* 1990;271:113–119.
96. Denison MS, Whitlock JP. Xenobiotic-inducible transcription of cytochrome P450 genes. *J Biol Chem* 1995;270:18175–18178.
97. Rangarajan PN, Padmanaban G. Regulation of cytochrome P450b/e gene expression by a heme- and phenobarbitone-modulated transcription factor. *Proc Natl Acad Sci USA* 1989;86:3963–3967.
98. Upadhya P, Venkateswara Rao M, Venkateswar V, Rangarajan PN, Padmanaban G. Identification and functional characterization of a *cis*-acting positive DNA element regulating CYP 2B1/B2 gene transcription in rat liver. *Nucleic Acids Res* 1992;20:557–562.
99. Ram N, Rao MV, Prabhu L, Nirodi CS, Sultana S, Vatsala PG, Padmanaban G. Characterization of a negative *cis*-acting DNA element regulating the transcription of CYP2B1/B2 gene in rat liver. *Arch Biochem Biophys* 1995;317:39–45.
100. Prabhu L, Upadhya P, Ram N, Nirodi CS, Sultana S, Vatsala PG, Mani SA, Rangarajan PN, Surolia

A, Padmanaban G. A model for the transcriptional regulation of the CYP2B1/B2 gene in rat liver. *Proc Natl Acad Sci USA* 1995;92:9628–9632.
101. Forrest LA, Shervington A, Phillips IR, Shephard EA. Regulation of cytochrome P4502B2 gene expression. *Biochem Soc Trans* 1994;22:125S.
102. Shephard EA, Forrest LA, Shervington A, Fernandez LM, Ciaramella G, Phillips IR. Interaction of proteins with a cytochrome P450 2B2 gene promoter: identification of two DNA sequences that bind proteins that are enriched or activated in response to phenobarbital. *DNA Cell Biol* 1994;13:793–804.
103. Ramsden R, Sommer KM, Omiecinski CJ. Phenobarbital induction and tissue-specific expression of the rat CYP2B2 gene in transgenic mice. *J Biol Chem* 1993;268:21722–21726.
104. Sidhu JS, Omiecinski CJ. cAMP-associated inhibition of phenobarbital-inducible cytochrome P450 gene expression in primary rat hepatocyte cultures. *J Biol Chem* 1995;270:12762–12773.

Toxicant–Receptor Interactions
Edited by Michael S. Denison and William G. Helferich

6

Antioxidant-Inducible Detoxifying Enzyme Gene Expression

Anil K. Jaiswal

Department of Pharmacology, Baylor College of Medicine, One Baylor Plaza, Houston, Texas, USA

Cellular exposure to antioxidants, xenobiotics, and drugs results in altered expression of genes encoding transcription factors that regulate the expression of drug-metabolizing enzymes. Cytochromes P450, P450 reductase, and other one-electron reducing enzymes metabolically activate xenobiotics and drugs to generate electrophiles and reactive oxygen species, resulting in electrophilic and oxidative stress leading to neoplasia (1). The detoxifying or chemopreventive enzymes, on the other hand, provide protection to the cells either by competing with the activating enzymes and preventing the formation of electrophiles and reactive oxygen species or by catalyzing their detoxification from the cells (2–6). The detoxifying enzymes include NAD(P)H:quinone oxidoreductases (NQO_1 and NQO_2), which catalyze two-electron reduction and detoxification of quinones (2–6); glutathione S-transferases (GSTs), which conjugate hydrophobic electrophiles and reactive oxygen species with glutathione (7–9); UDP-glucuronosyl-transferases (UDP-GT), which catalyze the conjugation of glucuronic acid with xenobiotics and drugs (10); epoxide hydrolase, which inactivates epoxides (11); γ-glutamylcysteine synthetase (γ-GCS), which plays a key role in the regulation of glutathione metabolism (12); and so on. The induction of detoxifying enzymes is one mechanism of critical importance in achieving chemoprevention.

Antioxidants are substances that delay or prevent the oxidation of cellular substrates and thus protect the cells against oxidative stress, cytotoxicity, mutagenicity, and carcinogenicity (2, 13–15). Several antioxidants (e.g., glutathione, superoxide dismutase, and catalase) act directly by scavenging the reactive oxygen species. On the other hand, others (2(3)-tert-butyl-4-hydroxy-anisole [BHA], butylated hydroxytoluene [BHT] and tert-butyl hydroquinone [t-BHQ]) work indirectly by coordinately increasing the expression of enzymes involved in elevation of cellular levels of antioxidants (e.g., γ-glutamyl cysteine synthetase) and detoxification of xenobiotics and drugs (e.g., NQOs, GSTs, UDP-GTs, epoxide hydrolases). The phenolic antioxidants (BHA and BHT) long have been known for their anticarcinogenic effects (13). Only recently, however, it has been found that anticarcinogenic properties of BHA and BHT are due to their capacity to increase the expression of detoxifying or chemopreventive enzymes (2). In addition to antioxidants, xenobiotics, tumor promoters, hydrogen peroxide, heavy metals, ionizing radiations, and UV light also increase the expression of various detoxifying enzyme genes (5, 6, 16, 17). The capacity of many diverse chemicals to block carcinogenesis correlates with their capacities to induce various detoxifying enzymes including NQOs and GSTs (2). Induction of NQO_1 and GST by "sulforaphane" from Saga broccoli blocks the formation of mammary tumors in Sprague-Dawley rats treated with single dose of 9,10-dimethyl-1,2-benzanthracene (18). This chapter discusses the mechanism of antioxidant-mediated coordinated induction of detoxifying enzyme genes including cis elements and trans-acting factors involved, and the mechanism of signal transduction from antioxidants and xenobiotics to the trans-acting factors that regulate the coordinated expression and induction of the various detoxifying enzyme genes.

Antioxidant-inducible genes encoding NQO_1, NQO_2, GST Ya, and GST P were cloned and sequenced (19–26). Among these genes, the regulation of NQO_1 and GST Ya gene expression and induction is best studied. The NQO_1 and GST Ya genes are expressed at higher levels in liver tumors and tumor cells, compared with normal liver and liver hepatocytes, and are induced coordinately in response to antioxidants and xenobiotics (2–4, 8, 9, 26).

Deletion mutagenesis studies of the human NQO_1 gene promoter identified several cis elements essential for expression and induction of the NQO_1 gene (4–6, 19). These elements include a 24–base pair fragment of the human antioxidant response element (hARE) that is required for both basal expression of the NQO_1 gene in liver tumor cells, compared with normal hepatocytes, and its induction by β-naphthoflavone (β-NF), BHA, t-BHQ, and hydrogen peroxide (27–29); a DNA fragment (between -780 and -365) required for 2,3,7,8-tetrachlorodibenzo-p-dioxin–induced expression of the NQO_1 gene (19); a basal element (region between -130 and -47) and an AP2 element (at nucleotide position -157) essential for cAMP-induced expression of the NQO_1 gene (30). hARE-like elements also have been found in the promoter regions of the rat NQO_1 gene (31), the human NQO_2 gene (21), the rat and mouse GST Ya subunit genes (9, 17, 32–36), and the rat GST P genes (9, 24, 25). The conservation of hARE in various detoxifying enzyme genes indicate that these genes may be coordinately regulated by a single mechanism involving hARE (17). Like human NQO_1 gene, the

NQO_2, GST Ya, and GST P genes also contain cis elements other than ARE, which contribute to their basal expression (9, 17, 21, 26). ARE elements in NQO_1, NQO_2, and GST genes, however, are the most important cis elements, which regulate a major portion of the basal expression and the induced expression in response to antioxidants and xenobiotics (17).

AREs

The AREs in various genes contain two AP1/AP1-like elements arranged in varying orientations separated by three or eight nucleotides, followed by a "GC" box (17, 21) (Fig. 1). The seven base pairs of the AP1 binding site (also known as *TPA response element* [TRE]) were characterized previously in the promoter regions of the collagenase

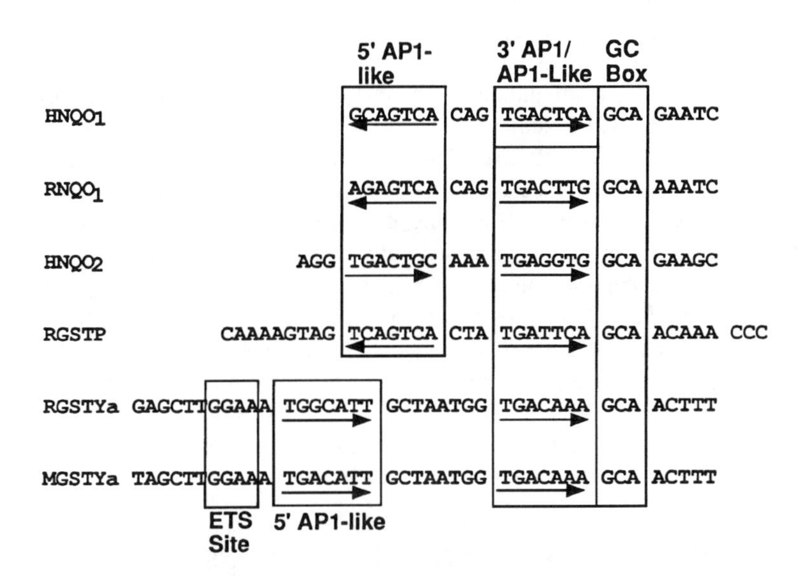

$HNQO_1$, Human NAD(P)H:Quinone $Oxidoreductase_1$ Gene ARE (hARE)
$RNQO_1$, Rat NAD(P)H:Quinone $Oxidoreductase_1$ Gene ARE (rARE)
$HNQO_2$, Human NAD(P)H:Quinone $Oxidoreductase_2$ Gene ARE (hARE)
RGSTP, Rat Glutathione S-transferase P Subunit Gene ARE (ARE)
RGSTYa, Rat Glutathione S-transferase Ya Subunit Gene ARE (ARE)
MGSTYa, Mouse Glutathione S-transferase Ya Subunit Gene ARE (EpRE)

FIG. 1. Alignment of antioxidant response elements (AREs) from six genes. The AP1 and AP1-like elements and their orientations are indicated. The 3′ AP1 site in the human NQO_1 gene ARE is a perfect consensus sequence and has been separated from AP1-like (imperfect consensus) elements in all other genes. The "GC" box and ETS binding sites are also shown in boxes. The rat NQO_1 and GST Ya gene AREs have been characterized to contain only one AP1-like element and "GC" box (5′puGTGACNNNGC3′). In these cases, additional sequences at 5′ and 3′ ends were included for alignment purposes.

and metallothionein genes, and it is required to activate expression of these genes in response to TPA (37). It may be noteworthy that human NQO_1 gene ARE is the only ARE that contains a perfect AP1 binding site (see Fig. 1). In all other genes, AREs contain imperfect (AP1-like) elements. hARE is a unique cis element, because it is this element rather than the TRE (a single AP1 element), that is responsive to antioxidants and xenobiotics (38). Mutations in the AP1/AP1-like elements and the GC box of the hARE revealed that the 3′ AP1 element is the most important for both high basal expression of the NQO_1 gene and its induction in response to β-NF and t-BHQ (38). The 5′ AP1-like element in the hARE is required for β-NF– and t-BHQ–induced response (38). The GC box is required for optimal expression and induction of the NQO_1 gene (38). The minimum ARE sequence required for expression and induction of rat GST Ya and rat NQO_1 genes has been determined as 5′puGTGACNNNGC3′, which contains only one AP1-like element (9, 34). Fine mutational analysis in the rat GST Ya gene ARE showed that only six base pairs, TGAC***GC, are essential for induction, and four out of these six base pairs (TGAC) also are required for basal expression (9, 34). These six base pairs are a part of the 3′ AP1 element and GC box that is highly conserved among the various AREs (see Fig. 1). This is in contrast to the reports on mouse GST Ya and rat GST P subunit and human NQO_1 gene AREs, which require presence of both of the AP1-like elements for their proper function (25, 36, 38). The GST P gene ARE containing two AP1-like elements exhibited a strong transcriptional enhancing activity in F9 embryonal carcinoma and HeLa cells (25). Mutation of one AP1-like element, however, severely affects the activity of the ARE in these cells.

In summary, the AREs in various genes usually contain two AP1 or AP1-like elements arranged in varying orientations generally separated by either three or eight nucleotides, followed by a GC box. AREs are unique cis elements because a single AP1 (TRE) element is nonresponsive to antioxidants and xenobiotics. The orientations and spacing between the two AP1 elements contained within various AREs may be important in determining the levels of basal and induced expression. It will be interesting to determine if these differences lead to significant changes in mechanisms of induction mediated by various AREs.

ARE–NUCLEAR PROTEIN INTERACTIONS

Nuclear trans-acting proteins bind to AREs on various genes and mediate basal and antioxidant- and xenobiotic-induced expression of detoxifying enzyme genes (28, 29, 38–44). However, a complete identification of nuclear proteins that bind to the ARE and upregulate ARE-mediated expression of detoxifying enzyme genes is still under investigation (17).

The products of the proto-oncogenes Jun and Fos bind to the AREs on human NQO_1 gene, mouse GST Ya subunit gene, and rat GST P subunit gene (27, 36, 38, 42, 43). Overexpression of c-Jun and c-Fos upregulates mouse GST Ya gene ARE-mediated expression and t-BHQ induction of chloramphenical acetyltransferase (CAT) gene in transfected F9 cells (36). On the other hand, the rat NQO_1 and GST Ya gene AREs did not bind to Jun and Fos proteins (9, 17, 39–41). These observations raised controversies regarding the role of Jun and Fos proteins in the regulation of ARE-mediated expression and induction of various detoxifying enzyme genes. Many reports suggested the involvement of nuclear proteins other than Jun and Fos in the regulation of ARE-mediated expression and induction of detoxifying enzyme genes (38–43, 45). Recently, Yoshioka et al. (44) reported the binding of c-Jun and Fra1 with mouse GST Ya subunit gene ARE. More recently, we found that overexpression of Jun with Fos or Fra1 in Hep-G2 cells negatively regulated the NQO_1 gene ARE-mediated CAT gene expression (6). We further found that repression of the ARE-mediated CAT gene expression in Hep-G2 cells was due to overexpression of Fos and Fra1 but not due to overexpression of Jun proteins (6). These results clearly suggested that nuclear proteins other than Jun and Fos regulate the hARE-mediated expression and induction of NQO_1 gene.

In summary, recent reports indicate that Jun and Fos proteins bind to the AREs from various detoxifying enzyme genes. There are also a number of arguments, however, suggesting that these proteins are not involved in regulating the expression and induction of these genes. These include that the ARE is a distinct element, compared with AP1 (TRE), even though it contains AP1/AP1-like elements, and it is ARE rather than AP1 that is responsive to antioxidants and xenobiotics (17, 38); Jun and Fos binding to the rat NQO_1 and GST Ya gene AREs have not been detected (9, 39–41); overexpression of Jun and Fos in Hep-G2 cells failed to upregulate human NQO_1 gene ARE-mediated expression of the CAT gene (6); and a positive correlation was not observed between the levels of expression of GST P and oncoproteins c-Jun and c-Fos in rat tissues and preneoplastic hepatic foci (45). Therefore, Jun and Fos binding to the AREs as observed in in-vitro band-shift assays may be due to the presence of AP1/AP1-like elements within the AREs. Nuclear proteins other than Jun and Fos may regulate the ARE-mediated expression and coordinated induction of various detoxifying enzyme genes. Further studies are required to identify these proteins.

SIGNAL TRANSDUCTION FROM ANTIOXIDANTS TO THE ARE FOR INCREASED EXPRESSION OF DETOXIFYING ENZYME GENES

The studies previously described also raise important issues regarding the mechanism of signal transduction from antioxidants and xenobiotics to the ARE. These include signal transduction from antioxidants and xenobiotics to the nuclear proteins that bind to the ARE; similar or different mechanisms with various classes of structurally different compounds; and differences and similarities in mechanisms among AREs from various genes. The nuclear proteins that bind to the ARE include jun/fos/fra and other unknown proteins. Among these proteins, the involvement of jun, fos, and fra1 in ARE-mediated regulation of detoxifying enzyme genes is questionable. We believe that nuclear proteins other than jun, fos, and fra1 positively regulate ARE-mediated expression and induction of various detoxifying enzyme genes. The identity of these nuclear proteins, however, remains unknown.

Antioxidants, xenobiotics, drugs, hydrogen peroxide, UV light, and heavy metals all induce the ARE-mediated expression of detoxifying enzyme genes. These are structurally different but may have one thing in common in that directly or indirectly they are capable of generating reactive oxygen species (46–48). The reactive oxygen has been implicated in the induction of a number of transcription factors in bacteria, yeast, and mammalian cells (49). These include oxyR and soxRS, which activate more than a dozen defense genes in bacteria (50–52); yAP1 and yAP2 in yeast, which regulate the expression of oxidative stress response genes superoxide dismutase, glutathione reductase, and glucose 6-phosphate dehydrogenase (53, 54); mac1, which activates catalase gene expression (55); and NF-kB and AP-1 in higher eukaryotic cells, which regulate the expression of several genes in immune response and antagonistic signals (56, 57). Two other eukaryotic transcription factors, serum response factor and hypoxia response factor 1, also have been suggested to be altered by reactive oxygen (49). Recently, it has been reported that quinone-mediated generation of hydroxyl radicals induces the ARE-mediated expression of mouse GST gene (58). This correlation remains unknown in case of AREs from other genes. Hydrogen peroxide, however, has been shown to induce ARE-mediated expression of rat GST Ya, rat NQO_1, and human NQO_1 genes (29, 34, 39) and may be an important intermediate in induction of these genes by antioxidant and xenobiotics. The role, if any, of electrophiles in the ARE-mediated expression and induction of detoxifying enzyme genes remains unknown.

The studies on the mechanism of signal transduction from antioxidants and xenobiotics are at a preliminary stage because of absence of information on nuclear proteins that bind to ARE and regulate the expression and induction of detoxifying enzyme genes. Several reports, however, indicate alterations in transcription and expression of Jun, Fos, and Fra proteins in response to antioxidants and xenobiotics (44, 59). It has been demonstrated that intracellular glutathione levels and increased expression of Jun and Fos nuclear proteins mediate induction of mouse GST Ya gene expression in response to xenobiotics and antioxidants (59, 60). In contrast, it has been reported that c-Fos expression is not induced by t-BHQ and that increased transcription of c-Jun and Fra1 may be responsible for activation of mouse GST Ya gene expression. The involvement of protein kinase C in the induction mechanism mediated by the mouse GST Ya subunit gene ARE was ruled out (36). The involvement of protein kinase C and tyrosine kinases in the transcriptional activation of the hARE-mediated NQO_1 gene expression by bifunctional and monofunctional inducers also had been ruled out (29). It is recommended, however, that the involvement of protein kinase C, tyrosine kinases, and other kinases should be further studied after complete identification of trans-acting proteins that mediate ARE-regulated detoxifying genes expression. Li and Jaiswal (29) also reported β-NF induced sulfhydryl modification of nuclear proteins that bind to the ARE and increase the expression of the NQO_1 gene. This is interesting because nuclear protein Ref-1, a major redox protein in HeLa cells, regulates the AP1 (c-Jun and c-Fos) proteins binding at the AP1 site by reducing

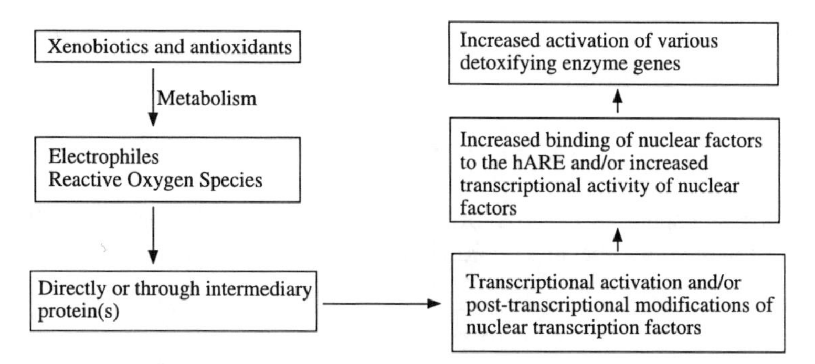

FIG. 2. Hypothetical model for mechanism of signal transduction from antioxidants and xenobiotics to the AREs.

the cysteine in the DNA binding domain of Jun and Fos proteins (61–63). The cysteine of c-Jun and c-Fos proteins that is regulated by Ref-1 is highly conserved in the DNA binding domains of other Jun- and Fos-related proteins including Jun-D, Jun-B, Fos-B, Fra1, and Fra2 (37). It is not clear, however, if Ref-1 is activated by antioxidants and xenobiotics or if it plays a role in ARE-mediated detoxifying-enzyme gene expression. A hypothetical model is presented in Fig. 2 based on what is known about ARE-mediated regulation of detoxifying-enzyme gene expression. Antioxidants and xenobiotics undergo metabolism to generate metabolites (electrophiles) and reactive oxygen species that may increase the transcription or expression of nuclear proteins or catalyze direct or indirect modification of nuclear proteins that bind to the ARE (see Fig. 2). The increased expression or modification of nuclear proteins results in increased binding of these proteins at the ARE and subsequently increased transcription of NQO_1 and other detoxifying-enzyme genes. The indirect modification of proteins may or may not involve redox factor Ref-1 and remains to be studied.

CONCLUSIONS AND FUTURE PERSPECTIVES

The studies on ARE-mediated regulation of expression and coordinated induction of the various detoxifying-enzyme genes are extremely important because of their role in detoxification and chemoprevention in cells. There is only limited information, however, available on trans-acting proteins that bind to the ARE and the mechanism of signal transduction from antioxidants and xenobiotics to these proteins. Pertinent questions concern complete identification of trans-acting proteins binding at the ARE, the possible involvement of redox proteins and other redox messengers, and the manner by which these messengers receive signals from external stimuli (antioxidants and xenobiotics) and transmit this information in the form of increased transcriptional levels of NQO_1, GST, and other detoxifying-enzyme genes. In addition, research into the types of naturally occurring and synthetic compounds that may act as potent antioxidants and inducers of the phase II enzyme genes through ARE should contribute to the development of chemoprevention and chemotherapeutics.

ACKNOWLEDGMENTS

This work was supported by NIH Grant # GM 47466 and Tobacco Council grant 3176A. We thank our colleague Dr. Kevin Ryder for helpful discussions.

REFERENCES

1. Nebert DW. Drug-metabolizing enzymes in ligand-modulated transcription. *Biochem Pharmacol* 1995;47:25–37.
2. Talalay P. Mechanisms of induction of enzymes that protect against chemical carcinogenesis. *Advances in Enzyme Regulation* 1989;28:149–159.
3. Riley RJ, Workman P. DT-diaphorase and cancer chemotherapy. *Biochem Pharmacol* 1992;43:1657–1669.
4. Belinsky M, Jaiswal AK. NAD(P)H:Quinone Oxidoreductase$_1$ (DT diaphorase) expression in normal and tumor tissues. *Cancer Metastasis Rev* 1993;12:103–117.
5. Joseph P, Xie T, Xu Y, Jaiswal AK. NAD(P)H:Quinone Oxidoreductase$_1$ (DT diaphorase): expression, regulation, and role in cancer. *Oncol Res* 1994;6:525–532.
6. Venugopal R, Joseph P, Jaiswal AK. Gene expression of DT-diaphorase in cancer cells. In: Cadenas E, Forman HJ, eds. *Oxidative Stress and Signal Transduction*. New York: Chapman and Hall, 1996, in press.
7. Pickett CB, Lu AYH. Glutathione S-transferases: gene structure, regulation and biological function. *Annu Rev Biochem* 1989;58:743–764.
8. Tsuchida S, Sato K. Glutathione transferases and cancer. *Crit Rev Biochem Mol Biol* 1992;27:337–384.
9. Rushmore TH, Pickett CB. Glutathione S-transferases, structure, regulation and therapeutic implications. *J Biol Chem* 1993;268:11475–11478.
10. Tephly T, Burchell B. UDP-glucuronosyl transferases: a family of detoxifying enzymes. *Trends Pharmacol Sci* 1990;11:276–279.
11. Oesch F, Gath I, Igarashi T, Glatt HR, Thomas H. Epoxide hydolases. In: Arinc E, Schenkman JB, Hodgson E, eds. *Molecular Aspects of Monooxygenases and Bioactivation of Toxic Compounds*. New York: Plenum Press, 1991;447–461.
12. Meister A, Anderson ME. Glutathione. *Annu Rev Biochem* 1983;52:711–760.
13. Wattenberg LW. Inhibition of carcinogenic and toxic effects of polycyclic aromatic hydrocarbons by phenolic antioxidants and ethoxyquin. *J Natl Cancer Inst* 1972;48:1425–1430.
14. Ulland BM, Weisburger JH, Yamamoto RS, Weisburger EK. Antioxidants and carcinogenesis: butylated hydroxytoluene, but not diphenyl-p-phenylenediamine, inhibits cancer induction by N-2-fluorenylacetamide and by N-hydroxy-N-2-fluorenylacetamide in rats. *Food Cosmet Toxicol* 1973;11:199–207.
15. Kahl R. Synthetic antioxidants: biochemical actions and interference with radiation, toxic compounds, chemical mutagens and chemical carcinogens. *Toxicol* 1984;33:185–228.
16. Prestera T, Holtzclaw WD, Zhang Y, Talalay, P. Chemical and molecular regulation of enzymes that detoxify carcinogens. *Proc Natl Acad Sci USA* 1993;90:2965–2969.
17. Jaiswal AK. Antioxidant response element. *Biochem Pharmacol* 1994;48:439–444.
18. Zhang Y, Kensler TW, Cho C, Posner GH, Talalay P. Anticarcinogenic activities of sulforaphane and structurally related synthetic norbornyl isothiocyanates. *Proc Natl Acad Sci USA* 1994;91:3147–3150.
19. Jaiswal AK. Human NAD(P)H:Quinone Oxidoreductase gene structure and induction by dioxin. *Biochemistry* 1991;30:10647–10653.
20. Bayney RM, Morton MR, Favreau LV, Pickett CB. Rat liver NAD(P)H:quinone reductase: regulation of quinone reductase gene expression by planar aromatic compounds and determination of the exon structure of the quinone reductase structural gene. *J Biol Chem* 1989;264:21793–21797.
21. Jaiswal AK. Human NAD(P)H:Quinone oxidoreductase 2: gene structure, activity, and tissue-specific expression. *J Biol Chem* 1994;269:14502–14508.

22. Telakowski-Hopkins CA, Rothkopf GS, Pickett CB. Structural analysis of a rat liver glutathione S-transferase Ya gene. *Proc Natl Acad Sci USA* 1986;83:9393–9397.
23. Telakowski-Hopkins CA, King RG, Pickett CB. Glutathione S-transferase Ya subunit gene: identification of regulatory elements required for basal level and inducible expression. *Proc Natl Acad Sci USA* 1985;85:1000–1004.
24. Sakai M, Okuda A, Muramatsu M. Multiple regulatory elements and phorbol 12-O-tetradecanoate 13-acetate responsiveness of the rat placental glutathione transferase gene. *Proc Natl Acad Sci USA* 1988;85:9456–9460.
25. Okuda A, Imagawa M, Maeda Y, Sakai M, Muramatsu M. Functional co-operativity between two TPA responsive elements in undifferentiated F9 embryonic stem cells. *EMBO J* 1990;9:1131–1135.
26. Daniel V. Glutathione S-transferases: gene structure and regulation of expression. *Crit Rev Biochem Mole Biol* 1993;28:173–207.
27. Li Y, Jaiswal AK. Regulation of human NAD(P)H:quinone oxidoreductase gene: role of AP1 binding site contained within human antioxidant response element. *J Biol Chem* 1992;267:15097–15104.
28. Li Y, Jaiswal AK. Identification of Jun-B as third member in human antioxidant response element-nuclear protein complex. *Biochem Biophys Res Commun* 1992;188:992–996.
29. Li Y, Jaiswal AK. Human antioxidant response element mediated regulation of type 1 NAD(P)H:Quinone Oxidoreductase gene expression: effect of sulfhydryl modifying agents. *Eur J Biochem* 1994;226:31–39.
30. Xie T and Jaiswal AK. AP-2 mediated regulation of human NAD(P)H:Quinone Oxidoreductase$_1$ (NQO$_1$) gene expression. *Biochem Pharm* 1996;51:771–778.
31. Favreau LV, Pickett CB. Transcriptional regulation of the rat NAD(P)H:quinone reductase gene. Identification of regulatory elements controlling basal level expression and inducible expression by planar aromatic compounds and phenolic antioxidants. *J Biol Chem* 1991;266:4556–4561.
32. Rushmore TH, King RG, Paulson KE, Pickett CB. Regulation of glutathione S-transferase Ya subunit gene expression: identification of a unique xenobiotic-responsive element controlling inducible expression by planar aromatic compounds. *Proc Natl Acad Sci USA* 1990;87:3826–3830.
33. Rushmore TH, Pickett CB. Transcriptional regulation of the rat glutathione S-transferase Ya subunit gene: characterization of a xenobiotic-responsive element controlling inducible expression by phenolic antioxidants. *J Biol Chem* 1990;265:14648–14653.
34. Rushmore TH, Morton MR, Pickett CB. The antioxidant responsive element: activation by oxidative stress and identification of the DNA consensus sequence required for functional activity. *J Biol Chem* 1991;266:11632–11639.
35. Friling RS, Bensimon A, Tichauer Y, Daniel V. Xenobiotic-inducible expression of murine glutathione S-transferase Ya subunit gene is controlled by an electrophile-responsive element. *Proc Natl Acad Sci USA* 1990;87:6258–6262.
36. Friling RS, Bergelson S, Daniel V. Two adjacent AP-1-like binding sites from the electrophile responsive element of the murine glutathione S-transferase Ya subunit gene. *Proc Natl Acad Sci USA* 1992;89:668–672.
37. Angel P, Karin M. The role of Jun, Fos and AP-1 complex in cell-proliferation and transformation. *Biochem Biophys Acta* 1991;1072:129–157.
38. Xie T, Belinsky M, Xu Y, Jaiswal AK. ARE- and TRE-mediated regulation of gene expression: response to xenobiotics and antioxidants. *J Biol Chem* 1995;270:6894–6900.
39. Favreau LV, Pickett CB. Transcriptional regulation of the rat NAD(P)H:quinone reductase gene: characterization of a DNA-protein interaction at the antioxidant responsive element and induction by 12-O-tetradecanoylphorbol 13-acetate. *J Biol Chem* 1993;268:19875–19881.
40. Nguyen T, Pickett CB. Regulation of rat glutathione-S-transferase Ya subunit gene expression: DNA-protein interaction at the antioxidant response element. *J Biol Chem* 1992;267:13535–13539.
41. Nguyen T, Rushmore TH, Pickett CB. Transcriptional regulation of a rat liver glutathione S-transferase Ya subunit gene. *J Biol Chem* 1994;269:13656–13662.
42. Diccianni MB, Imagawa M, Muramatsu M. The dyad palindromic glutathione transferase P enhancer binds multiple factors including AP1. *Nucleic Acids Res* 1992;20:5153–5158.
43. Oridate N, Nishi S, Inuyama Y, Sakai M. Jun and Fos related gene products bind to and modulate the GPE I, a strong enhancer element of the rat glutathione transferase P gene. *Biochem Biophys Acta* 1994;1219:499–504.
44. Yoshioka K, Deng T, Cavigelli M, Karin M. Antitumor promotion by phenolic antioxidants: inhibition of AP-1 activity through induction of Fra expression. *Proc Natl Acad Sci USA* 1995;92: 4972–4976.

45. Suzuki S, Satoh K, Nakano H, Hatayama I, Sato K, Tsuchida S. Lack of correlated expression between the glutathione S-transferase P-form and the oncogene products c-Jun and c-Fos in rat tissues and preneoplastic hepatic foci. *Carcinogenesis* 1995;16:567–571.
46. Halliwell B, Gutteridge JMC, eds. *Free Radicals in Biology and Medicine*, ed 2. Oxford: Clarendon Press, 1989.
47. De Long MJ, Santamaria AB, Talalay P. Role of cytochrome P_1-450 in the induction of NAD(P)H:quinone reductase in a murine hepatoma cell line and its mutants. *Carcinogenesis* 1987;8:1549–1553.
48. Talalay P, De Long MJ, Prochaska HJ. Molecular mechanisms in protection against carcinogenesis. In: Cory JG, Szentivanyi A, eds. *Cancer Biology and Therapeutics*. New York: Plenum Press, 1987;197–216.
49. Pahl HL, Baeuerle PA. Oxygen and the control of gene expression. *BioEssays* 1994;16:497–502.
50. Demple B, Amabile-Cuevas CF. Redox redux: the control of oxidative stress responses. *Cell* 1991;67:837–839.
51. Li Z, Demple B. SoxS, an activator of superoxide stress genes in Escherichia coli: purification and interaction with DNA. *J Biol Chem* 1994;269:18371–18377.
52. Toledano MB, Kullik I, Trinh F, Baird PT, Schneider TD, Storz G. Redox-dependent shift of OxyR-DNA contacts along an extended DNA-binding site: a mechanism for differential promoter selection. *Cell* 1994;78:897–909.
53. Zitomer RS, Lowry CV. Regulation of gene expression by oxygen in *Saccharomyces cerevisiae. Microbiol Rev* 1992;56:1–11.
54. Kuge S, Jones N. YAP1 dependent activation of TRX2 is essential for the response of *Saccharomyces cerevisiae* to oxidative stress by hydroperoxides. *EMBO J* 1994;13:655–664.
55. Jungmann J, Reins HA, Lee J, Romeo A, Hassett R, Kosman D, Jentsch S. Mac-1, a nuclear regulatory protein related to Cu-dependent transcription factors is involved in Cu/Fe utilization and stress resistance in yeast. *EMBO J* 1993;12:5051–5056.
56. Schreck R, Baeuerle PA. Assessing oxygen radicals as mediators in activation of inducible eukaryotic transcription factor NF-kB. *Methods Enzymol* 1994;234:151–163.
57. Meyer M, Schreck R, Baeuerle P. H_2O_2 and antioxidants have opposite effects on activation of NF-kB and AP-1 in intact cells: AP-1 as secondary antioxidant-responsive factor. *EMBO J* 1993;12:2005–2015.
58. Pinkus R, Weiner LM, Daniel V. Role of quinone-mediated generation of hydroxyl radicals in the induction of glutathione S-transferase gene expression. *Biochem* 1995;34:81–88.
59. Bergelson S, Pinkus R, Daniel V. Induction of AP1 (Fos/Jun) by chemical agents mediates activation of glutathione S-transferase and quinone reductase gene expression. *Oncogene* 1994;9:565–571.
60. Bergelson S, Pinkus R, Daniel V. Intracellular glutathione levels regulate Fos/Jun induction and activation of glutathione S-transferase gene expression. *Cancer Res* 1994;54:36–40.
61. Abate C, Patel L, Rauscher III FJ, Curran T. Redox regulation of Fos and Jun DNA-binding activity *in vitro. Science* 1990;249:1157–1161.
62. Xanthoudakis S, Curran T. Identification and characterization of Ref-1, a nuclear protein that facilitates AP-1 DNA binding activity. *EMBO J* 1992;11:653–665.
63. Xanthoudakis S, Miao G, Wang F, Pan YE, Curran T. Redox activation of Fos-Jun DNA binding activity is mediated by a DNA repair enzyme. *EMBO J* 1992;11:323–3335.

Toxicant–Receptor Interactions
Edited by Michael S. Denison and William G. Helferich

7

Mechanism of Heavy Metal–Inducible Gene Transcription

Fuminori Otsuka

Department of Environmental Toxicology, Faculty of Pharmaceutical Sciences, Teikyo University, Sagamiko, Kanagawa, Japan

Shinji Koizumi

Department of Experimental Toxicology, National Institute of Industrial Health, Nagao, Kawasaki, Japan

Heavy metals such as Zn, Fe, and Cu are indispensable for almost all living organisms. The intrinsic chemical nature of each of these essential metals is utilized in fundamental cellular functions, such as the respiratory electron-transfer system and a number of enzyme reactions. Maintenance of the functional protein structure is another role of metals, and the Zn finger structure, for instance, now is known to be an important module of the DNA-binding domain in a certain group of transcription factors. These essential metals can be toxic, however, if present in excessive amounts as well as if deficient. For example, transition metals, such as Cu and Fe, are thought to give rise to oxidative stress in vivo, through the generation of hydroxyl radicals by a Fenton-like reaction (1). On the other hand, nonessential heavy metals, such as Cd and Hg, are believed to exert their toxic actions mainly by interactions with cellular macromolecules, leading to the diverse toxic manifestations through modulation of protein functions and DNA damage (2). Because so many cellular components can interact with heavy metals, however, it often is difficult to specify the target molecules as well as the primary effects of the heavy metals. Zn ions in the finger-loop structure of many transcription factors are proposed to be one of the potential targets of metal toxicity (3). Accordingly, unfavorable modulations of the gene expression may partly underlie the divergent toxic effects of heavy metals. Although our current knowledge of such modulations is quite limited, genes induced by heavy metals have been investigated, such as those for metallothionein (MT), heme oxygenase, and heat-shock proteins (HSPs), which are assumed to be involved in the cellular defense systems against heavy metal toxicity. In particular, expression of the metallothionein genes is investigated as a model for elucidating the molecular mechanism of cellular responses to heavy metals. This chapter focuses on the factors involved in the metal-regulated transcriptional activation of metallothionein genes. In addition, it refers to genes now known to be induced by heavy metals and to the putative mechanisms of their metal-dependent induction.

REGULATION OF THE MT GENES

General Aspects of MTs

MTs are a family of short polypeptides with the remarkable feature of a high cysteine content. Most of cysteines in the MT molecule are arranged in Cys-Xaa-Cys and Cys-Xaa-Xaa-Cys configurations (Xaa indicating an amino acid other than Cys) that can form a metal thiolate cluster (4–6). Since the first discovery of MT in equine kidneys, the MT proteins or genes have been purified or cloned from various species. MTs are known to be widely distributed from cyanobacteria to humans.

Mammalian MTs are composed of approximately 60 amino acids and are characterized by a lack of aromatic amino acids. Usually, there are two subgroups of MT (I and II) with slightly different amino-acid sequences, but whose functional differences still remain ambiguous. Structural analyses of MTs have shown that cysteines of MTs form two metal thiolate clusters: a C-terminal half α domain and an N-terminal half β domain (5–8). Four- and three-divalent metals such as Zn and Cd can bind to the α and β domains in a tetrahedral configuration, respectively. In the case of Cu, six ions are trigonally coordinated in each domain. Inter- and intramolecular exchanges of heavy metals in these domains can occur rapidly (9), which is probably important for the physiologic functions of MTs, such as homeostatic maintenance of essential metals and regulation of protein functions by donating and eliminating the essential metals.

MTs are directly induced by various heavy metals, and cis-acting elements responsive to heavy metals have been identified in the upstream region of the MT genes.

One of the biological functions of MTs is assumed to be prevention of metal toxicity by binding the heavy metals. This has been suggested strongly by the relationship between heavy-metal resistance and preinduction or overexpression of MTs in experimental animals and cultured cells. Recently, MT gene null mice were shown to be susceptible to cadmium intoxication as judged by hepatic poisoning (10). On the other hand, a wide variety of nonmetallic reagents, which includes several cytokines [interleukin 1 (11), interleukin 6 (12), and interferon (13)] and glucocorticoid hormone, has been reported to induce MTs in vitro or in vivo (5, 6). The hepatic MT induction also is observed in association with inflammatory conditions (14) and X radiation (15), suggesting that MT is a stress-inducible protein. Although the precise mechanism of the induction by these stimuli is still unclear except for glucocorticoid hormone, they have attracted much attention in relation to oxidative stress. Recently, an upstream region responsive to hydroxyperoxide has been located in the upstream of the mouse MT-I gene (16). Furthermore, from the finding that the sulfhydryl group of MTs can act as a radical scavenger in vitro (17), it is possible that MTs are involved in the defense line against oxidative stress. Such a potential function of MTs as well as the putative functions in the heavy-metal metabolism in cellular development, differentiation, and proliferation now are being investigated actively.

Regulation of the Yeast MT Gene

The induction of mammalian MTs by heavy metals is regulated mainly at the level of transcription (18, 19). This transcriptional activation is mediated by the cis-acting element located in the upstream region of the MT genes and the trans-acting factors that bind to the element, as is the case with other inducible genes. Before describing the MT induction in mammalian cells, it is worth referring to the MT induction system of yeast cells, because the well-established metalloregulating system in this unicellular organism shows an example for the mechanism by which eukaryotes sense the intracellular concentrations of heavy metals and maintain metal homeostasis. In yeast cells, MT is induced by a single-component signal transduction system, in which a metal-sensor molecule itself acts as a transcription factor for the MT gene.

The MT protein of *Saccharomyces cerevisiae* is encoded by the *CUP1* gene that determines the Cu-resistant phenotype of yeast cells (20). Although the locations of cysteine residues in the CUP1 protein are only distantly related to those in the mammalian MTs (21, 22), the functions of these proteins are similar: both protect cells from metal toxicity by binding metal ions (23). The remarkable difference between mammalian and yeast cells is that CUP1 is induced exclusively by Cu(I) and Ag(I) (21, 24), whereas the mammalian MTs can be induced by a variety of monovalent and divalent heavy metals.

In the upstream region of the *CUP1* gene, the Cu responsiveness resides in the region from −105 to −148 relative to the transcription start site (24, 25). This region, designated *UAScup1* (UASc), can confer Cu responsiveness to a heterologous promoter fused downstream (24). A Cu-responsive transcription factor, ACE1/CUP2 (for convenience, the term ACE1 is used hereafter) was isolated as a gene that can complement the recessive phenotype of a mutant cell line with a defect in the ability to induce CUP1 (26, 27). The ACE1 protein is constitutively expressed in yeast cells and is distributed in the nucleus (28).

ACE1 apparently can be divided into two domains (24, 28, 29). The amino-terminal half of the protein involves 12 cysteine residues and is rich in basic amino acids. This domain by itself, expressed as a recombinant protein, can bind to UASc. On the other hand, the C-terminal half domain is free from cysteine residues and is highly acidic. ACE1, lacking the C-terminal half domain cannot induce *CUP1* gene transcription if introduced into an *ACE1* deletion mutant, suggesting that the C-terminal half domain is involved in transcriptional activation. The functions of these two domains were confirmed by exchanging them with the corresponding domains of the yeast GAL4 transcription factor (30). In addition, ACE1 can interact with general transcription factors, including mammalian TFIID, and may be necessary for both the formation and maintenance of the transcription initiation complex (31, 32).

The most striking feature of ACE1 as a metalloregulatory transcription factor resides in its N-terminal DNA binding domain. Cysteine residues in this domain are arranged in the Cys-Xaa-Cys and Cys-Xaa-Xaa-Cys configurations, which resemble those of the MT protein itself (24, 28, 29). This indicates that the DNA-binding domain also involves a Cu(I)-binding ability. The DNA binding domain translated in vitro was partially protected from trypsin digestion by preincubation with Cu(I) and Ag(I), but not by incubation with other heavy metals (24). Fürst et al. (24) hypothesized that Cu(I) binding provokes a structural change of the N-terminal domain to form an MT-like Cu(I)–thiolate cluster, thereby conferring DNA-binding ability to ACE1 (24). This hypothesis has been supported by various structural analyses (33–35). On the other hand, the functional importance of individual cysteine residues in the DNA-binding domain has been demonstrated by site-directed mutagenesis (30). Eleven out of twelve cysteines in this domain have been found to be essential for Cu-induced structural change of ACE1. Furthermore, mutations of two basic amino acids, Arg^{94} and Lys^{53}, have been found to give rise to a decrease in the capacity of DNA binding without changing the Cu-binding ability. These lines of evidence reveal another aspect of ACE1: The amino acids necessary for the two functions, DNA binding and Cu binding, are interdigitated in a small domain of a single transcription factor.

Binding of Cu(I) ions to ACE1 occurs highly cooperatively (36). This implies that subtle changes in the intracellular Cu concentration are sufficient for the structural change of ACE1, thus leading to the induction of CUP1. On the other hand, the CUP1 protein negatively regulates the expression of its own gene (37). This is probably due to a decrease in intracellular concentrations of free Cu(I) ions as a result of sequestration by CUP1, because primate MT without any structural homology to CUP1 also shows the same effect as CUP1 if expressed under the control of UASc in a *CUP1*-deletion mutant strain (23). Consequently, the intracellular homeostasis of Cu(I) would be maintained by the Cu(I)-sensor molecule ACE1.

It is noteworthy that the *SOD1* gene encoding Cu-, Zn-superoxide dismutase, which is an important enzyme involved in the prevention of oxidative stress, is another target gene of ACE1 (38, 39). The upstream region of the gene contains one ACE1 binding site and the Cu-dependent SOD1 induction is not observed in an *ACE1* deletion mutant. Therefore, CUP1 and SOD1 are co-induced by Cu ions under the regulation of the same Cu-regulatory transcription factor.

Regulation of Mammalian MT Genes

Metal Responsive Element

Compared with the upstream region of the *CUP1* gene of yeast cells, those of the mammalian MT genes usually are more complex, containing various target sequences of transcription factors, such as SP1 (40–42), AP1 (40, 43, 44), AP2 (45, 46), AP4 (47), MLTF (42, 48), and the glucocorticoid hormone receptor (49), although not all these sequences are present in each MT gene upstream region. Nucleotide sequences necessary for the metal responsiveness of the MT genes were analyzed by deletion mapping of the mouse MT-I (50, 51) and human MT-IIA genes (49). In the upstream regulatory region of the mouse MT-I gene, the metal responsiveness resides in two regions: the proximal (around −50) and distal (around −120) regions relative to the transcription initiation site. A short 12–base pair nucleotide sequence in the proximal region was selected as a candidate for metal-responsive element (MRE) (50, 51), and the function was confirmed by demonstrating the ability of a synthetic oligonucleotide containing this sequence to confer metal responsiveness to a heterologous promoter (50). Similar, but not identical, sequences have been shown to be present in the upstream region of various MT genes, and the consensus sequence was deduced to be TGCRCNCGGCC, in which *R* is a purine and *N* is any nucleotide (52). The first seven nucleotides are particularly conserved, and this is termed the *MRE core sequence*. In fact, mutations in the core sequence have been found to destroy metal-dependent transcriptional activation (53). Usually, MRE involves a GC-rich sequence adjacent to the core, which resembles the SP1 binding site. The MRE sequence is totally different from that of the ACE1 binding site (5′-TC$[T]_{4-6}$GCTG-3′) in UASc in yeast cells (54), suggesting that transcription factors that bind to MRE are not related to ACE1.

MREs in each of the upstream regions of the MT genes (52) are named in the order of proximity from the transcription start site, such as MREa, MREb, and so on. Using MREa of the mouse MT-I gene, Searle et al. (55) indicated that multiple MREs are required for more efficient induction by metals, but both the location and orientation of the elements are less important. Moreover, multimeric MREd of the mouse MT-I gene can activate transcription in response to heavy metals even if located downstream of a reporter gene (56). This indicates that MRE can act as a metal-inducible enhancer. Each MRE, however, is not equally functional. Among the MREs of the mouse MT-I gene, for instance, MREd is the most potent, MREe is not functional, and other MREs have a marginal effect (52). Because of the potency of MREd of the mouse MT-I gene (mMREd), it often has been used in various studies.

MRE-Binding Factors

As described in the previous sections, the yeast *CUP1* gene was found to be controlled by the positive regulator ACE1. Séguin et al. (57) suggested that the regulation of mammalian MT genes involves a positive regulatory mechanism by means of an in-vivo competition assay. Consistent with this, in-vivo footprinting analyses of the upstream regions of the mouse MT-I (42) and rat MT-I genes (41) have revealed the protection of multiple MREs in association with heavy-metal treatment.

In order to obtain a heavy metal–responsive transcription factor, sequence-specific MRE binding proteins have been searched by means of various DNA binding assays, such as mobility-shift assay, UV cross-linking analyses, South-western blotting, and footprinting. To date, more than 10 MRE binding factors have been reported; characteristics of most of them were reviewed in (58). It should be noted that there are factors whose binding to MRE does not require the addition of exogenous heavy metals (59–65), as well as factors responsive to Cd or Zn (56, 66–69). Furthermore, a factor that can bind to MRE in response to multiple species of heavy metals has not yet been discovered. Factors responsive to Cd and Zn are candidates for metal regulatory factors, and Zn-responsive factors are described later in this chapter. Other factors, which are not responsive to any heavy metals, include those with unique characteristics: MREBP (62, 64), MRE-BF1 and MRE-BF2 (63). MREBP dissociates from MRE in the presence of high concentrations of various heavy metals in vitro. MRE-BF1 disappears and MRE-BF2 appears after incubation of cells in the presence of heavy metals.

To circumvent the limitation of the biochemical methods, genetic approaches using yeast cells recently have been used to isolate the cDNA coding for MRE-binding proteins (70, 71). Such an approach is possible because there is no yeast factor that binds to mMREd, and the methods include the utilization of yeast cells transformed with a vector carrying an appropriate reporter gene driven by a promoter containing mMREd in a heavy metal–dependent manner. Xu (70) reported the isolation of mouse factor MafY that can complement the *gal*$^-$ phenotype of mutant cells by activating bacterial *galK* gene via multimeric mMREd. The MafY cDNA encodes a small protein consisting of 99 amino acids at most, which contains one Zn finger motif. The transactivation function of MafY is not dependent on heavy metals but is dependent on the carbon source in the culture medium, suggesting modulation of the MafY function by a specific yeast metabolism. In addition, binding of MafY to MRE has not been demonstrated. Inouye et al. (71) reported the isolation of another mouse cDNA, M96, using the one-hybrid system, in which mouse cDNA clones are expressed as fusion proteins with the transactivation domain of the yeast GAL4 transcription factor and are screened by the activity of a reporter gene fused downstream of multimeric mMREd (71). M96 encodes a 45-kDa protein with four Zn fingers that exhibit homology to the trithorax proteins. The glutathione S-transferase (GST)-M96 fusion protein expressed in Escherichia coli can bind to mMREd in a sequence-specific manner, even in the absence of heavy metals. If the recombinant protein was prepared from *E. coli* cultured without the addition of exogenous Zn, however, it lost

MRE-binding activity, suggesting the requirement of Zn for the functional form of M96.

At present, there are no data suggesting a relationship between the two cDNA clones and other MRE-binding factors, and a conclusive mechanism cannot be illustrated from such divergence of MRE binding factors. As described subsequently, however, much progress has been made recently in studies on Zn-responsive MRE binding factors.

MTF-1 and Related Factors

MTF-1 was discovered in HeLa cell nuclear extracts by mobility shift assay using mMREd as a probe (56). Factor binding to mMREd is dependent on the addition of exogenous Zn in vitro. MTF-1 requires not only the core sequence of mMREd but also the flanking GC-rich sequence for its binding. The cDNA coding for MTF-1 has been isolated by expression cloning from mouse libraries using multimeric mMREd as a probe (72). The cDNA thus obtained is 2550 base pairs in length and contains an open reading frame encoding a protein of 72.5 kDa. The recombinant proteins expressed in primate and mouse cells can bind to MRE in a sequence-specific manner, and their partial proteolytic digests show the same pattern of retarded bands as that of endogenous MTF-1 if analyzed in mobility-shift assay, indicating that the cloned cDNA encodes MTF-1. The predicted amino-acid sequence of MTF-1 exhibits domain structures typical of various transcription factors. Specifically, the N-terminal domain contains six tandem TFIIIA-type Zn finger structures known as a DNA-binding motif, and the C-terminal half involves the acidic, proline-rich, and serine/threonine-rich domains presumed to be responsible for transcriptional activation. The Zn-finger domain by itself can bind to MRE, and the three C-terminal domains have been demonstrated to be functional by estimating the activity of chimeric proteins with the DNA binding domain of the yeast GAL4 transcription factor (73). MTF-1 is expressed constitutively and is encoded by a single copy gene (72). Using the cDNA of mouse MTF-1 (mMTF-1), human MTF-1 (hMTF-1), was isolated (74). Compared with mMTF-1, the human homologue has 78 additional amino acids at the C terminus, but they exhibit 93% identity in the predicted amino acid sequence. Despite such high homology, there is a difference between these two factors in the capacity of the Zn-dependent transcriptional activation via MRE. Overexpressed hMTF-1 activates transcription in response to Zn (74), whereas mMTF-1 is less responsive to Zn (72, 75). Comparison of various chimeric proteins between mMTF-1 and hMTF-1 shows that this difference does not reside in the C-terminal extension of hMTF-1 but is located in either the acidic or the proline-rich region (73).

The importance of MTF-1 in heavy metal–dependent transcription was demonstrated by antisense RNA expression (75) and the gene knockout experiment (76). Palmiter (75) established a BHK cell line with the bacterial β-galactosidase gene (β-*gal*) integrated stably into a chromosome, whose expression was controlled by upstream multimeric mMREd. Overexpression of mMTF-1 in this cell line increased the expression of *β-gal*, whereas expression of antisense MTF-1 RNA suppressed the induction of the gene by Zn, indicating the importance of mMTF-1 in the MRE-mediated transcriptional activation. Heuchel et al. established embryonic stem cells, both of whose alleles of the mMTF-1 gene were disrupted (76). In the null embryonic stem cells, reporter gene expression mediated by MRE as well as that of the endogenous MT-I and -II genes was not induced by treatment with a variety of heavy metals, including Zn, Cd, Cu, Hg, Ni, and Pb. Expression of MTF-1 in these cells restored the Zn-dependent transcription via MRE. These lines of evidence indicate that MTF-1 is an essential factor in the heavy metal–dependent transcriptional activation via MRE.

Zinc activated protein (ZAP) (68) and zinc regulatory factor (ZRF) (69) are MRE-binding factors found in nuclear extracts prepared from rat liver and HeLa cells, respectively. Although, these two factors and MTF-1 are similar in their requirement for Zn ions for binding to MRE, ZRF is different from ZAP and MTF-1 in its requirement for a few nucleotides upstream of the MRE core for DNA binding (69, 77). ZRF was purified from HeLa cell nuclear extracts to near homogeneity as a 116-kDa protein (78). The amino-acid sequences of several ZRF fragments were identical with those present in the N-terminal half of hMTF-1. This indicates that ZRF and MTF-1 are the same or at least closely related factors.

Possible Mechanisms of Heavy Metal–Dependent Transcriptional Activation of Mammalian MT Genes

As described in the previous sections, the yeast factor ACE1 contains the MT-like cysteine cluster structure that points to a molecular mechanism for sensing metal ions. On the other hand, the heavy metal–related structure of the mammalian factor MTF-1 is only the classical Zn fingers (72, 74), and not the cysteine cluster structure. Given that the heavy metal–dependent transcriptional activation via MRE cannot occur without MTF-1, how do various metal signals converge upon MTF-1? Palmiter (75) proposed the hypothesis that various heavy metal signals are converted into a Zn signal; specifically, Zn ions act as a second messenger. In this model, Zn ions that are weakly associated with various proteins form a Zn reservoir, and the concentration of free Zn ions increases by replacing Zn in the reservoir with other heavy metals such as Cd.

A simplified mechanism is that MTF-1 itself acts as a Zn sensor (76). Generally, Zn ions in the finger structure are thought to stabilize the functional protein structure (79), and it is unknown whether Zn can regulate the DNA-binding activity of transcription factors. Heuchel et al. (76) proposed the possibility that MTF-1 acts as a Zn sensor through its Zn finger structure. To assess this possibility, Radke et al. (73) prepared a chimeric protein between the N-terminal domain of MTF-1 containing the Zn fingers and the transcription-activation domain of viral protein VP16. The protein indeed can activate transcription in a Zn-dependent manner, but the extent of Zn responsiveness is less than that of intact MTF-1. The authors have suggested that the full Zn responsiveness results from collaboration between different functional domains of MTF-1.

Another possibility is that the DNA binding activity of MTF-1 is suppressed by an inhibitor molecule and that heavy-metal signals liberate MTF-1 from the inhibitor to activate transcription (74, 75). In this case, the putative inhibitor molecule may be the real metal sensor. This mechanism is similar to that of activation of NF-kB, whose activity is suppressed by the formation of a complex with I-kB in cytosol (80).

If multiple MRE binding factors are taken into consideration, the transcriptional activation via MRE seems to be more complex, and each MRE-binding factor possibly cooperates with MTF-1 to augment efficient metal-dependent transcription. Furthermore, it also is possible that a suppressor-type MRE-binding factor is involved in the mechanism; such a factor can suppress transcription via MRE and is replaced by MTF-1 if MT expression is required. MREBP (64) and MRE-BF1 (63) may be such factors, because their binding to MRE seems to be negatively regulated by heavy metals. Clearly, isolation of the genes encoding various MRE-binding factors and reconstitution of the heavy-metal dependent transcription in vitro are necessary to clarify the molecular mechanism of the heavy metal–dependent MT induction in mammalian cells.

REGULATION OF OTHER METAL-INDUCIBLE GENES

Because MRE acts as a metal-inducible enhancer, genes that have MRE in the upstream region could be induced by heavy metals. Actually, the presence of MRE-like sequences has been indicated in the upstream regions of several genes; for example, those encoding acute-phase proteins (α_1-acid glycoprotein and C-reactive protein) (81), interleukin 8 (82), estrogen receptor (83), heme oxygenase (HO) (84, 85), and αB-crystallin (86). Of these genes, only the gene encoding α1-acid glycoprotein has been suggested to be regulated by MRE (81). To date, HO and HSPs have been relatively well investigated as heavy metal-inducible genes, but their induction seems to be independent of MRE, as described subsequently.

HO

An isozyme of HO, HO-1, is known to be induced by various stimuli, including heat shock and heavy metals (Cd, Co) as well as its substrate heme (87–90). HO is a rate-limiting enzyme in the heme catabolism and catalyzes the cleavage of heme, producing biliverdin that subsequently is converted into bilirubin by biliverdin reductase (91). Because bilirubin has been shown to function as an antioxidant (92), it is conceivable that the induction of HO-1 by heavy metals is involved in the protection system against the oxidative stress generated by metals.

Induction of HO-1 by heavy metals is regulated at the level of transcription (87, 93). Recent analyses of the upstream regions of human and mouse HO-1 genes, however, have suggested that the mechanisms underlying HO-1 induction by heavy metals are different between these two species (93, 94). By deletion analysis of the 5′ upstream region of the mouse HO-1 gene, Alam and Zhining (95) have located the Cd-responsive region at about 4000 base pairs upstream from the transcription start site. The Cd responsiveness resides in the 268–base pair fragment containing two copies each of the AP1 and the C/EBP binding sites. Site-directed mutagenesis and deletion analyses showed that the AP1 binding sites are more important for Cd responsiveness than the C/EBP binding sites, although both are necessary for full enhancer activity. The fragment also is responsive to Zn and Hg.

On the other hand, Takeda et al. (93) obtained completely different results for the human HO-1 gene. In their case, the Cd-responsive region also is located at about 4000 base pairs from the transcription start site. Further dissection of this region, however, indicated that a 10–base pair sequence, TGCTAGATTT, is required for Cd responsiveness. Moreover, the formation of two complexes of nuclear proteins and a synthetic probe with the Cd-responsive sequence was demonstrated by mobility-shift assay. The Cd-responsive element thus determined is quite different from MRE consensus and is not found in human MT-gene upstream regions. It is noted that this element is not responsive to Zn (96), whereas MRE can respond to a variety of heavy metals. At present, the discrepancy between the mouse and human HO-1 genes remains elusive. In any case, the Cd-dependent transcriptional activation of the HO-1 gene appears to be mediated by a different mechanism from that of the MT genes.

HSPs

Induction of various HSPs by heavy metals has been demonstrated (86, 97–99). HSPs are induced by various kinds of environmental stress in addition to heat shock (100, 101). As some HSPs are known to facilitate the proper folding of proteins, their induction by heavy metals implies protection of proteins from metal toxicity or repair of damaged proteins. The HSP genes have multiple heat-shock elements in the upstream region, which is the target sequence of the heat shock factor (HSF) (102). HSF is constitutively expressed as an inactive form in mammalian cells and is activated by various stresses to form a trimer, acquiring binding ability to the heat shock element. It has been reported that the active DNA-binding form of HSF increases in association with the treatment of heavy metals (103–106). Therefore, induction of HSPs by heavy metals may be due to the activation of HSF through some indirect effects of metals, such as generating abnormal proteins, which can trigger the induction of HSPs (107).

Other Genes

It is reasonable to postulate that the induction of genes described previously is involved in the cellular defense systems against metal toxicity. On the other hand, there have been reports on the heavy metal–inducible genes that possibly lead to harmful effects on normal cellular functions. Toxicologically, such genes are important for investigating the pathogenicity of heavy metals. Firstly, the proto-oncogenes, c-Myc

(108, 109), c-Jun (108, 109) and c-Fos (109–112), have been reported to be induced by heavy metals in various cell species. Although the mechanism of the induction of these proto-oncogenes is still controversial, stabilization of mRNA through inhibition of protein synthesis by heavy metals has been suggested to contribute, at least in part, to the induction (110, 112). The activation of these proto-oncogenes is reminiscent of the mitogen mimetic action of divalent heavy metals (release of Ca ions and increase in inositol polyphosphate formation) (113) and often has been discussed in the context of metal carcinogenesis. In fact, it has been suggested that Cd induction of c-Myc mRNA in NRK-49F cells is mediated by signal-transduction pathways involving protein kinase C (109). Also, tetradecanoyl phorbol acetate–inducible genes, including *TIS8* (*egr1*) and *TIS11*, were shown to be induced by Cd or Zn in Swiss 3T3 cells (111). Further information is necessary, however, to clarify the relation to metal carcinogenesis as well as the induction mechanism. Secondly, Horiguchi et al. (82) has reported that interleukin 8 is induced by Cd at both the mRNA and protein levels in human peripheral mononuclear cells (82). Because interleukin 8 is a neutrophil chemotactic and activating factor, infiltration of neutrophils observed with severe renal damage caused by Cd can be explained. In this case, reactive oxygen intermediates generated by Cd was implicated as the cause of interleukin 8 induction. Thirdly, induction of estrogen-regulated genes (progesterone receptor, cathepsin D, and pS2) in a human breast-cancer cell line by Cd has been reported (83). Such estrogen-mimetic effects were not observed with Zn and were attributed to mediation by the estrogen receptor. Although the implication of this phenomenon in vivo is ambiguous, the author has pointed out the possibility of breast cancer occurring via derangement of cell proliferation by Cd.

CONCLUSIONS AND FUTURE PERSPECTIVES

Although our knowledge of metal-inducible genes, particularly on the mechanism of transcriptional activation, remains limited, some of them are thought to be involved in the cellular defense system against heavy-metal toxicity. In this context, investigation of the genes induced by heavy metals is worthwhile because their induction may reflect the toxic actions of heavy metals. Of these genes, the MT genes have been well investigated in terms of the mechanism of transcriptional activation by heavy metals at the molecular level. Recent progress in the studies on transcription factors for the MT genes has revealed that the mechanism of heavy metal–dependent transcriptional activation in mammalian cells, is not simply analogous to that in yeast cells, despite the similarity of the biological functions of MTs in each organism. Clarification of the induction system of MTs is a critical key for understanding the mechanism of cellular responses to heavy metals in mammalian cells. Nevertheless, MRE-mediated transcriptional activation is not the only mechanism by which mammalian cells respond to heavy metals, and other direct or indirect pathways to reach various transcription factors can be considered. Divergent mechanisms of metal-dependent gene expression might correspond to the differential effects of heavy metals. Clarification of such mechanisms will shed light on unknown cellular defense systems and their networks organized to protect cells from the toxicity of heavy metals.

REFERENCES

1. Halliwell B, Gutteridge JM. Oxygen toxicity, oxygen radicals, transition metals and disease. *Biochem J* 1984;219:1–14.
2. Clarkson TW. Effects—general principles underlying the toxic action of metals. In: Friberg L, Nordberg GF, Vouk VB, eds. *Handbook on Toxicology of Metals.* Amsterdam: Elsevier, 1979;99–117.
3. Sunderman FJ, Barber AM. Finger-loops, oncogenes, and metals: Claude Passmore Brown memorial lecture. *Ann Clin Lab Sci* 1988;18:267–288.
4. Hamer DH. Metallothionein. *Annu Rev Biochem* 1986;55:913–951.
5. Kägi JHR, Kojima Y, eds. *Metallothionein II.* Basel: Birkhäuser Verlag, 1987.
6. Suzuki KT, Imura N, Kimura M, eds. *Metallothionein III.* Basel: Birkhäuser Verlag, 1993.
7. Braun W, Wagner G, Wörgötter E, Vasak M, Kägi JH, Wüthrich K. Polypeptide fold in the two metal clusters of metallothionein-2 by nuclear magnetic resonance in solution. *J Mol Biol* 1986;187:125–129.
8. Furey WF, Robbins AH, Clancy LL, Winge DR, Wang BC, Stout CD. Crystal structure of Cd, Zn metallothionein. *Science* 1986;231:704–710.
9. Otvos JD, Liu X, Shen G, Basti M. Dynamic aspects of metallothionein structure. In: Suzuki KT, Imura N, Kimura M, eds. *Metallothionein III.* Basel: Birkhäuser Verlag, 1993;57–74
10. Masters BA, Kelly EJ, Quaife CJ, Brinster RL, Palmiter RD. Targeted disruption of metallothionein I and II genes increases sensitivity to cadmium. *Proc Natl Acad Sci USA* 1994;91:584–588.
11. Karin M, Imbra RJ, Heguy A, Wong G. Interleukin 1 regulates human metallothionein gene expression. *Mol Cell Biol* 1985;5:2866–2869.
12. Schroeder JJ, Cousins RJ. Interleukin 6 regulates metallothionein gene expression and zinc metabolism in hepatocyte monolayer cultures. *Proc Natl Acad Sci USA* 1990;87:3137–3141.
13. Friedman RL, Stark GR. alpha-Interferon-induced transcription of HLA and metallothionein genes containing homologous upstream sequences. *Nature* 1985;314:637–639.
14. Sobocinski PZ, Canterbury WJ. Hepatic metallothionein induction in inflammation. *Ann N Y Acad Sci* 1982;389:354–367.
15. Matsubara J, Shida T, Ishioka K, Egawa S, Inada T, Machida K. Protective effect of zinc against lethality in irradiated mice. *Environ Res* 1986;41:558–567.
16. Dalton T, Palmiter RD, Andrews GK. Transcriptional induction of the mouse metallothionein-I gene in hydrogen peroxide-treated Hepa cells involves a composite major late transcription factor/antioxidant response element and metal response promoter elements. *Nucleic Acids Res* 1994;22:5016–5023.
17. Thornalley PJ, Vasak M. Possible role for metallothionein in protection against radiation-induced oxidative stress: kinetics and mechanism of its reaction with superoxide and hydroxyl radicals. *Biochim Biophys Acta* 1985;827:36–44.
18. Durnam DM, Palmiter RD. Transcriptional regulation of the mouse metallothionein-I gene by heavy metals. *J Biol Chem* 1981;256:5712–5716.
19. Karin M, Andersen RD, Herschman HR. Induction of metallothionein mRNA in HeLa cells by dexamethasone and by heavy metals. *Eur J Biochem* 1981;118:527–531.
20. Fogel S, Welch JW. Tandem gene amplification mediates copper resistance in yeast. *Proc Natl Acad Sci USA* 1982;79:5342–5346.
21. Butt TR, Sternberg EJ, Gorman JA, Clark P, Hamer D, Rosenberg M, Crooke ST. Copper metallothionein of yeast, structure of the gene, and regulation of expression. *Proc Natl Acad Sci USA* 1984;81:3332–3336.
22. Winge DR, Nielson KB, Gray WR, Hamer DH. Yeast metallothionein: sequence and metal-binding properties. *J Biol Chem* 1985;260:14464–14470.
23. Thiele DJ, Walling MJ, Hamer DH. Mammalian metallothionein is functional in yeast. *Science* 1986;231:854–856.
24. Fürst P, Hu S, Hackett R, Hamer D. Copper activates metallothionein gene transcription by altering the conformation of a specific DNA binding protein. *Cell* 1988;55:705–717.
25. Thiele DJ, Hamer DH. Tandemly duplicated upstream control sequences mediate copper-induced transcription of the *Saccharomyces cerevisiae* copper-metallothionein gene. *Mol Cell Biol* 1986;6:1158–1163.

26. Thiele DJ. ACE1 regulates expression of the *Saccharomyces cerevisiae* metallothionein gene. *Mol Cell Biol* 1988;8:2745–2752.
27. Welch J, Fogel S, Buchman C, Karin M. The CUP2 gene product regulates the expression of the CUP1 gene, coding for yeast metallothionein. *EMBO J* 1989;8:255–260.
28. Szczypka MS, Thiele DJ. A cysteine-rich nuclear protein activates yeast metallothionein gene transcription. *Mol Cell Biol* 1989;9:421–429.
29. Buchman C, Skroch P, Welch J, Fogel S, Karin M. The CUP2 gene product, regulator of yeast metallothionein expression, is a copper-activated DNA-binding protein. *Mol Cell Biol* 1989;9:4091–4095.
30. Hu S, Fürst P, Hamer D. The DNA and Cu binding functions of ACE1 are interdigitated within a single domain. *New Biol* 1990;2:544–555.
31. Culotta VC, Hsu T, Hu S, Fürst P, Hamer D. Copper and the ACE1 regulatory protein reversibly induce yeast metallothionein gene transcription in a mouse extract. *Proc Natl Acad Sci USA* 1989;86: 8377–8381.
32. Kambadur R, Culotta V, Hamer D. Cloned yeast and mammalian transcription factor TFIID gene products support basal but not activated metallothionein gene transcription. *Proc Natl Acad Sci USA* 1990;87:9168–9172.
33. Casas FJ, Hu S, Hamer D, Karpel RL. Spectroscopic characterization of the copper(I)-thiolate cluster in the DNA-binding domain of yeast ACE1 transcription factor. *FEBS Lett* 1991;281:205–208.
34. Dameron CT, Winge DR, George GN, Sansone M, Hu S, Hamer D. A copper-thiolate polynuclear cluster in the ACE1 transcription factor. *Proc Natl Acad Sci USA* 1991;88:6127–6131.
35. Nakagawa KH, Inoue C, Hedman B, Karin M, Tullius TD, Hodgson KO. Evidence from EXAFS for a copper cluster in the metalloregulatory protein CUP2 from yeast. *J Am Chem Soc* 1991;113:3621–3623.
36. Fürst P, Hamer D. Cooperative activation of a eukaryotic transcription factor: interaction between Cu(I) and yeast ACE1 protein. *Proc Natl Acad Sci USA* 1989;86:5267–5271.
37. Hamer DH, Thiele DJ, Lemontt JE. Function and autoregulation of yeast copperthionein. *Science* 1985;228:685–690.
38. Carri MT, Galiazzo F, Ciriolo MR, Rotilio G. Evidence for co-regulation of Cu, Zn superoxide dismutase and metallothionein gene expression in yeast through transcriptional control by copper via the ACE 1 factor. *FEBS Lett* 1991;278:263–266.
39. Gralla EB, Thiele DJ, Silar P, Valentine JS. ACE1, a copper-dependent transcription factor, activates expression of the yeast copper, zinc superoxide dismutase gene. *Proc Natl Acad Sci USA* 1991;88:8558–8562.
40. Lee W, Haslinger A, Karin M, Tjian R. Activation of transcription by two factors that bind promoter and enhancer sequences of the human metallothionein gene and SV40. *Nature* 1987;325:368–372.
41. Andersen RD, Taplitz SJ, Wong S, Bristol G, Larkin B, Herschman HR. Metal-dependent binding of a factor *in vivo* to the metal-responsive elements of the metallothionein 1 gene promoter. *Mol Cell Biol* 1987;7:3574–3581.
42. Mueller PR, Salser SJ, Wold B. Constitutive and metal-inducible protein:DNA interactions at the mouse metallothionein I promoter examined by *in vivo* and *in vitro* footprinting. *Genes Dev* 1988;2:412–427.
43. Lee W, Mitchell P, Tjian R. Purified transcription factor AP-1 interacts with TPA-inducible enhancer elements. *Cell* 1987;49:741–752.
44. Angel P, Imagawa M, Chiu R, and Karin M. Phorbol ester-inducible genes contain a common cis element recognized by a TPA-modulated trans-acting factor. *Cell* 1987;49:729–739.
45. Mitchell PJ, Wang C, Tjian R. Positive and negative regulation of transcription in vitro: enhancer-binding protein AP-2 is inhibited by SV40 T antigen. *Cell* 1987;50:847–861.
46. Imagawa M, Chiu R, Karin M. Transcription factor AP-2 mediates induction by two different signal-transduction pathways: protein kinase C and cAMP. *Cell* 1987;51:251–260.
47. Mermod N, Williams TJ, Tjian R. Enhancer binding factors AP-4 and AP-1 act in concert to activate SV40 late transcription *in vitro*. *Nature* 1988;332:557–561.
48. Carthew RW, Chodosh LA, Sharp PA. The major late transcription factor binds to and activates the mouse metallothionein I promoter. *Genes Dev* 1987;1:973–980.
49. Karin M, Haslinger A, Holtgreve H, Rechards DI, Krauter P, Westphal HM, Beato M. Characterization of DNA sequences through which cadmium and glucocorticoid hormones induce human metallothionein-IIA gene. *Nature* 1984;308:513–519.
50. Stuart GW, Searle PF, Chen HY, Brinster RL, Palmiter RD. A 12-base-pair DNA motif that is repeated several times in metallothionein gene promoters confers metal regulation to a heterologous gene. *Proc Natl Acad Sci USA* 1984;81:7318–7322.

51. Carter AD, Felber BK, Walling M, Jubier M-F, Schmidt CJ, Hamer DH. Duplicated heavy metal control sequences of the mouse metallothionein-I gene. *Proc Natl Acad Sci USA* 1984;81:7392–7396.
52. Stuart GW, Searle PF, Palmiter RD. Identification of multiple metal regulatory elements in mouse metallothionein-I promoter by assaying synthetic sequences. *Nature* 1985;317:828–831.
53. Culotta VC, Hamer DH. Fine mapping of a mouse metallothionein gene metal response element. *Mol Cell Biol* 1989;9:1376–1380.
54. Evans CF, Engelke DR, Thiele DJ. ACE1 transcription factor produced in Escherichia coli binds multiple regions within yeast metallothionein upstream activation sequences. *Mol Cell Biol* 1990;10:426–429.
55. Searle PF, Stuart GW, Palmiter RD. Building a metal-responsive promoter with synthetic regulatory elements. *Mol Cell Biol* 1985;5:1480–1489.
56. Westin G, Schaffner W. A zinc-responsive factor interacts with a metal-regulated enhancer element (MRE) of the mouse metallothionein-I gene. *EMBO J* 1988;7:3763–3770.
57. Séguin C, Felber BK, Carter AD, Hamer DH. Competition for cellular factors that activate metallothionein gene transcription. *Nature* 1984;312:781–785.
58. Koizumi S, Otsuka F. Factors involved in the transcriptional regulation of metallothionein genes. In: Suzuki K, Imura N, Kimura M, eds. *Metallothionein III*. Basel: Birkhäuser, 1994;457–474.
59. Imbert J, Zafarullah M, Culotta VC, Gedamu L, Hamer D. Transcription factor MBF-I interacts with metal regulatory elements of higher eucaryotic metallothionein genes. *Mol Cell Biol* 1989;9:5315–5323.
60. Labbé S, Prévost J, Remondelli P, Leone A, Séguin C. A nuclear factor binds to the metal regulatory elements of the mouse gene encoding metallothionein-I. *Nucleic Acids Res* 1991;19:4225–4231.
61. Séguin C. A nuclear factor requires Zn^{2+} to bind a regulatory MRE element of the mouse gene encoding metallothionein-1. *Gene* 1991;97:295–300.
62. Koizumi S, Otsuka F, Yamada H. A nuclear factor that interacts with metal responsive elements of a human metallothionein gene. *Chem Biol Interact* 1991;80:145–157.
63. Czupryn M, Brown WE, Vallee BL. Zinc rapidly induces a metal response element-binding factor. *Proc Natl Acad Sci USA* 1992;89:10395–10399.
64. Koizumi S, Suzuki K, Otsuka F. A nuclear factor that recognizes the metal-responsive elements of human metallothionein IIA gene. *J Biol Chem* 1992;267:18659–18664.
65. Labbé S, Larouche L, Mailhot D, Séguin C. Purification of mouse MEP-1, a nuclear protein which binds to the metal regulatory elements of genes encoding metallothionein. *Nucleic Acids Res* 1993;21:1549–1554.
66. Séguin C, Prévost J. Detection of a nuclear protein that interacts with a metal regulatory element of the mouse metallothionein I gene. *Nucleic Acids Res* 1988;16:10547–10560.
67. Andersen RD, Taplitz SJ, Oberbauer AM, Calame KL, Herschman HR. Metal-dependent binding of a nuclear factor to the rat metallothionein-I promoter. *Nucleic Acids Res* 1990;18:6049–6055.
68. Searle PF. Zinc dependent binding of a liver nuclear factor to metal response element MRE-a of the mouse metallothionein-I gene and variant sequences. *Nucleic Acids Res* 1990;18:4683–4690.
69. Koizumi S, Yamada H, Suzuki K, Otsuka F. Zinc-specific activation of a HeLa cell nuclear protein which interacts with a metal responsive element of the human metallothionein-IIA gene. *Eur J Biochem* 1992;210:555–560.
70. Xu C. cDNA cloning of a mouse factor that activates transcription from a metal response element of the mouse metallothionein-I gene in yeast. *DNA Cell Biol* 1993;12:517–525.
71. Inouye C, Remondelli P, Karin M, Elledge S. Isolation of a cDNA encoding a metal response element binding protein using a novel expression cloning procedure: the one hybrid system. *DNA Cell Biol* 1994;13:731–742.
72. Radtke F, Heuchel R, Georgiev O, Hergersberg M, Garglio M, Dembic Z, Schaffner W. Cloned transcription factor MTF-1 activates the mouse metallothionein I promoter. *EMBO J* 1993;12:1355–1362.
73. Radtke F, Georgiev O, Müller HP, Brugnera E, Schaffner W. Functional domains of the heavy metal-responsive transcription regulator MTF-1. *Nucleic Acids Res* 1995;23:2277–2286.
74. Brugnera E, Georgiev O, Radtke F, Heuchel R, Baker E, Sutherland GR, Schaffner W. Cloning, chromosomal mapping and characterization of the human metal-regulatory transcription factor MTF-1. *Nucleic Acids Res* 1994;22:3167–3173.
75. Palmiter RD. Regulation of metallothionein genes by heavy metals appears to be mediated by a zinc-sensitive inhibitor that interacts with a constitutively active transcription factor, MTF-1. *Proc Natl Acad Sci USA* 1994;91:1219–1223.

76. Heuchel R, Radtke F, Georgiev O, Stark G, Aguet M, Schaffner W. The transcription factor MTF-1 is essential for basal and heavy metal-induced metallothionein gene expression. *EMBO J* 1994;13:2870–2875.
77. Koizumi S, Otsuka F. Transcriptional regulation of the metallothionein gene: metal-responsive element and zinc regulatory factor. In: Sarkar B, ed. *Genetic Response to Metals.* New York: Mercel Dekker, 1995;397–410.
78. Otsuka F, Iwamatsu A, Suzuki K, Ohsawa M, Hamer DH, Koizumi S. Purification and characterization of a protein that binds to metal responsive elements of the human metallothionein IIA gene. *J Biol Chem* 1994;269:23700–23707.
79. Klug A, Rhodes D. Zinc fingers: a novel protein fold for nucleic acid recognition. *Cold Spring Harbor Symposia Quantitative Biology* 1987;52:473–482.
80. Baeuerle PA, Baltimore D. IkB: a specific inhibitor of the NF-kB transcription factor. *Science* 1988;242:540–546.
81. Yiangou M, Ge X, Carter KC, Papaconstantinou J. Induction of several acute-phase protein genes by heavy metals: a new class of metal-responsive genes. *Biochemistry* 1991;30:3798–3806.
82. Horiguchi H, Mukaida N, Okamoto S, Teranishi H, Kasuya M, Matsushima K. Cadmium induces interleukin-8 production in human peripheral blood mononuclear cells with the concomitant generation of superoxide radicals. *Lymphokine Cytokine Res* 1993;12:421–428.
83. Garcia-Morales P, Saceda M, Kenney N, Kim N, Salomon, DS, Gottardis MM, Solomon HB, Sholler PF, Jordan VC, Martin MB. Effect of cadmium on estrogen receptor levels and estrogen-induced responses in human breast cancer cells. *J Biol Chem* 1994;269:16896–16901.
84. Müller RT, Taguchi H, Shibahara S. Nucleotide sequence and organization of the rat heme oxygenase gene. *J Biol Chem* 1987;262:6795–6802.
85. Alam J, Cai J, Smith A. Isolation and characterization of the mouse heme oxygenase-1 gene: distal 5′ sequences are required for induction by heme or heavy metals. *J Biol Chem* 1994;269:1001–1009.
86. Head MW, Corbin E, Goldman JE. Coordinate and independent regulation of αB-crystallin and hsp27 expression in response to physiological stress. *J Cell Physiol* 1994;159:41–50.
87. Alam J, Shibahara S, Smith A. Transcriptional activation of the heme oxygenase gene by heme and cadmium in mouse hepatoma cells. *J Biol Chem* 1989;264:6371–6375.
88. Lin JH, Villalon P, Martasek P, Abraham NG. Regulation of heme oxygenase gene expression by cobalt in rat liver and kidney. *Eur J Biochem* 1990;192:577–582.
89. Mitani K, Fujita H, Sassa S, Kappas A. Activation of heme oxygenase and heat shock protein 70 genes by stress in human hepatoma cells. *Biochim Biophys Res Commun* 1990;166:1429–1434.
90. Tyrrell RM, Applegate LA, Tromvoukis Y. The proximal promoter region of the human heme oxygenase gene contains elements involved in stimulation of transcriptional activity by a variety of agents including oxidants. *Carcinogenesis* 1993;14:761–765.
91. Shibahara H. Heme oxygenase—Regulation of and physiological implication in heme catabolism. In: Fujita H, ed. *Regulation of Heme Protein Synthesis.* Dayton: AlphaMed Press, 1994;103–116.
92. Stocker R, Yamamoto Y, McDonagh AF, Glazer AN, Ames BN. Bilirubin is an antioxidant of possible physiological importance. *Science* 1987;235:1043–1046.
93. Takeda K, Ishizawa S, Sato M, Yoshida T, Shibahara S. Identification of a cis-acting element that is responsible for cadmium-mediated induction of the human heme oxygenase gene. *J Biol Chem* 1994;269:22858–22867.
94. Alam J. Multiple elements within the 5′ distal enhancer of the mouse heme oxygenase-1 gene mediate induction by heavy metals. *J Biol Chem* 1994;269:25049–25056.
95. Alam J, Zhining D. Distal AP-1 binding sites mediate basal level enhancement and TPA induction of the mouse heme oxygenase-1 gene. *J Biol Chem* 1992;267:21894–21900.
96. Takeda K, Fujita H, Shibahara S. Differential control of the metal-mediated activation of the human heme oxygenase-1 and metallothionein IIA genes. *Biochem Biophys Res Commun* 1995;207:160–167.
97. Levinson W, Oppermann H, Jackson J. Transition series metals and sulfhydryl reagents induce the synthesis of four proteins in eukaryotic cells. *Biochim Biophys Acta* 1980;606:170–180.
98. Heikkila JJ, Browder LW, Gedamu L, Nickells RW, Schultz GA. Heat-shock gene expression in animal embryonic systems. *Can J Genet Cytol* 1986;28:1093–1105.
99. Hiranuma K, Hirata K, Abe T, Hirano T, Matsuno K, Hirano H, Suzuki K, Higashi K. Induction of mitochondrial chaperonin, hsp60, by cadmium in human hepatoma cells. *Biochem Biophys Res Commun* 1993;194:531–536.
100. Burdon RH. Heat shock and the heat shock proteins. *Biochem J* 1986;240:313–324.
101. Lindquist S. The heat-shock response. *Ann Rev Biochem* 1986;55:1151–1191.

102. Morimoto RI, Sarge KD, Abravaya K. Transcriptional regulation of heat shock genes: a paradigm for inducible genomic responses. *J Biol Chem* 1992;267:21987–21990.
103. Mosser DD, Theodorakis NG, Morimoto RI. Coordinate changes in heat shock element-binding activity and HSP70 gene transcription rates in human cells. *Mol Cell Biol* 1988;8:4736–4744.
104. Sarge KD, Murphy SP, Morimoto RI. Activation of heat shock gene transcription by heat shock factor 1 involves oligomerization, acquisition of DNA-binding activity, and nuclear localization and can occur in the absence of stress. *Mol Cell Biol* 1993;13:1392–1407.
105. Hatayama T, Asai Y, Wakatsuki T, Kitamura T, Imahara H. Regulation of hsp70 synthesis induced by cupric sulfate and zinc sulfate in thermotolerant HeLa cells. *J Biochem (Tokyo)* 1993;114:592–597.
106. Abe T, Konishi T, Katoh T, Hirano H, Matsukuma K, Kashimura M, and Higashi K. Induction of heat shock 70 mRNA by cadmium is mediated by glutathione suppressive and non-suppressive triggers. *Biochem Biophys Acta* 1994;1201:29–36.
107. Ananthan J, Goldberg AL, Voellmy R. Abnormal proteins serve as eukaryotic stress signals and trigger the activation of heat shock genes. *Science* 1986;232:522–524.
108. Jin P, Ringertz NR. Cadmium induces transcription of proto-oncogenes c-jun and c-myc in Rat L6 myoblasts. *J Biol Chem* 1990;265:14061–14064.
109. Tang N, Enger MD. Cd^{2+}-induced c-myc mRNA accumulation in NRK-49F cells is blocked by the protein kinase inhibitor H7 but not by HA1004, indicating that protein kinase C is a mediator of the response. *Toxicology* 1993;81:155–164.
110. Andrew GK, Harding MA, Calvet JP, Adamson ED. The heat shock response in HeLa cells is accompanied by elevated expression of the c-fos proto-oncogene. *Mol Cell Biol* 1987;7:3452–3458.
111. Epner DE, Herschman HR. Heavy metals induce expression of the TPA-inducible sequence (TIS) genes. *J Cell Physiol* 1991;148:68–74.
112. Taylor GA, Blackshear PJ. Zinc inhibits turnover of labile mRNA in intact cells. *J Cell Physiol* 1995;162:378–387.
113. Smith JB, Dwyer SD, Smith L. Cadmium evokes inositol polyphosphate formation and calcium mobilization: evidence for a cell surface receptor that cadmium stimulates and zinc antagonizes. *J Biol Chem* 1989;264:7115–7118.

PART III

Xenobiotics and Second Messenger Systems

Toxicant–Receptor Interactions
Edited by Michael S. Denison and William G. Helferich

8

Modulation of Protein Kinases by Xenobiotics

Burra V. Madhukar

Department of Pediatrics and Human Development, Michigan State University, East Lansing, Michigan, USA

Cellular homeostasis is maintained through a series of complex biochemical processes that are tightly controlled both spatially and temporally. Perturbation of any one or more of these biochemical processes conceivably can lead to the failure of this homeostasis, with deleterious consequences to the cells, the tissue of which the affected cells are a part, and finally the organism comprising the affected tissues. Such perturbations are recognized in many disease states such as cancer. In some instances exposure to exogenous agents such as toxic environmental chemicals, natural toxins, or UV light may perturb cellular homeostasis through disruption of regulatory components that are involved in this process. This is considered a toxic response that leads to failure of a specific organ function. It is important, therefore, to understand the complex biochemical processes taking part in cell growth and proliferation and how foreign chemicals interfere with these biochemical pathways and elicit a toxic response. This chapter limits its discussion to these biochemical pathways regulated by protein kinases, enzymes that phosphorylate cellular proteins by transferring the phosphate group from ATP to target proteins of the cell to alter their function. It long has been recognized that kinase phosphorylation of cellular proteins plays a major role in cell growth and proliferation and in the function of hormones and growth factors (1–5). In many cases, phosphorylation activates the function of a protein; however, there are a few examples in which phosphorylation inactivates a protein so as to release an associated factor that produces the response. The known protein kinases can be grouped broadly into two major classes: the tyrosine kinases and serine/threonine kinases. The former class of kinases phosphorylates proteins at tyrosine amino-acid residues and the latter at serine/threonine residues. There are some exceptions to this classification that are discussed later in this chapter. The substrate proteins, however, may be phosphorylated by either of these two kinase types at the appropriate residue. Although phosphorylation of proteins is a covalent modification, the process is reversible; that is, phosphorylated proteins are dephosphorylated by enzymes termed *protein phosphatases*. Protein phosphorylation by kinases and dephosphorylation by phosphatases play a central role in many cellular processes and in the transmission of extracellular signals to the interior of the cell for growth control. Perturbations in the regulatory mechanisms of protein kinases have been known to be caused by synthetic and natural chemicals. Such a modification by exogenous chemicals may lead to loss of growth control, such as that which occurs during carcinogenesis.

TOXIC CHEMICALS AND HUMAN HEALTH

Many toxic effects have been known to result from exposure to foreign chemicals or xenobiotics. Several natural and synthetic chemicals are particularly important in this regard. From a toxicologic standpoint a number of synthetic chemicals are major concerns because of their effects on human health. For example, several organochlorine pesticides have been extensively used in agriculture to boost food production and protect from vector-borne diseases such as malaria or plague. Although the use of these chemicals helped improve the social and economic status of our society, some of these chemicals have persisted in the environment for many years after their last application. A good example of such persistent xenobiotic chemicals is 2,2-bis(p-chlorophenyl)-1,1,1-trichloroethane (p,p′-DDT). The use of this pesticide resulted in a tremendous increase in crop production and eradicated vector-borne diseases such as malaria, but the environmental stability of DDT and related pesticides led to their persistence, bioaccumulation, and biomagnification within the food chain. Exposure to such chemicals through the food chain endangered certain wildlife populations and threatened human health with increased cancer risk, effects on reproduction, and other effects. Besides pesticides, a variety of other industrial chemicals also have been extensively used and discharged into the environment. These include chlorinated solvents such as chloroform, carbon tetrachloride, and methylene chloride. These chlorinated aliphatic compounds are highly toxic and have been linked to cancers and other effects. The greatest concern, however, lies with the halogenated aromatic hydrocarbon (HAH)-type xenobiotics. The reason for this is that they are highly lipophilic and resistant to environmental and biologic biodegradation and thus tend to accumulate in the adipose tissue of the body. As such, these chemicals biomagnify through the food chain. With humans at the upper end of the food chain, there is a high potential that the concentrations of environmentally stable xenobiotics may reach levels at which they can produce toxic effects in humans. A legitimate question to ask is, what effects should we be concerned with? A survey of the toxicologic effects of many HAHs reveals that these agents are able to induce long-term effects that might impact on reproduction, neuronal development, and carcinogenesis. All of the long-term toxic effects of xenobiotics conceivably could involve alterations in gene expression. This review discusses how xenobiotics alter gene expression through their effects on protein kinases and the downstream components of these cell-signaling pathways. Much work in this area has been done with non-HAH agents such as the phorbol ester tumor promoters, and these also are discussed.

HAH-type environmental chemicals, such as the polychlorinated dioxins, dibenzofurans, and biphenyls (PCBs), are widespread ubiquitous environmental chemicals. 2,3,7,8-tetrachlorodibenzo-p-dioxin (TCDD), the prototypical and most potent HAH, is perhaps the most toxic chemical ever synthesized by man (6–8). Although the coplanar PCBs, polychlorinated dibenzofurans, and some polychlorinated dibenzo-p-dioxins are known to evoke a spectrum of toxic effects similar to TCDD, they do so with significantly reduced potencies. The LD_{50} values for TCDD in various experimental animals are widely variable, with guinea pig being the most sensitive species ($LD_{50} = 0.1\ \mu g/kg$ body weight) and hamster is the least sensitive ($LD_{50} = 5000\ \mu g/kg$). Among wildlife and domestic animals there also is a considerable degree of variation in relative sensitivities to lethality produced by TCDD (6). Despite of intensive research efforts, the mechanisms of toxicity of TCDD and related chemicals remains unclear. At the cellular, biochemical, and molecular levels, however, considerable work has been carried out by various investigators that has provided insight into many aspects of the mechanism of action of TCDD (9, 10). One of the initial regulatory events in TCDD-induced toxicity is the specific high-affinity binding of TCDD to a cytosolic receptor protein termed the aryl hydrocarbon receptor (AhR). Initial identification and characterization of the AhR, its ligand-binding specificity, and its relationship to the toxicity of TCDD and related HAHs was first described by Poland and coworkers (11, 12). The AhR functions in a manner that is somewhat analogous to that described for steroid-hormone receptors (see chapter 1 for a detailed description of the AhR and its mechanism of action). Questions remain, however, as to whether toxic actions of such compounds are solely mediated through receptor interactions or whether some toxic effects are mediated through AhR-independent mechanisms.

In the late 1980s our group started investigations into the effects of TCDD exposure in vivo in animal models and in vitro in cell-culture systems (13–15). A surprising finding was the action of TCDD on the development of mouse embryos. We noted that TCDD exposure of mouse embryos in utero and neonatal mice via milk induced early eyelid opening and tooth eruption, effects similar to those seen in neonatal rats and mice given epidermal growth factor (EGF) (16, 17). The implication of this finding was enormous, because our results suggested that dioxin-like chemicals may interfere with normal developmental and tissue differentiation processes involving interaction with cell-signaling pathways used by growth factors and hormones. This key observation has led to subsequent work by many other investigators on the interactions of dioxin-like xenobiotics with growth-factor and growth factor–receptor pathways. Based on our findings, we proposed that dioxin-like xenobiotics might cause downregulation of growth-factor binding to their receptors through an activation of the tyrosine kinase activity of the EGF receptor (16, 18, 19). How such activation occurs is still unclear; however, it is possible that dioxin might produce this effect by inducing the activity of a second protein kinase that, in turn, activates the EGF receptor and subsequently causes its internalization. Alternatively, dioxin may induce the synthesis of transforming growth factor α (TGF-α) in the hepatocyte, which subsequently activates the EGF receptor in an autocrine fashion and stimulates receptor internalization. Following our reports, others have shown increased transcriptional activity of the TGF-α gene in vivo and in vitro following exposure to dioxin (20–22), providing support for the latter hypothesis.

PROTEIN KINASES AND GROWTH CONTROL

Protein kinases are enzymes that phosphorylate cellular proteins at either serine/threonine or tyrosine residues. Although the majority (99%) of cellular protein phosphorylations occur at serine/threonine residues, tyrosine phosphorylation by protein kinases has drawn attention since the discovery that certain growth factors and oncogenes have tyrosine-kinase activity and that the phosphorylation of tyrosine residues of cellular proteins may be a critical determinant in cell proliferation and oncogenesis (23, 24). In the following discussion attention is focused on certain serine/threonine protein kinases and tyrosine kinases, their role in the regulation of gene expression, and their modulation by xenobiotic chemicals.

Tyrosine Kinases

Many growth factors such as EGF and fibroblast growth factor (FGF) bind with cell-surface receptors and transduce signals that eventually result in mitogenesis and cell proliferation. The binding of the growth factor to its cognate receptor activates the receptor by a conformational change and subsequent autophsophorylation and enables it to function as a kinase to transfer a phosphate group (from ATP) to tyrosine residues of cellular proteins. Under normal conditions, tyrosine phosphorylation accounts for about 1% of total phosphorylation activities in the cells. After it was discovered that growth-factor receptors function as tyrosine kinases there was a heightened interest

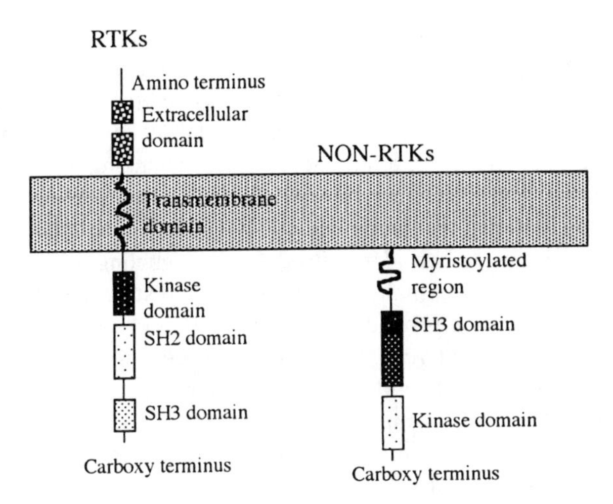

FIG. 1. Diagrammatic representation of the structure of receptor tyrosine kinases (RTKs) and nonreceptor tyrosine kinases (NRTKs). See text for details.

in the cellular proteins that are phosphorylated by these kinases and the mechanisms of these interactions. Although there are a large number of distinct growth factors, the receptors of most of them share some structural similarities (Fig. 1), including an extracellular ligand-binding domain, a transmembrane domain, and a cytosolic kinase domain (25–27). How does the interaction of the growth factor with its receptor activate the latter? In general, after a growth factor (ligand) binds to its receptor on the cell membrane, the receptor is internalized and dimerizes with neighboring receptor. Immediately following dimerization the receptors are autophosphorylated on tyrosine residues. In some cases it is thought that dimerized receptors phosphorylate each other on tyrosine residues (28, 29). Autophosphorylation of the receptor induces its kinase activity, which then can phosphorylate target (substrate) proteins on the tyrosine residues. Different receptors have tyrosine residues at various sites in the kinase domain that become phosphorylated on ligand binding. Transduction of the signal from this point on is performed by proteins referred to as *signaling molecules*. Until recently, the sequence of events following receptor activation and leading to the phosphorylation of the substrate proteins has remained a "black box." Research in this area has intensified during the past few years, and as a result we now have a better understanding of the downstream participants of the signaling cascade. Once the tyrosine residue becomes phosphorylated, the activated receptor recruits certain cytosolic proteins to the membrane as well as phosphorylating membrane-associated proteins on tyrosine. This recruitment and phosphorylation of proteins sets signal transduction into motion and involves activation and inactivation of a number of downstream kinases that are discussed subsequently. In studies with platelet-derived growth factor β (PDGF-β) it first became evident that certain proteins are recruited by the activated receptor. Activated PDGF-β receptor binds with the enzymes phosphatidylinositol 3

(PI3) kinase, phospholipase C γ (PLC-γ) and the GTPase activating protein (GAP). Using synthetic peptides it has been demonstrated that signaling molecules such as PI3 kinase bind to specific sites on the receptor tyrosine kinases (activated receptor, RTKs) [reviewed in (30)]. Without going into molecular details of such specificity, it is important to bear in mind that the amino-acid sequence flanking the phosphotyrosine residues of the receptor determines the specific interaction between the signaling molecules and the activated receptor.

An important question to consider is just how the signaling molecules recognize and bind with the RTKs. Examination of the amino-acid structure of these receptor-binding signaling molecules reveals the presence of short stretches of 100 amino acids, termed *SH2 domains* (*src homology domains* after they were first identified in the retroviral src protein). The SH2 domains of crk oncogene directly binds with tyrosine phosphorylated proteins in transformed cells (31–33). Other SH2 domain–containing proteins that bind with the activated RTKs are PI3 kinase, PLC-γ, and GAP (34–37). The domains of the RTKs to which each of the SH2-containing signaling proteins bind, however, are different. This has been demonstrated by mutation and deletion-analysis studies. For example, mutation of Tyr-751 of human PDGF receptor abolishes its recruitment of PI3 kinase without affecting the binding of PLC-γ or GAP (38, 39). In the case of EGF and PDGF receptors, other proteins designated as "adapters/linkers" that consist almost entirely of SH2 and SH3 domains also may be recruited through phosphorylation of the tyrosine (40–42). GRB-2 is an adapter protein that contains two SH3 domains flanked by one SH2 domain (43). Tyrosine phosphorylation of GRB-2, enhances its association with the guanine nucleotide releasing factor through another protein, m-sos (the mammalian homolog of the *Drosophila* protein son of sevenless). This interaction activates the ras protein from its GDP-bound state to GTP-bound form, an important molecular switch for signal transduction and downstream transmission to the nucleus to control cell growth (44–46). Another protein that is important in ras signaling pathway is SHC protein. SHC interacts with activated EGF receptor and other RTKs and binds with GRB-2, providing an alternate link between growth factor receptors and the ras signaling pathway (43, 47, 48). A general scheme of the signal transduction pathway linking the components of the growth factor receptor–initiated signaling is presented in Fig. 2.

Obviously not all receptors use the same set of molecules for signal transduction. The nature of the signaling molecules employed by each type of activated receptors depends on the ability of each receptor to provide recognition sites for the molecules. The ultimate goal of this series of events is the alteration of gene expression, which results in cell growth or differentiation events.

Nonreceptor Tyrosine Kinases

Although the above discussion focused on the tyrosine-kinase activity associated with growth factor receptors, a second group of tyrosine kinases, the nonreceptor tyrosine kinases (NRTKs) that are cytosolic and do not possess extracellular ligand-binding

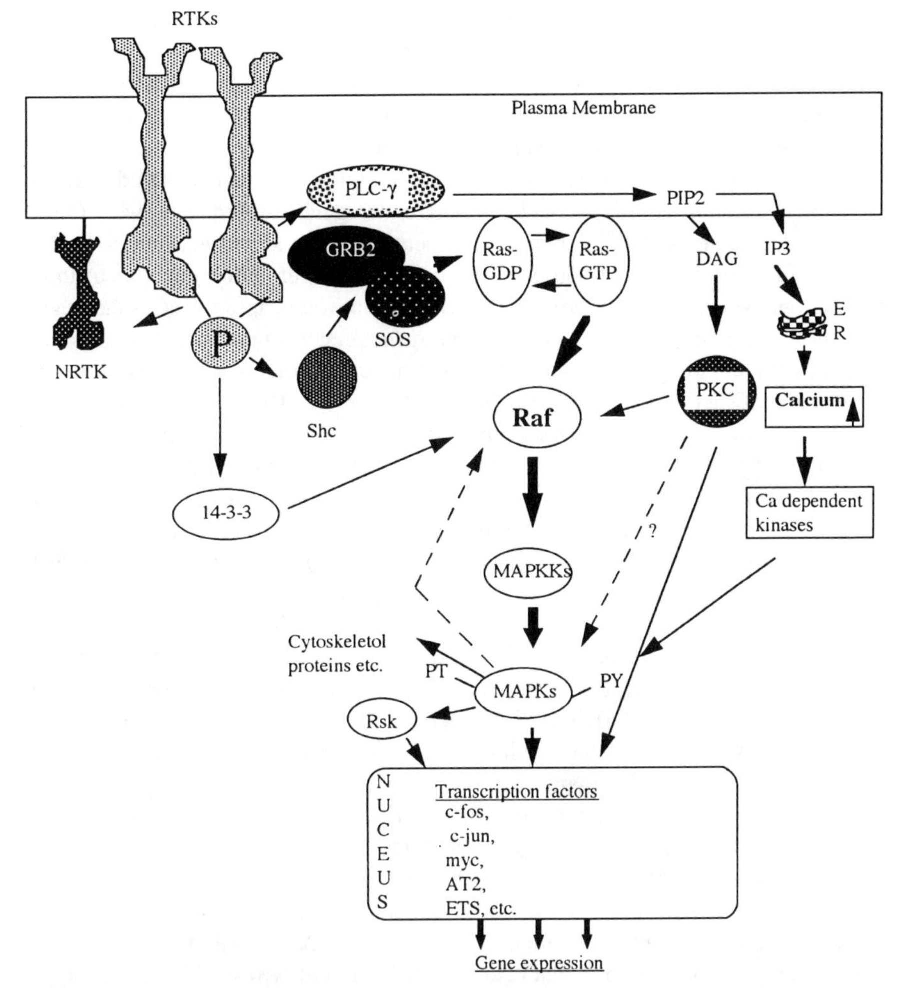

FIG. 2. Tyrosine kinase–mediated signaling cascade and regulation of gene expression. Growth-factor binding to the cell surface receptors initiates the cascade of events that transducers the extracellular signals to the cell nucleus to induce gene expression. Nonreceptor tyrosine kinases (NRTKs) also can be activated by activated receptor tyrosine kinases (RTKs). Both types of tyrosine kinases recruit adapter proteins containing SH domains such as GRB2 to activate ras protein. As a result, a series of downstream kinases are activated. The activated mitogen activated protein kinases (MAPK) translocate to the nucleus to induce transcription, factor activities. One of the substrates for activated tyrosine kinases in phospholipase C-γ (PLC-γ), which extends the signal transduction through another pathway, via the formation of two second messengers, DAG and IP3. In turn these two second messengers combine with the MAPK pathway to regulate gene expression. See text for additional details.

domain, also is important in signal transduction cascade. Certain oncogene products, c-src for example, belong to this category of protein tyrosine kinases.

Oncogenes are cellular genes that when mutated or overexpressed lead to neoplastic (cancer) growth. The normal cellular counterparts of oncogenes are termed *proto-oncogenes*, the controlled expression of which is central for normal cell proliferation. Oncogenes were first described as RNA virus (retrovirus)–encoded genes that produced tumor in birds and rodents. These oncogenes subsequently were found to be mutated versions of cellular genes just mentioned. Some oncogenes encode proteins that function as growth factors; others code for growth-factor receptors. A third group of oncogenes codes for proteins that function as transcription factors that regulate gene expression. The fourth group of oncogenes codes for proteins that transduce growth-factor signals and includes oncogenes encoding proteins with kinase activity. A prototype of such genes is the cellular oncogene, c-src. The viral homolog of this gene is v-src (first described as *Rous sarcoma virus protein*). Structurally (see Fig. 2) this family of proteins contains a kinase domain over a stretch of 300 amino acids that also is present in RTKs discussed above. In addition, these proteins contain a region within their amino terminus for the addition of myristic acid (which facilitates membrane anchoring) as well as two additional domains, SH2 and SH3 (49, 50). How, then, is the tyrosine kinase activity of these proteins stimulated and regulated? Although the nonreceptor tyrosine kinase proteins have no ligand-binding domain, it has been shown that their kinase activity is induced by interaction with activated RTKs and other cellular kinases (50, 51). In murine fibroblasts, for example, PDGF activates c-src kinase activity (52, 53). The kinase activity of c-src also may be regulated by protein–protein interactions that result in the phosphorylation of Tyr-527 residue that lies outside of the kinase domain in the carboxy terminus of the c-src protein. The importance of regulating the phosphorylation at this site has been demonstrated. In studies with polyoma middle T antigen–transforming I protein, it has been found that the 58-kDa middle T protein complexes with c-src by associating with it in the vicinity of Tyr-527 and prevents phosphorylation and thereby the activity of c-src (54, 55). A second possible mechanism by which c-src can be activated is by binding of the SH2 domain of a protein to Tyr-527 that results in activation of tyrosine kinase activity as in the case of the crk oncogene (31, 56). The Crk oncogene protein is associated with the viral gag protein as gag-crk complex isolated from the avian sarcoma virus strain CT10 (56). Molecular and biochemical analysis revealed that gag-crk associates with tyrosine phosphorylated proteins such as c-src via the SH2 domain and activate the latter (57, 58). Thus, the activation of the nonreceptor protein kinases by RTKs either directly or indirectly and by other proteins implies that protein–protein interactions are a central theme of signal-transduction scheme.

Regulation of Protein Tyrosine Kinases by Xenobiotics

In assessing the effects of xenobiotics on gene regulation one must consider the possible mechanisms by which xenobiotics alter gene expression. The first possibility to consider is that the xenobiotics may bind with a cell surface or soluble receptor and thus mimic a hormone or a growth factor, thereby eliciting a similar spectrum of biochemical modulations and biologic activity. Estrogenic chemicals such as the pesticides p,p′-DDT and β-HCH (hexachlorocyclohexane) possess estrogenic activity by binding with the estrogen receptor (59–62). Treatment of the responsive cells such as the human breast cancer cells MCF-7 with estrogen mimics causes estrogen-receptor transformation and nuclear association (63). Although in these instances the receptors are not kinases, they illustrate the ability of xenobiotics to modulate cellular functions by mimicking endogenous regulators. The other possibility is that the xenobiotic exposure leads to the production of growth factors that then interact with their receptors and transduce a biologic signal. In fact, our group was the first to report that treatment of rats or mice with TCDD decreased ^{125}I-EGF binding to the plasma membrane receptors but enhanced the phosphorylation of EGF receptors (15). We also showed that treatment of neonatal mice with TCDD caused the developmental changes described previously (16). Examination of the phosphorylation state of EGF receptor immunoprecipitated from the liver of TCDD-treated animals showed increased tyrosine phosphorylation of the EGF receptor (15, 19). These observations provided new insights into the mechanism of action of dioxin and isosteric compounds and their ability to modulate growth factor– and growth factor receptor–mediated signaling. A third possibility is that the xenobiotics may directly bind to and activate the modulating protein kinases, as is the case with the phorbol ester, 12-O-tetradecanoylphorbol-13-acetate (TPA), a potent activator of protein kinase C (PKC). TPA has structural similarity to the endogenous activator of PKC, namely diacylglycerol (DAG), and thus is able to bind to and activate PKC. Thus, in this instance, the kinase functions both as a receptor and as a protein kinase.

SERINE/THREONINE PROTEIN KINASES

Serine/threonine phosphorylations account for nearly 99% of total phosphoproteins. The majority of protein kinases in the cell phosphorylate proteins at serine/threonine residues. The major groups of serine/threonine kinases are divided on the basis of the molecular requirement for activation. For example, c-AMP–dependent protein kinase (PKA) requires c-AMP for activation, whereas the calcium/calmodulin protein kinase has requirement for calcium and calmodulin, and so forth. One of the most prominent of serine/threonine kinases is the calcium- and phospholipid-dependent PKC and one of the most potent activators of this enzyme is TPA. The discussion here focuses on PKC and related serine/threonine kinases. We mention only briefly the participation of PKA in the regulation of other serine/threonine kinases during signal transduction.

PKC originally was described as a calcium- and phospholipid-dependent kinase that phosphorylates substrate proteins at serine and threonine residues (64–66). Castagna and associates (67) first described activation of this enzyme in response to TPA and implicated its activation in the tumor-promoting potential of TPA. The endogenous activator of this enzyme is DAG, a membrane lipid that is produced by hydrolysis of the phosphatidyl inositol bisphosphate (PIP2) by activated PLC-γ. The other product of this reaction is inositol triphosphate (IP3). Together, DAG and IP3 form a potent second-messenger system that regulates a series of biochemical cascades critically important for cell function and proliferation. DAG binds to and activates PKC, thereby increasing its association with the plasma membrane in the presence of physiologic concentrations of calcium. IP3, the second product of PIP2 hydrolysis, binds with a receptor of the endoplasmic reticulum and causes the release of free calcium from the endoplasmic reticulum. The transient increase in calcium levels is sufficient to trigger a number of calcium-dependent responses through activation of several calcium-dependent kinases such as the calcium/calmodulin–dependent kinase (68). Obviously the cell must regulate the production and degradation of such important molecules rigorously under normal circumstances to avoid unregulated turning on of the signaling cascades. DAG quickly is metabolized to phosphatidic acid to be utilized subsequently for the generation of PIP2 or arachidonic acid (69). Calcium that is released from the endoplasmic reticulum either is pumped out of the cells by the activated plasma membrane–associated calcium pump or sequestered by mitochondria or the endoplasmic reticulum (70, 71). The transient nature of the signals suggests that they are further transduced by other components of the transduction pathway in order to alter gene expression because, in the final analysis, altered gene expression requires that a responsive gene be transcribed into a protein. PKC activated by DAG becomes membrane-associated where its kinase activity is enhanced, resulting in the phosphorylation of a number of cellular proteins. TPA and other tumor-promoting phorbol esters have structural similarity to DAG. Many of the physiologic effects of DAG are a result of activation of PKC that, in turn, phosphorylates cellular substrate proteins that have a role in cell proliferation and differentiation. Because DAG is rapidly metabolized, the activation state of PKC also is transient, functioning only to transduce signals to the downstream elements. The transient nature of DAG may be a mechanism by which this endogenous PKC activator may effectively utilize cytosolic pools of PKC. According to this reasoning, DAG activates only a small pool of total available cytosolic PKC and translocates it to the membrane. The rapid metabolism of DAG then prevents further translocation of PKC to the membrane. While the activated pool of PKC participates in signal transduction events, the cytosolic PKC still is available for the newly produced DAG. One of the reasons TPA has more sustained effect on PKC is that it is rather slowly metabolized in cells. Thus, although DAG activation of PKC may have transient effect on substrate proteins, TPA effects may persist for a longer period of time, resulting in sustained effects on downstream elements.

Multiple Isozymes of PKC

Initially it was thought that PKC was a single enzyme. Nishizuka and associates isolated three fractions from rat brain with similar cofactor requirements and designated them as *PKC I, II,* and *III.* Subsequently these have been redesignated as *alpha, beta,* and *gamma*. Following this characterization, other investigators also have reported that PKC is a family of multiple isozymes. Pioneering work from the laboratories of Nishizuka, Parker, and others have identified several isozymes with different requirements for activation (72–77). The 11 isozymes presently known can be divided broadly into three major groups, the first of which is the "conventional" PKCs [alpha, beta (βI, βII), and gamma] that require calcium and phospholipid for activation. The enzymatic assays for measuring PKC activity in the cells and its translocation was based originally on these isozymes, in which one measures the activity of protein kinase by determining the substrate (usually histone or peptides derived from the EGF-receptor sequence or myelin basic protein) (78, 79). The second group of PKC isozymes are the "novel" PKCs that have no calcium requirement for activation; PKC δ, ε, η, θ, and μ isoforms belong to this group. Finally, the third group of PKC isozymes is exemplified by PKC ζ, λ, and τ, which are referred to as *atypical PKCs* because these isozymes require neither calcium nor DAG for activation. It should be noted that not all of the isozymes are expressed in all types of cells and tissue. There is a clear tissue-specific and developmental stage-specific expression of these isozymes that suggests that the various isozymes have specific functions [for a more detailed discussion please refer to (72–75)]. The brain, however, expresses abundant levels of the major PKC isozymes, namely α, β, γ, and δ and is a major tissue source for the isolation and purification of PKCs. Structurally, the conventional PKCs contain an N-terminal regulatory half and a C-terminal catalytic domain that, in turn, contain four conserved regions, C1 through C4, and five variable regions, V1 through V5 (80). The N-terminal regions of all members contain at least one Zn-finger region upstream of which there is a sequence of amino acids that functions as a pseudosubstrate to regulate the kinase activity of PKC.

PKC isozymes now are recognized as pivotal regulators of many cellular functions as well as regulators of cell growth and differentiation. Inappropriate activation can, therefore, have serious consequences on cell growth and homeostasis. A major role for PKC has been implicated from studies on the tumor promoter TPA. The exact nature of the endogenous substrates has not been definitively identified, but a number of potential candidate proteins for phosphorylation by PKCs have been identified. One of the most prominent among them is the myrystoylated alanine rich c kinase substrate (MARCKS) protein, which is specifically phosphorylated by activated PKC (81–84). MARCKS is a membrane-associated protein that if phosphorylated by PKC translocates to the cytosol and binds with the calmodulin (85). Other endogenous proteins whose activity is modulated by activation of PKC include transcription factors such as AP1, and cytoskeletal proteins such as vinculin (86, 87). The phosphorylation of nuclear proteins such as AP1 suggests that PKC in its active state must translocate to the nucleus. Indeed, it has been shown that such nuclear localization of PKC does occur (88, 89). A number of hormone receptors show increased phosphorylation in response to PKC activation by TPA or other activators. Although phosphorylation may be important in receptor transformation (activation), inhibition of kinase activity, at least in some cases, did not prevent receptor activation.

Activation of Protein Kinase C by Xenobiotics

As discussed earlier, the phorbol ester, TPA, is by far the most potent activator of PKC, although a number of other non–phorbol ester tumor promoters also activate PKC (90–92). There have been very few studies on PKC activation by environmental xenobiotics such as the HAHs, and initial observations in in-vitro systems did not find activation of PKC by these agents (93–96). Bombick et al. (97) reported PKC translocation in hepatocytes of rats treated with TCDD. This activation could be an indirect effect of TCDD on other cellular components. Indeed, Matsumura and associates have reported that TCDD induced the activity of two cellular oncogenes, namely src and ras (97, 98). Activation of these oncogenes is known to activate the phosphatidylinositol bisphosphate pathway, thus leading to the activation of PKC through increased DAG production (99). The failure of TCDD to activate PKC in cell cultures suggests that activation of the AhR alone was insufficient to activate PKC. More recent studies from our laboratory have suggested that certain PCBs, notably the coplanar-type PCBs, caused translocation of a limited number of PKC isozymes (alpha and delta isozymes) from the cytosol to the membrane in WB F-344 rat liver epithelial cells and C3H10T1/2 mouse embryo cells (100, 101). Thus, the effect of HAHs on PKC activation does not seem to be universal.

The activation of protein kinases by environmental toxicants such as PCBs might explain why many of these compounds can alter gene regulation and thus play a potential role in carcinogenesis in addition to their spectrum of toxic effects. Because protein phosphorylations profoundly alter the function of the proteins, the activation of kinases by xenobiotics is expected to alter the function of specific types of cells and then the tissue that contains the altered cell types. What are the mechanisms by which xenobiotics alter kinase activities? Taking the tumor promoter TPA as an example, one can consider that xenobiotic agents can substitute for an endogenous modulator of a kinase and mimic the effects of the endogenous chemical. TPA activates PKC predominantly by translocating the cytosolic from to the plasma membrane. This transforms PKC into an active state by releasing the kinase domain from the pseudosubstrate region of PKC. The formation of activated PKC then leads to the phosphorylation of cellular proteins, some of which are considered to be central to cell proliferation and differentiation. Thus, the pleiotrophic or differential effects of kinase-activating xenobiotics can be explained by the nature of proteins that are affected in a given cell. TPA and similar tumor promoters are known to alter gene expression through activation of PKC isozymes that, in turn, phosphorylate and activate transcription factors to induce gene expression. One of the transcription factors that is activated by TPA treatment of cells is the transcription factor AP1. This nuclear protein is a complex of two cellular oncogenes, c-fos and c-jun. Because the primary effect of TPA on cells is activation of PKC, it is conceivable that the functional activation of AP1 might be related to activation of PKC. This implies that either c-fos or c-jun are substrates for phosphorylation by PKC, as has been shown in a number of cell systems (102–104). The activated AP1 will then bind with enhancer elements termed *TPA response elements* (TREs) in the promoter regions of genes to initiate transcription of the genes. Karin and coworkers (105–108) have reported that the expression of human collagenase gene was induced by TPA, and this induction was associated with the presence of activated AP1. Molecular analysis of the promoter region of the collagenase gene revealed that it contained a short sequence of nucleotides (a TRE), to which AP1 was bound. Although the exact nature of the many other genes regulated by AP1 is not known, studies with reporter genes such as chloramphenicol acetyltransferase and luciferase containing upstream TPA recognition sites have shown that PKC activation is indeed involved in TPA-induced transcriptional activation (109, 110).

SIGNAL TRANSDUCTION BY CYTOSOLIC KINASES AND ras PROTEIN

The previous discussion focused on kinases that initiate cell signaling pathways. In order for the cellular signals to be transduced into the nucleus and culminate in altered gene expression, the cell must be equipped with mechanisms that efficiently carry out this function. This is accomplished by a series of kinases and proteins that are closely associated with them. As we note in the course of the following discussion, the modular nature of this cascade is a terrific adaptive mechanism for transduction of the signals to the cell's nucleus. This means that there is a seeming redundancy in the signaling molecules that carry these signals generated at the cell surface. It also is important to remember that aberrations in these signaling components are at the very root of unrestrained growth such as that occurs during carcinogenesis.

Mitogen-Activated Protein Kinases (MAPKs)

The activation of RTKs and NRTKs leads to rapid tyrosine phosphorylation of membrane proteins such as PI3 kinase and PLC-γ with enzyme activity as well other adapter proteins such as GRB2, SOS (vertebrate homolog of son of sevenless gene product), and shc. Activation of PLC-γ, in turn, leads to the hydrolysis of PIP2 that generates DAG and IP3, the former of which activates PKC, a serine/threonine kinase as discussed previously. A second consequence of activated RTKs is the activation of the adapter proteins such as GRB2 that contain SH2 and SH3 domains. Tyrosine phosphorylation of these proteins enables them to activate the ras ($p^{21\,ras}$) protein from its GDP-bound form to GTP-bound active form. Activated ras recruits a serine/threonine kinase, raf-1, from the cytosol to the plasma membrane. Activated raf-1, in turn, phosphorylates and activates downstream kinases, specifically the MAPKs and MAPKKs (Fig. 2).

c-Raf-1 is a 70- to 74-kDa cytoplasmic protein with intrinsic kinase activity encoded by c-raf-1 gene, the cellular homolog of v-raf, the transforming gene of the murine sarcoma virus 3611 (111). The c-raf-1 gene is expressed in all cell types. Two other raf-1 related genes, A-raf and B-raf, appear to be restricted to specific cell types (111, 112). Structurally, the 74-kDa raf protein consists of two functional domains, the amino-terminal regulatory domain and a carboxy-terminal kinase domain

(113, 114). The amino-terminal domain contains two conserved regions, CR1 and CR2, and the carboxy terminal contains the third conserved region, CR3. There is some indication that binding of c-raf to the membrane is stabilized by its association with another protein, the 14-3-3 protein [a homolog of PKC (115)]. How does Raf-1 protein become activated? Treatment of cells with growth factors such as PDGF and fibroblast growth factor results in rapid hyperphosphorylation of Raf-1 on tyrosine residues that causes a shift in electrophoretic mobili ty of the protein (116). It is not certain whether Raf-1 also is autophosphorylated on membrane association. A second possible mode of Raf-1 activation is by serine/threonine phosphorylation. Indeed, it has been shown that treatment of murine fibroblasts with TPA enhances Raf-1 phosphorylation, suggesting that Raf-1 is a substrate for phosphorylation by PKC (117). Thus, PKC may be considered a kinase kinase for Raf-1. Raf-1 protein has also been reported to be phosphorylated by c-AMP-dependent protein kinase, PKA (118). Raf-1 phosphorylation by PKA that occurs at the serine 43 residue in the amino terminal region of Raf-1, however, inhibits its activity, suggesting that PKA phosphorylation of this protein may play a feedback regulatory role to uncouple ras–Raf interactions.

Activation and membrane association of Raf-1 enables it to function as a serine/threonine kinase. Raf-1 is not the only protein kinase that functions as an upstream regulator of MAP- and ERK (extracellular signal related kinase)-regulated kinase (MEK) and MAPKs. Other proteins with MAPKKK activity have been identified (119–121). The multiplicity of these MAPKKKs (MEKKs) suggests the diversity of cell-signaling cascades and the necessity to equip the cells with more than a single member within each kinase group.

MAPKKs

An immediate downstream substrate for c-raf kinase is MAPKK (also called MEK or ERK kinase). MEK is a serine/tyrosine kinase that phosphorylates MAPKs on threonine (or serine) and tyrosine residues and thus functions as a dual-specific kinase. Similar to MAPKKKs, multiple forms of MAPKKs exist. At least six members of the MEK protein kinases are known so far (122). As expected, each member of the MEK group shows some specificity toward the downstream MAPKs. Each MEK acts specifically on one or two MAPKs (123). Although Raf-1 and other MAPKKKs phosphorylate MEKs exclusively on Ser 218 and Ser 222, MEKs also are phosphorylated by MAPKs on both serine and threonine. The function of the latter phosphorylation may be a feedback signaling.

MAPKs

The signals generated at the cell surface are transduced via the ras-raf cascade and relayed to MAPKs, the final component of the kinase pathway, from which the signals are propagated through phosphorylation of both cytosolic and nuclear proteins and result in gene expression and altered cell function. MAPK was first identified by Ray and Sturgill (124) and Hoshii et al. (125) as a kinase that phosphorylated microtubule-associated protein 2 in studies on insulin sensitive serine/threonine kinase of 3T3 LI preadipocytes. Subsequently, Kreb's group observed that the myelin basic protein was phosphorylated by a meiosis-activated protein kinase in frog and starfish oocytes (126). In 1991, Bolton and coworkers (127) cloned and characterized of two members of this group, 44-kDa and 42-kDa proteins that were designated as *ERK1* and *ERK2*, respectively. The terms MAPK or ERKs now are used to describe this group of protein kinases. The presently known MAPKs can be grouped into four subfamilies based on their structural similarities and regulation: ERK, JNK-SAPK, p38, and ERK3 [reviewed in (128)]. The first three subfamily members require phosphorylation on both tyrosine and threonine residues for activation, while the fourth group lacks the dual phosphorylation motif. In addition, it appears that members of the MAPK family are evolutionarily highly conserved. Homologous proteins with kinase activity were reported to be active in the mating pheromone signal transduction pathway of the budding yeast *Sacchromyces cerevisiae* and in the fission yeast *S. pombe* [reviewed in (129)].

MAPK1 and MAPK2 are widely distributed in all cells and tissue. As mentioned, MAPKs are active only when phosphorylated on both threonine (or serine) and tyrosine residues. This has been confirmed in studies using phosphatase 2A, a selective serine/threonine phosphatase, and tyrosine-specific phosphatases such as CD45 (130–133). The importance of the threonine and tyrosine phosphorylation sites on the two MAPKs, ERK1 and ERK2, also has been demonstrated using mutants that lack these sites. Such mutant proteins produced in bacteria failed to be phosphorylated by purified MEK in vitro (134).

As mentioned previously, MAPKs are a part of the signal-transduction cascade activated by RTKs as well as NRTKs and transduced by ras protein. The ras-raf-MEK pathway, however, is not the only mechanism involved in the activation of MAPKs. MAPKs ERK1 and ERK2 were shown to be phosphorylated in cells treated with the phorbol ester TPA (125, 135, 136). Because TPA is an activator of PKC, it is implied that ERKs are substrates for PKC. It may be, however, that PKC-induced phosphorylation of MAPKs may be indirect possibly through PKC activation of Raf-1 kinase. This is supported by the fact PKC is a serine/threonine kinase and is not known to function as a dual-specific kinase. Thus, PKC may phosphorylate Raf-1 on serine/threonine, leading to the phosphorylation downstream of MEK and MAPK. It is noteworthy that PKC also is a multigene family of enzymes, and it is possible that only specific isozymes may participate in the activation of Raf-1 and MAPKs. Another ras-independent pathway that activates MAPKs may be via the G-protein–coupled agonists that signal via the serpentine receptor (137). Other possible mechanisms of MAPK activation include v-src protein, pp60-induced phosphatidyl choline hydrolysis by phospholipase D, and PI3 kinase–induced production of phosphatidylinositol triphosphate that directly activates a PKC isoform, PKC-ζ (138–141).

At this point, it is pertinent to consider the nature of substrates that are phosphorylated by MAPK and their physiologic importance in cell-growth regulation and differentiation. As is discussed subsequently, activation of the MAPK signal-transduction pathway initiated by growth factors may not necessarily result in mitogenesis, but under certain circumstances may induce differentiation of the cells. A number of proteins have been identified as substrates for activated MAPKs. A major portion of the information regarding these substrates was gained from studies on ERK1 and ERK2 MAPKs. A MAPK substrate, the bovine Myelin basic protein (MBP), contains the consensus recognition sequence for MAPK phosphorylation was Pro-Xaa-Ser/Thr-Pro, where *Xaa* is a basic or neutral amino acid and there is a critical requirement for the last propyl residue. There are a large number of proteins with such sequence (142). Among the proteins identified as substrates for MAPKs are ribosomal S6 kinase (rsk), the MAPK-activated protein kinase 2 of skeletal muscles, and heat-shock proteins of 25 kDa and 27 kDa (142). Certain nuclear proteins that function as transcription factors have been identified as potential substrates for phosphorylation by MAPKs. This implies that activated MAPKs are translocated to the nucleus. This, indeed, has been shown by immunofluorescence studies to occur in growth factor–treated cells (143, 144). Among the nuclear proteins that are targets of MAPKs are the transcription factors, AP1, c-Elk, c-myc, Tal-1 and E12 (145). The phosphorylation of cytosolic phospholipase A2 by ERKs also has been reported, which suggests the pivotal role that these MAPKs play in the regulation of second-messenger functions.

MAPKs also have been shown to phosphorylate upstream components such as the Raf-1 MEK of the MAPK cascade, as well as receptor tyrosyl kinases that lie further upstream (138, 146, 147). Although the importance of these reverse interaction is not clear, it may suggest a feedback regulation of the ras-Raf-MEK-MAPK pathway by MAPKs.

As mentioned previously, the MAPK signal-transduction pathway may have a major role in cell proliferation because the pathway is regulated and activated by growth-factor receptors that possess tyrosine kinase activity. Although this is true for growth factors such as EGF and PDGF, other growth factors may utilize the same pathway that ultimately results in differentiation. In PC-12 cells, a rat pheochromocytoma cell line, EGF induces MAPK activity and mitogenesis. On the other hand, in cells treated with nerve growth factor the same MAPKs are activated but result in nerve growth factor–induced differentiation (148–150). To explain such a paradox one must consider the duration of activation of MAPKs in cells treated with these two growth factors. MAPK activation by mitogenic growth factors is rapid and transient, whereas MAPK activation in cells treated with nerve growth factor is sustained for a much longer time. This suggests that duration of MAPK activation is a critical determinant in the decision for proliferation or differentiation. In addition, the final result may be dependent on a particular cell type (151).

Although we know that many environmental chemicals alter gene expression and promote neoplastic growth of cancer-initiated cells, we do not know the detailed mechanisms of their effects. In light of the fact that the tumor promoter TPA activates the MAPK pathway, it would be interesting to investigate whether environmental chemicals such as the HAHs modulate this signal-transduction pathway. Unpublished preliminary work from our laboratory demonstrates that certain PCBs activate ERK 1 and ERK 2 in a rat-liver epithelial cell line. Both TPA and TGF-α also activate these two MAPKs. The activation of MAPKs by TGF-α, however, was much more rapid but transient (lasting only 1 hour after treatment) compared with that observed in PCB-treated cells. These preliminary results certainly point to the potential of environmental toxicants such as the PCBs to interact with signal-transduction pathways to modulate gene expression.

CONCLUSIONS AND FUTURE PERSPECTIVES

During the past decade our knowledge of the cell-signaling pathways has grown significantly, and many of the "black boxes" in the signal-transduction cascade pathway

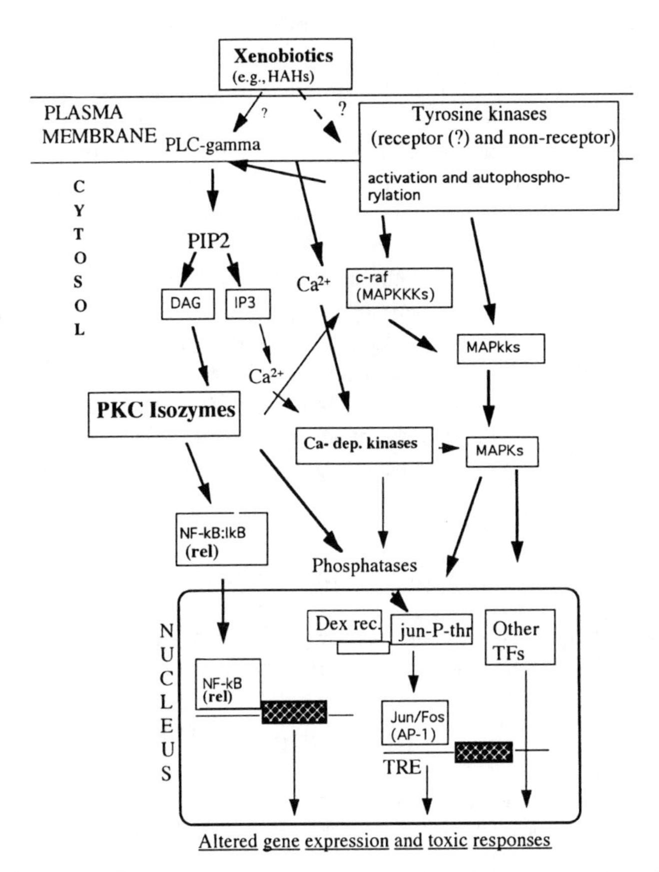

FIG. 3. A model for modulation of gene expression by xenobiotic chemicals through activation of protein tyrosine kinase signaling pathways. Xenobiotic chemicals such as the halogenated aromatic hydrocarbons (HAHs) can activate the same pathways used by growth-factor receptor kinases to modulate gene expression. This can be accomplished through activation of protein kinase C isozymes and increase in intracellular calcium. PKC isozymes can phosphorylate upstream of MAP kinases kinases such as raf-1, thus activating the MAPK pathway. A transient increase in intracellular calcium can activate calcium-dependent protein kinases, such as the calcium/calmodulin kinases, that can phosphorylate MAPKs. PKC also may be involved in the activation of transcription factors such as AP-1 and NF-kB that control gene expression. See Fig. 2 for abbreviations and text.

have been uncovered. We now have a better understanding of how extracellular signals such as hormones or growth factors interacting with their receptors on the cell surface or cytoplasm transduce these signals to the nucleus, where these signals are deciphered and alter gene expression. Such an alteration in gene expression can result in cell proliferation or cell differentiation depending on the specific cell type and the type of signaling systems that are affected. Alternatively, it may result in an altered physiology of the cells. It is important to remember that in disease states such as cancer there is a subversion of one or more of these signaling cascades. Despite of the significant progress in our understanding of the biochemical and cellular aspects of cell-signaling cascades we have not been able to apply this knowledge to understand, except in the case of a few chemicals such as TPA or TCDD, how toxic environmental chemicals interact with the signaling systems and elicit specific toxic responses. We do know that TPA and TCDD evoke pleiotrophic biochemical and toxic effects through binding to receptors or receptor-like proteins, thereby altering transcriptional programs of the cells. Many environmental chemicals are suspected to be carcinogenic not by virtue of their mutagenic effect but because they are able to promote the growth of carcinogen-initiated cancer cell growth. Such an effect is possible only if those chemicals are able to alter gene regulation by interfering with the normal function of cell-signaling mechanisms. It is surprising that we seem to know very little of such interactions of environmental chemicals. Based on our laboratory observations, chemicals such as PCBs can potentially interact with these cell-signaling pathways to alter gene expression (Fig. 3), and the same may be true with many other persistent environmental toxicants. It is thus imperative that we investigate the interactions of such chemicals with cellular-signaling cascades and the involvement and importance of these interactions in cell-, tissue-, and organ-specific toxicity.

ACKNOWLEDGMENT

Preparation of this review and some of the preliminary work described herein were supported by NIEHS Superfund grant no. ES-04911.

REFERENCES

1. Greengard P. Phosphorylated proteins as physiological effectors. *Science* 1978;199:146–152.
2. Shenolikar S. Protein phosphorylation: hormones, drugs and bioregulation. *FASEB J* 1988;2:2753–2764.
3. Shenolikar S. Control of cell function by reversible protein phosphorylation. *J Cyclic Nucl Prot Phosphoryl* 1987;11:531–541.
4. Hemmings HC, Nairn AC, McGuinnes TL, Huganir RL, Greengard P. Role of protein phosphorylation in neuronal signal transduction. *FASEB J* 1989;3:1583–1592.
5. Krebs EG. The enzymology of control by protein phosphorylation. *Enzymes* 1986;XVII:3–20.
6. Poland A, Kende AS. 2,3,7,8-tetrachlorodibenzo-p-dioxin: environmental contaminant and molecular probe. *Fed Proc* 1976;35:2404–2411.
7. Schwetz BA, Norris JM, Sparschu GL, Rowe UK, Gehring PJ, Emerson JL, Gerbig CG. Toxicology of chlorinated dibenzo-p-dioxins. *Environ Health Perspect* 1973;5:87–99.

8. Kociba RJ, Keyes DG, Beyer JE, Carreon RM, Wade CE, Dittenber DA, Kalnins RP, Franson LE, Park CN, Barnard SD, Hummel RA, Humiston CG. Results of a two-year chronic toxicity and oncogenicity study of 2,3,7,8-tetrachlorodibenzo-p-dioxin in rats. *Toxicol Appl Pharmacol* 1978;46:279–303.
9. Whitlock JP, Jr. Mechanistic aspects of dioxin action. *Chem Res Toxicol* 1993;6:754–763.
10. Greenlee WF, Sutter TR, Marcus C. Molecular basis of dioxin actions on rodent and human target tissue. In: Spitzer HL, Slaga TJ, Greenlee WF, McClain RM, eds. *Receptor-Mediated Biological Process: Implications for Evaluating Carcinogenesis*. New York:Wiley-Liss 1994;47–57.
11. Poland A, Glover E, Robinson JR, Nebert DW. Genetic expression of aryl hydrocarbon hydroxylase activity: induction of monooxygenase activities and cytochrome P1-450 formation by 2,3,7,8-tetrachlorodibenzo-p-dioxin in mice genetically "nonresponsive" to other aromatic hydrocarbons. *J Biol Chem* 1974;249:5599–5606.
12. Poland A, Glover E, Kende AS. Stereospecific, high affinity binding of 2,3,7,8-tetrachlorodibenzo-p-dioxin by hepatic cytosol: evidence that the binding species is receptor for induction of aryl hydrocarbon hydroxylase. *J Biol Chem* 1976;251:4936–4946.
13. Brewster DW, Madhukar BV, Matsumura F. Influence of 2,3,7,8-TCDD on the protein composition of the plasma membrane of the hepatic cells from the rat. *Biochem Biophys Res Commun* 1983;107;68–74.
14. Bombick DW, Matsumura F, Madhukar BV. TCDD (2,3,7,8-tetrachlorodibenzo-p-dioxin) causes reduction in the low density lipoprotein (LDL) receptor activities in the hepatic plasma membrane of the guinea pig and rat. *Biochem Biophys Res Commun* 1985;127:296–302.
15. Madhukar BV, Ebner K, Matsumura F, Bombick DW, Brewster DW, Kawamoto T. 2,3,7,8-tetrachlorodibenzo-p-dioxin causes an increase in protein kinases associated with epidermal growth factor receptor in the hepatic plasma membrane. *J Biochem Toxicol* 1988;3:261–277.
16. Madhukar BV, Brewster DW, Matsumura F. Effects of *in vivo* administered 2,3,7,8-tetrachlorodibenzo-p-dioxin on receptor binding of epidermal growth factor in the hepatic plasma membrane of the rat, guinea pig, mouse and hamster. *Proc Natl Acad Sci USA* 1984;81:548–554.
17. Cohen S. Isolation of a mouse submaxillary gland protein accelerating incisor eruption and eyelid openingin the new born animal. *J Biol Chem* 1962;237:4297–4304.
18. Matsumura F, Brewster DW, Bombick DW, Madhukar BV. Studies on the molecular basis of TCDD-caused changes in proteins associated with the liver plasma membrane. In: Rappe C, Choudhary G, Keith L, eds. *Chlorinated Dioxins and Dibenzofurans in Perspective.* Chelsea: Lewis Publishers, 1986;243–268.
19. Kawamoto T, Matsumura F, Madhukar BV, Bombick DW. Effects of TCDD on the EGF receptor binding of XB mouse keratinizing epithelial cells. *J Biochem Toxicol* 1989;4:173–182.
20. Hudson G, Toscano WA, Jr, Greenlee WF. 2,3,7,8-tetrachlorodibenzo-p-dioxin (TCDD) modulates epidermal growth factor (EGF) binding to basal cells from a human keratinocyte cell line. *Toxicol Appl Pharmacol* 1986;82:481–492.
21. Choi EJ, Toscano DG, Ryan JA, Riede N, Toscano WA, Jr. Dioxin induces transforming growth factor-α in human keratinocytes. *J Biol Chem* 1991;266:9591–9597.
22. Gaido KW, Maness SC, Leonard LS, Greenlee WF. 2,3,7,8-tetrachlorodibenzo-p-dioxin-dependent regulation of transforming growth factor-α and -$\beta 2$ expression in a human keratinocyte cell line involves both transcriptional and posttranscriptional control. *J Biol Chem* 1992;267:24591–24595.
23. Hunter T, Cooper JA. Protein-tyrosine kinases. *Ann Rev Biochem* 1985;54:897–930.
24. Comoglio PM, Di Renzo MF, Gaudino G, Ponzetto C, Prat M. Tyrosine kinase and control of cell proliferation. *Am J Resp Dis* 1990;142:S16–S19.
25. Hanks SK, Quinn AM, Hunter T. The protein kinase family: conserved features and deduced phylogeny of the catalytic domains. *Science* 1988;241:42–52.
26. Williams LT. Signal transduction by platelet derived growth factor receptor. *Science* 1989;243:1564–1570.
27. Ullrich A, Schlessigner J. Signal transduction by receptors with tyrosine kinase activity. *Cell* 1990;61:203–212.
28. Bellot F, Crumley G, Kaplow JM, Schlessinger J, Jaye M, Dionne CA. Ligand-induced transphosphorylation between different FGF receptors. *EMBO J* 1991;10:2849–2854.
29. Schlessinger J, Ullrich A. Growth factor signaling by receptor tyrosine kinases. *Neuron* 1992;9:383–391.
30. Fantl WJ, Johnson DE, Williams LT. Signalling by receptor tyrosine kinases. *Ann Rev Biochem* 1993;62:453–481.

31. Matsuda M, Mayer BJ, Fukui Y, Hanafusa H. Binding of transforming protein, p47gag-crk, to a broad range of phosphotyrosine containing proteins. *Science* 1990;248:1537–1539.
32. Mayer BJ, Hanafusa H. Mutagenic analysis of the v-crk oncogene: requirement for SH2 and SH3 domains and correlation between increased cellular phosphotyrosine and transformation. *J Virol* 1990;64:3581–3589.
33. McGalde CJ, Ellis C, Reedijk M, Anderson D, Mbamalu G, Reith AD, Panayotou G, End P, Bernstein A, Kazlauskas A. SH2 domains of the p85α subunit of phosphatidylinositol 3-kinase binding to growth factor receptors. *Mol Cell Biol* 1992;12:991–997.
34. Myers MG, Becker JM, Sun XJ, Shoelson S, Hu P, Schlessinger J, Yoakim M, Schaffhausen B, White MF. IRS-1 activates phosphatidylinositol 3′-kinase by associating with src homology 2 domains of p85. *Proc Natl Acad Sci USA* 1992;89:10350–10354.
35. Mohammadi M, Honeggar AM, Rotin D, Fischer R, Bellot F, Li W, Dionne CA, Jaye M, Rubinstein M, Schlessinger J. A tyrosine phosphorylated carboxy-terminal peptide of the fibroblast growth factor receptor (Flg) is a binding site for the SH2 domain of phospholipase C-γ 1. *Mol Cell Biol* 1991;11:5068–5078.
36. Morrison DK, Kaplan DR, Rhee SG, Williams LT. Platelet-derived growth factor receptor (PDGFR)-dependent association of phospholipase C-γ with PDGF receptor complex. *Mol Cell Biol* 1990;10:2359–2366.
37. Merengere LEM, Pawson T. Identification of residues in GTP-ase activating protein Src homology 2 domains that control binding to tyrosine phosphorylated growth factor receptors and p62. *J Biol Chem* 1992;267:22779–22786.
38. Coughlin SR, Escobedo JA, Williams LT. Role of phosphatidylinositol kinase in PDGF receptor signal transduction. *Science* 1989;243:1191–1194.
39. Klazuskas A, Ellis C, Pawson T, Cooper JA. Binding of GAP to activated PDGF receptors. *Science* 1990;247:1578–1581.
40. Lehmann J, Reithmuller G, Johnson J. Nck, a melanoma cDNA encoding a cytoplasmic protein consisting of the src homology units SH2 and SH3. *Nucleic Acid Res* 1990;18:1048–1060.
41. Clark SG, Stern MJ, Horvitz HR. C. elegans cell signalling gene Sem-5 encodes a protein with SH2. *Nature* 1992;356:340–344.
42. Mayer BJ, Hamaguchi M, Hanafusa H. A novel oncogene with structural similarity to phospholipase C. *Nature* 1988;332:272–275.
43. Lowenstein EJ, Daly RJ, Batzer AG, Li W, Margolis B, Lammers R, Ullrich A, Skolnik EY, Bar-Sagi D, Schlessinger J. The SH2 and SH3 domain-containing protein Grb2 links receptor tyrosine kinases to ras signaling. *Cell* 1992;70:431–442.
44. Egan SE, Gidding BW, Brooks MW, Buday L, Sizeland AM, Weinberg RA. Association of Sos Ras exchange protein with Grb2 is implicated in tyrosine kinase signal transduction and transformation. *Nature* 1993;363:45–51.
45. Bonfini L, Karlovich CA, Dasgupta C, Banerjee U. The son of sevenless gene product: a putative repressor of Ras. *Science* 1992;255:603–606.
46. Schlessinger J. SH2/SH3 signaling proteins. *Curr Opin Genet Dev* 1994;4:25–29.
47. Pawson T, Schlessinger J. SH2 and SH3 domains. *Curr Biol* 1992;3:434–442.
48. Rozakis-Adcock M, Fernley R, Wade J, Pawson T, Bowtell D. The SH2 and SH3 domains of mammalian Grb2 couple the EGF receptor to the Ras activator mSos1. *Nature* 1993;363:83–85.
49. Cross FR, Garber EA, Pellman D, Hanafusa H. A short sequence in the $p60^{src}$N terminus is required for $p60^{src}$ myristylation and membrane association and for cell transformation. *Mol Cell Biol* 1985;4:1834–1842.
50. Gutkind JS, Lacal PM, Robbins KC. Thrombin-dependent association of phosphatidylinositol 3-kinase with $pp60^{src}$ and $p59^{fyn}$ in human platelets. *Mol Cell Biol* 1990;10:3806–3809.
51. Golden A, Brugge JS. Thrombin treatment induces rapid changes in tyrosine phosphorylation in platelets. *Proc Natl Acad Sci USA* 1989;86:901–905.
52. Gould KL, Hunter T. Platelet-derived growth factor induces multisite phosphorylation of pp60c-src and increases its protein tyrosine kinase activity. *Mol Cell Biol* 1988;8:3345–3356.
53. Kypta RM, Goldberg Y, Ulug ET, Courtneidge SA. Association between the PDGF receptor and members of the src family of tyrosine kinases. *Cell* 1990;62:481–492.
54. Courtneidge SA. Activation of the pp60c-src kinase by middle T antigen binding or by dephosphorylation. *EMBO J* 1985;4:1471–1477.
55. Courtneidge SA, Smith AE. Polyoma virus transforming protein associates with the product of the c-src cellular gene. *Nature* 1983;303:435–439.

56. Mayer BJ, Hamaguchi M, Hanafusa H. A novel viral oncogene with structural similarity to phosphlipase C. *Nature* 1988;332:272–275.
57. Mayer BJ, Hanafusa H. Mutagenic analysis of the v-crk oncogene: requirement for SH2 and SH3 domains and correlation between increaed cellular phosphotyrosine and transformation. *J Virol* 1990;64:3581–3589.
58. Matsuda M, Mayer BJ, Hanafusa H. Identification of domains of the v-crk oncogene product sufficient for association with phosphotyrosine-containing proteins. *Mol Cell Biol* 1991;1:1607–1613.
59. Robison AK, Schmidt WA, Stancel GM. Estrogenic activity of DDT: estrogen-receptor profiles and the responses of individual uterine cell types following o,p′-DDT administration. *J Toxicol Environ Health* 1985: 16:493–508.
60. Mason RR, Schulte GJ. Interaction of o,p′-DDT with the estrogen-binding protein (EBP) of DMBA-induced rat mammary tumors. *Res Commun Chem Pathol Pharmacol* 1981;33:119–128.
61. Bitman J, Cell HS, Harris SJ, Fries GF. Estrogenic activity of o,p′-DDT in the mammalian uterus and avian oviduct. *Science* 1968;62:371–372.
62. Coosen R, van Velsen FL. Effects of β-isomer of hexachlorocyclohexane on estrogen-sensitive human mammary tumor cells. *Toxicol Appl Pharmacol* 1989;101:310–318.
63. Steinmetz R, Young PCM, Caparell-Grant A, Gize EA, Madhukar BV, Ben-Jonathan N, Bigsby RM. Novel estrogenic action of the pesticide residue β-hexachlorocyclohexane in human breast cancer cells. *Cancer Res* 1996;56:5403–5409.
64. Inoue M, Kishimoto A Takai Y, Nishizuka Y. Studies on a cyclic nucleotide-dependent protein kinase and its proenzyme in mammalian tissues. *J Biol Chem* 1977;252:7610–7616.
65. Takai Y, Kishimoto A, Inoue M, Nishizuka Y. Studies on cyclic nucleotide-independent protein kinase and its proenzyme in mammalian tissue: I. Purification and its characterization of an active enzyme from bovine cerebellum. *J Biol Chem* 1977;252:7603–7609.
66. Takai Y, Kishimoto A, Iwasa Y, Kawahara Y, Mori T, Nishizuka Y. Calcium-dependent activation of a multifunctional protein kinase by membrane phospholipids. *J Biol Chem* 1979;254:3692–3695.
67. Castagna M, Takai Y, Kaibuchi K, Sano K, Kikkawa U, Nishizuka Y. Direct activation of calcium-activated, phospholipid-dependent protein kinase by tumor-promoting phorbol esters. *J Biol Chem* 1982;257:7847–7851.
68. Jacob R. Calcium oscillations in electrically non-excitable cells. *Biochim Biophys Acta* 1990;1052:427–438.
69. Khan WA, Blobe GC, Hannun YA. Arachidonic acid and free fatty acids as second messengers and the role of protein kinase C. *Cell Signal* 1995;7:171–184.
70. Meldrum E, Parker PJ, Carozzi A. The PtdIns-PLC superfamily and signal transduction. *Biochim Biophys Acta* 1991;1092:49–71.
71. Berridge MJ, Irvine RF. Inositol phosphates and cell signaling. *Nature* 1989;341:197–205.
72. Nishizuka Y. The role of protein kinase C in cell surface signal transduction and tumor promotion. *Nature* 1984;308:693–695.
73. Nishizuka Y. The molecular heterogeneity of protein kinase C and its implications for cellular regulation. *Nature* 1988;334:661–665.
74. Nishizuka Y. Studies and prospectives of the protein kinase C family for cellular regulation. *Cancer* 1989;63:1892–1903.
75. Ono Y, Kurokawa T, Kawahara K, Nishimura O, Muramoto R, Igarashi K, Sugino Y, Kikkawa U, Ogita K, Nishizuka Y. Cloning of rat brain protein kinase C complementary DNA. *FEBS Lett* 1986;203:111–115.
76. Ono Y, Fujii T, Ogita K, Kikkawa U, Igarashi K, Nishizuka, Y. Identification of three additional members of rat protien kinase C family: δ, ε and ζ subspecies. *FEBS Lett* 1987;226:125–128.
77. Parker PJ, Kour G, Marais RM, Mitchell F, Pears CJ, Schaap D, Stabel S, Webster C. Protein kinase C: a family affair. *Mol Cell Endocrinol* 1989;65:1–11.
78. Kraft AS, Anderson WB. Phorbol esters increase the amount of $Ca2+$, phospholipid dependent protein kinase associated with plasma membrane. *Nature* 1983;301:621–624.
79. Hannun YA, Loomis CR, Bell RM. Activation of protein kinase C by Triton X-100 mixed micelles containing diacylglycerol and phosphatidyl serine. *J Biol Chem* 1985;260:10039–10043.
80. Hug H, Sarre TF. Protein kinase C isozymes: divergence in signal transduction? *Biochem J* 1993;291:329–343.
81. Aderem A. The MARCKS brothers: a family of protein kinase C substrates. *Cell* 1992;71:713–716.

82. Hyatt SL, Liao L, Aderem A, Nairn AC, Jaken S. Correlation between protein kinase C binding proteins and substrates in REF52 cells. *Cell Growth Regul* 1994;5:495–502.
83. Brooks SF, Herget T, Broad S, Rozengurt E. The expression of 80K/MARCKS, a major substrate of protein kinase C (PKC), is down-regulated through both PKC-dependent and -independent pathways: effects of bombesin, platelet-derived growth factor, and cAMP. *J Biol Chem* 1992;267:14212–14218.
84. Herget T, Brooks SF, Broad S, Rozengurt E. Relationship between the major protein kinase C substrates acidic 80-kDa protein-kinase-C substrate (80k) and myristoylated alanine-rich C-kinase substrate (MARCKS): members of a gene family or equivalent genes in different species. *Eur J Biochem* 1992;209:7–14.
85. Graff JM, Young TN, Johnson JD, Blackshear PJ. Phosphorylation-regulated calmodulin binding to a prominent cellular substrate for protein kinase C. *J Biol Chem* 1989;264:21818–21823.
86. Azzi A, Boscoboinik D, Hensey C. The protein kinase C family. *Eur J Biochem* 1992;208:547–557.
87. Nishizuka Y. Protein kinase C and lipid signaling for sustained cellular responses. *FASEB J* 1995;9:484–496.
88. Buchner K, Lindschau C, Hucho F. Nuclear localization of protein kinase Cα and its association with nuclear components in Neuro-2a neuroblastoma cells. *FEBS Lett* 1997;406:61–65.
89. Trubiani O, Rana RA, Stuppia L, Di Primio R. Nuclear translocation of βII PKC isoenzyme in phorbol ester-stimulated KM-3 pre-B human leukemia cells. *Exp Cell Res* 1995;221:172–178.
90. Blumberg PM. *In vitro* studies on the mode of action of the phorbol esters, potent tumor promoters: *Part I. CRC Crit Rev Toxicol* 1980;8:153–197.
91. Ashendel CL, Staller JM, Boutwell RK. Protein kinase activity associated with a phorbol ester receptor purified from mouse brain. *Cancer Res* 1983;43:4333–4337.
92. Castagna M, Takai Y, Kaibuchi K, Sano K, Kikkawa U, Nishizuka Y. Direct activation of calcium-activated, phospholipid-dependent protein kinase by tumor-promoting phorbol esters. *J Biol Chem* 1982;257:7847–7851.
93. Kramer CM, Sando JJ, Holsapple MP. Lack of direct effect of 2,3,7,8-tetrachlorodibenzo-p-dioxin (TCDD) on protein kinase C activity in EL4 cells. *Biochem Biophys Res Commun* 1986;140:267–272.
94. Moser GJ, Meyer SA, Smart RC. The chlorinated pesticide mirex is a novel nonphorbol ester-type tumor promoter in mouse skin. *Cancer Res* 1992;52:631–636.
95. Moser GJ, Smart RC. Hepatic tumor-promoting chlorinated hydrocarbons stimulate protein kinase C activity. *Carcinogenesis* 1989;10:851–856.
96. Rotenberg SA, Weinstein IB. Two polychlorinated hydrocarbons cause phospholipid-dependent protein kinase C activation *in vitro* in the absence of calcium. *Mol Carcinogen* 1991;4:477–481.
97. Bombick DW, Madhukar BV, Brewster DW, Matsumura F. TCDD (2,3,7,8-tetrachlorodibenzo-p-dioxin) cause increases in protein kinases especially protein kinase C in the hepatic plasma membrane of the guinea pig and rat. *Biochem Biophys Res Commun* 1985;127;296–302.
98. Bombick DW, Matsumura F. 2,3,7,8-tetrachlorodibenzo-p-dioxin causes elevation of the levels of the protein tyrosine kinase 60src. *J Biochem Toxicol* 1987;2:141–154.
99. Jankun J, Matsumura F, Kaneko H, Trosko JE, Pellicer A, Greenberg AH. Plasmid-aided insertion of MMTV-LTR and ras DNAs to NIH 3T3 fibroblast cell makes them responsive to 2,3,7,8-TCDD causing overexpression of p21ras and downregulation of EGF receptor. *Mol Toxicol* 1989;2:177–186.
100. Madhukar BV, Little RA, Klaunig JE, Trosko JE. Activation of protein kinase C alpha (PKC-α) and inhibition of intercellular communication by polychlorinated biphenyls (PCB) in rat liver epithelial cells. *Toxicologist* 1995;15:28.
101. Beavens JS, Little RA, Madhukar BV. Altered subcellular distribution of protein kinase C (PKC) isozymes in C3H10T1/2 mouse embryo cells exposed to halogenated aromatic hydrocarbons (HAHs). *Toxicologist* 1996;30:289.
102. Hunter T, Karin M. The regulation of transcription by phosphorylation. *Cell* 1992;70:375–387.
103. Karin M, Liu ZG, Zhandi E. AP-1 function and regulation. *Curr Opin Cell Biol* 1997;9:240–246.
104. Angel P, Hattori, Smeal T, Karin M. The jun proto-oncogene is positively autoregulated by its product, Jun/AP-1. *Cell* 1988;55:875–885.
105. Angel P, Imagawa, M, Chiu R, Stein B, Imbra RJ, Rahmsdorf HJ, Jonat C, Herrlich P, Karin M. Phorbol ester-inducible genes contain a common cis element recognized by TPA-modulated trans-acting factor. *Cell* 1987;49:729–739.
106. Angel P, Allegratto EA, Okino ST, Hattori K, Boyle WJ, Hunter T, Karin M. Oncogene jun encodes a sequence specific trans-activator similar to AP-1. *Nature* 1988;332:166–171.

107. Chiu R, Boyle WJ, Meek J, Smeal T, Hunter T, Karin M. The c-fos protein interacts with c-jun/AP-1 to stimulate transcription of AP-1 responsive genes. *Cell* 1988;54:541–552.
108. Boyle WJ, Smeal T, Defize LHK, Angel P, Woodgett JR, Karin M, Hunter T. Activation of protein kinase C decreases phsosphorylation of c-jun at sites that negatively regulates its DNA-binding activity. *Cell* 1991;64:573–584.
109. Watts RG, Ben-Ari ET, Bernstein LR, Birrer MJ, Winterstein D, Wendel E, Colburn NH. c-jun and multistate carcinogenesis: Association of overexpression of introduced c-jun with progression toward a neoplastic endpoint in mouse JB6 cells sensitive to tumor promoter-induced transformation. *Mol Carcinogen* 1995;13:27–36.
110. Madhukar BV, Little RA, Beavens JS. Non-coplanar polychlorinated biphenyls (PCBs) induce AP-1 transcription factor mediated expression of luciferase activity. *Toxicologist* 1996;30(1, Pt. II):290.
111. Morrison DK, Heidecker G, Rap UR, Copeland TD. Identification of the major phosphorylation sites of the Raf-1 kinase. *J Biol Chem* 1993;268:17309–17316.
112. Rapp UR, Heidecker G, Huleihel M, Cleveland JL, Choi WC, Pawson T, Ihle JN, Anderson WB. raf Family serine/threonine protein kinases in mitogen signal transduction. *Cold Spring Harbor Symposia of Quantitative Biology* 1988;53:173–184.
113. Morrison, DK, Cutler RE, Jr. The complexity of Raf-1 regulation. *Curr Opin Cell Biol* 1997;9:174–179.
114. Heidecker G, Huleihel M, Cleveland JL, Kolch W, Beck TW, Lloyd P, Pawson T, Rapp UR. Mutational activation of c-raf-1 and definition of the minimal transforming sequence. *Mol Cell Biol* 1990;10:2503–2512.
115. Fantl WJ, Muslin AJ, Kikuchi A, Martin JA, MacNicol AM, Gross RW, Williams LT. Activation of Raf-1 by 14-3-3 proteins. *Nature* 1994;371:612–613.
116. Morrison DK, Kaplan DR, Rapp UR, Roberts TM. Signal transduction from membrane to cytoplasm: growth factors and membrane-bound oncogene products increase Raf-1 phosphorylation and associated protein kinase activity. *Proc Natl Acad Sci USA* 1988;85:8855–8859.
117. Kolch W, Heidecker G, Kochs G, Hummel R, Vahidi H, Mischak H, Finkenzeller G, Marme D, Rapp UR. Protein kinase Cα activates Raf-1 by direct phosphorylation. *Nature* 1993;364:249–252.
118. Hafner S, Adler HS, Mischak H, Janosch P, Heidecker G, Wolfman A, Pippig S, Lohse M, Ueffing M, Kolch W. Mechanism of inhibition of Raf-1 by protein kinase A. *Mol Cell Biol* 1994;14:6696–6703.
119. Lange-Carter CA, Pleiman CM, Gardner AM, Blumer KJ, Johnson GL. A divergence in the MAP kinase regulatory network defined by MEK kinase and Raf. *Science* 1993;360:315–319.
120. Blank JL, Gerwisn P, Elliott EM, Sather S, Johnson GL. Molecular cloning of mitogen-activated protein/ERK kinase kinases (MEKK) 2 and 3. *J Biol Chem* 1996;271:5361–5368.
121. Han J, Lee JD, Jiang Y, Li Z, Feng L, Ulevitch RJ. Characterization of the structure and function of a novel MAP kinase kinase (MKK6). *J Biol Chem* 1996;271:2886–2891.
122. Waskeiwicz AJ, Cooper JA. Mitogen and stress response pathways: MAP kinase cascades and phosphatase regulation in mammalian and yeast. *Curr Opin Cell Biol* 1995;7:798–805.
123. Brunet A, Poussegur J. Identification of MAP kinase domains by redirecting stress signals into growth factor responses. *Science* 1996;272:1652–1655.
124. Ray LB, Sturgill TW. Rapid stimulation by insulin of a serine/threonine kinase in 3T3-LI adipocytes that phosphorylates microtubule-associated protein 2 *in vitro*. *Proc Natl Acad Sci USA* 1987;84:15072–1506.
125. Hoshii M, Nishida E, Sakai H. Activation of a Ca2+-inhibitable protein kinase that phosphorylates microtubule-associated protein 2 *in vitro* by growth factors, phorbol ester and serum in quiescent cultured human fibroblasts. *J Biol Chem* 1988;263:5396–5401.
126. Cicirelli M, PelechSL, Krebs EG. Activation of multiple kinases during the burst in protein phosphorylation that precedes the first meiotic division in Xenopus oocytes. *J Biol Chem* 1988;263:2009–2019.
127. Bolton TG, Nye SH, Robbins DJ, IP, NY, Rodziejewska E, Morgenbesses SD, DePenho RA, Panayotatos N, Cobb MH, Yancopoulos GD. Erks, a family of protein-serine/threonine kinases that are activated and tyrosine phosphorylated in response to insulin and NGF. *Cell* 1991;65:663–675.
128. Robinson MJ, Cobb MH. Mitogen-activated protein kinase pathways. *Curr Opin Cell Biol* 1997; 9:180–186.
129. Robbins DJ, Zhen E, Cheng M, Xu S, Ebert D, Cobb MH. MAP kinases ERK1 and ERK2: pleiotropic enzymes in a ubiquitous signaling network. *Adv Cancer Res* 1994;63:93–116.
130. Anderson NG, Maller JL, Tonks NK, Sturgill TW. Requirement for integration of signals from two distinct phosphorylation pathways for activation of MAP kinase. *Nature* 1990;343:651–653.

131. Ahn NG, Seger R, Bratlien RL, Diltz CD, Tonks NK, Krebs EG. Multiple components in an epidermal growth factor-stimulated protein kinase cascade: *in vitro* activation of a myelin basic protein/microtubule-associated protein 2 kinase. *J Biol Chem* 1991;266:4220–4227.
132. Bolton TG, Cobb MH. Identification of multiple extracellular signal-regulated kinases (ERKs) with antipeptide antibodies. *Cell Reg* 1991;2:357–371.
133. Alessi DR, Smythe C, Keyse SM. The human CL100 gene encodes a Tyr/Thr-protein phosphatase which potently and specifically inactivates MAP kinase and suppresses its activation by oncogenic ras in Xenopus oocyte extracts. *Oncogene* 1993;8:2015–2020.
134. Posada J, Cooper JA. Requirements for phosphorylation of MAP kinase during meiosis in Xenopus oocytes. *Science* 1992;255:212–215.
135. Yamaguchi K, Ogita K, Nakamura S, Nishizuka Y. The protein kinase C isoforms leading to MAP-kinase activation in CHO cells. *Biochem Biophys Res Commun* 1995;210:639–647.
136. Young SW, Dickens M, Tavare JM. Activation of mitogen-activated protein kinase by protein kinase C isotypes alpha, beta I, and gamma, but not by epsilon. *FEBS Lett* 1996;384:181–184.
137. Winitz S, Russell M, Quian N-X, Dwyer L, Johnson GL. Involvement of Ras and Raf in G1-coupled receptor activation of mitogen-activated protein (MAP) kinase kinase and MAP kinase. *J Biol Chem* 1993;268:19196–19199.
138. Song J, Pfeffer LM, Foster DA. v-src Increases diacylglycerol levels via a type D phospholipase-mediated hydrolysis of phosphotidylcholine. *Mol Cell Biol* 1991;11:4903–4908.
139. Duronio V, Welham M, Abraham S, Dryden P, Schrader JW. p21ras Activation via hemopoietin receptors and c-kit requires tyrosine kinase activity but not tyrosine phosphorylation of GAP. *Proc Natl Acad Sci USA* 1992;89:1587–1591.
140. Nakanishi H, Brewer KA, Exton JH. Activation of the ζ isozyme of protein kinase C by phosphatidylinositol 3,4,5-triphosphate. *J Biol Chem* 1993;268:13–16.
141. Downes CP, Carter AN. Phosphoinositide 3-kinase: a new effector in signal transduction? *Cell Signal* 1991;3:501–513.
142. Pelech SL, CHarest DL, Mordret GP, Siow YL, Palaty C, Campbell D, Charlton L, Samiei M, Sanghera JS. Networking with mitogen-activated protein kinases. *Mol Cell Biochem* 1993;127/128:157–169.
143. Lenormand P, Sardet C, Pages G, L'Allemain G, Brunet A, Pouyssegur J. Growth factors induce nuclear translocation of MAP kinases (p42mapk and p44mapk) but not of their activator MAP kinase kinase (p45mapkk) in fibroblasts. *J Cell Biol* 1993;122:1079–1088.
144. Lamy F, Wilkin F, Baptist M, Posada J, Roger PP, Dumont JE. Phosphorylation of mitogen-activated protein kinases is involved in the epidermal growth factor and phorbol ester, but not in the thyrotropin/cAMP, thyroid mitogenic pathway. *J Biol Chem* 1993;268:8398–8401.
145. Blenis J. Signal transduction via the MAP kinases: proceed at your own RSK. *Proc Natl Acad Sci USA* 1993;90:5889–5892.
146. Brunet A, Pages G, Pouyssegur J. Growth factor-stimulated MAP kinase induces rapid retrophosphorylation and inhibition of MAP kinase Kinase (MEK1). *FEBS Lett* 1994;346:299–303.
147. Anderson NG, Li P, Marsden LA, Williams N, Roberts TM, Sturgill TW. Raf-1 is a potential substrate for mitogen-activated protein kinase *in vivo* . *Biochem J* 1991;277:573–576.
148. Mark MD, Liu, Y, Wong ST, Hinds TR, Storm DR. Stimulation of neurite outgrowth in PC12 cells by EGF and KCl depolarization: A Ca2+-independent phenomenon. *J Cell Biol* 1995;130:701–710.
149. Cowley S, Paterson H, Kemp P, Marshall CJ. Activation of MAP kinase kinase is necessary and sufficient for PC12 differentiation and for transformation of NIH 3T3 cells. *Cell* 1994;77:841–852.
150. Sano M, Nishiyama K, Kitajima S. A nerve growth factor-dependent protein kinase that phosphorylates microtubule-associated proteins *in vitro*: possible involvement of its activity in the outgrowth of neurites from PC12 cells. *J Neurochem* 1990;55:427–435.
151. Schamel WWA, Dick TP. Signal transduction: specificity of growth factors explained by parallel distribution processing. *Med Hypotheses* 1996;47:249–255.

Toxicant–Receptor Interactions
Edited by Michael S. Denison and William G. Helferich

9

Cellular Calcium Regulation and Signaling

Isaac N. Pessah and Patty W. Wong

Department of Molecular Biosciences, School of Veterinary Medicine and Center for Environmental Health Sciences, University of California, Davis, California, USA

GLOBAL CALCIUM HOMEOSTASIS: EXTRACELLULAR CALCIUM SIGNALING

In most living cells, both extracellular and intracellular calcium levels ($[Ca^{2+}]_o$ and $[Ca^{2+}]_i$, respectively) are highly regulated, often with the expense of energy. Normally $[Ca^{2+}]_o$ is buffered at approximately 1 to 2 mM (1). In addition to supplying Ca^{2+} necessary for intracellular signaling and regulation of several cell functions described herein, $[Ca^{2+}]_o$ also plays a role as a first messenger important to the integrity of several structural (e.g., bone matrix) and functional (e.g., blood clotting) processes. The extremely stable $[Ca^{2+}]_o$ level is attributed to homeostatic mechanisms that consist of Ca^{2+}-sensory cells (e.g., parathyroid cells) and Ca^{2+}-responsive tissues (e.g., kidney, bowel, and bone). The "global" $[Ca^{2+}]_o$ balance in a mammal is regulated through concerted changes in disposition, mobilization, and excretion of calcium (in the free and bound forms) by these systems (2–4).

Recent cloning and characterization of a plasmalemmal Ca^{2+}-sensing receptor from bovine parathyroid has provided insight into the role of Ca^{2+} as first messenger (2). The cDNA encoding the Ca^{2+}-sensing receptor reveals similar sequence homology with metabotropic glutamate receptors (2, 5). Amino acid–sequence analysis of the Ca^{2+}-sensing receptor suggests a large N-terminus extracellular domain that contains Ca^{2+} binding sites and seven transmembrane domains, characteristic of the G protein–coupled receptor superfamily. Ca^{2+}-sensory cells are highly sensitive to minute alteration of $[Ca^{2+}]_o$ in circulation (6). In this respect, Ca^{2+}-sensory cells and their plasmalemmal Ca^{2+}-sensing receptor also are important in fine-tuning $[Ca^{2+}]_o$ in specific body compartments such as the thick ascending limb and the collecting duct of the nephron. Genetic defects involving plasmalemmal Ca^{2+}-sensing receptors have been shown to underlie hypercalcemia or hypocalcemia in affected individuals (7, 8). Recent studies have shown the differential expression of plasmalemmal Ca^{2+}-sensing receptors among various cell types and tissues (9, 10) and could reflect specialization in function. In parathyroid, Ca^{2+} acts as mediator for the release of parathyroid hormone and calcitonin. The delicate Ca^{2+} sensing system also is important in kidney, because 1 mM Ca^{2+} almost is saturating with normal phosphate and oxalate levels in urinary tract. The system functions to prevent precipitation of salts by closely adjusting the amount of Ca^{2+} in urine. Another example of $[Ca^{2+}]_o$ as a first messenger can be found in skin (11). Maturation of keratinocytes from basal layer requires the Ca^{2+} gradient across the different skin cell layers and provides an important signal for differentiation.

In neuronal tissues, membrane depolarization triggers influx of Ca^{2+} into pre- and postsynaptic cells. Local $[Ca^{2+}]_o$ can be reduced dramatically during neuronal activity under normal or pathophysiologic conditions (12–14). Plasmalemmal Ca^{2+}-sensing receptors have been shown to differentially localize at nerve terminals throughout rat brain, suggesting a possible role in regulating neurotransmitter release in response to changing $[Ca^{2+}]_o$ at the synaptic cleft (3). Ca^{2+}-sensing receptors also have been demonstrated to alter the activity of voltage-gated Ca^{2+} channels in the presynaptic cleft and N-Methyl-D-aspartate-type glutamate receptors at postsynaptic sites. In hippocampus, Ca^{2+}-sensing receptors have been suggested to be involved in induction of long-term potentiation (LTP). Instead of sensing global changes in circulating $[Ca^{2+}]_o$, Ca^{2+}-sensing receptors of the central nervous system appear to function as local indicators of changes in the microenvironment of the synapse, which results from neuronal activities.

Extracellular Ca^{2+} binding to plasmalemmal receptors mediates signal transduction through a G protein–coupled pathway (6, 15), thereby functioning as first messenger to modulate cellular activities in response to systemic changes in $[Ca^{2+}]_o$ (e.g., parathyroid and kidney) as well as changes in microenvironments (e.g., central nervous system).

CALCIUM AS SECOND MESSENGER: INTRACELLULAR CALCIUM SIGNALING

In order to provide a rapid and precise signaling system, a messenger should bind to its target molecule with sufficient affinity to confer specificity. Electrostatic attraction between signaling messenger and target molecule (usually protein) provides a rapid and reversible means of communication between the cell surface and internal components of the cell. Calcium, a divalent cation, binds to proteins through electrostatic interaction, having higher affinity than that of monovalent cations (K^+ and Na^+). The ionic radius of Ca^{2+} compared with other physiologic ions (Mg^{2+} has a smaller ionic radius than Ca^{2+}) permits binding to protein sites without creating excessive steric constraints to the protein backbone (16). The empty outermost *d*-electron shell of Ca^{2+} enables it to form a complex with multiple coordination numbers, thereby permitting greater flexibility in the types of interactions with electron donating moieties found on proteins. A large number of proteins have evolved tertiary structures that enhance the binding affinity for Ca^{2+} (e.g., EF hands) and through which Ca^{2+} confers regulation of structure and function.

Depending on cell type, intracellular calcium concentration ($[Ca^{2+}]_i$) is buffered at 10 to 100 nM at rest (17, 18). The extremely low $[Ca^{2+}]_i$ is tightly regulated by Ca^{2+} extrusion through plasmalemma (17–20), sequestration within intracellular stores (endoplasmic/sarcoplasmic reticulum [ER/SR] and mitochondria), and binding to cytosolic proteins that in essence act to buffer free Ca^{2+} in the cytosol. Hence, a greater than 10,000-fold Ca^{2+} gradient is maintained across the plasmalemma at rest. Ca^{2+} at high concentration tends to complex with inorganic phosphate and precipitates. Therefore, low $[Ca^{2+}]_i$ is required in order to maintain a significant level of inorganic phosphate in regeneration of ATP, the most important fuel molecule and cofactor for protein phosphorylation. More important, the extremely high Ca^{2+} gradient across the plasmalemma is essential for the role of Ca^{2+} as second messenger (21). On stimulation, many cells respond by increasing $[Ca^{2+}]_i$ towards micromolar level. The rapid rise of $[Ca^{2+}]_i$ can be attributed to either Ca^{2+} influx through Ca^{2+} channel on plasmalemma (21) or Ca^{2+} mobilization from intracellular stores (22).

At the surface of nonexcitable cells, ligands that activate receptors coupled to GTP-binding proteins or receptor tyrosine kinases can mobilize Ca^{2+} from intracellular stores by activating phospholipase C β or phospholipase C γ, respectively (Fig. 1). Phospholipase C hydrolysis of phosphatidyl inositol 4,5-bisphosphate (an integral phospholipid found at the inner leaflet of the plasma membrane) produces inositol 1,4,5-triphosphate (IP3) and 1,2-diacylglycerol (DAG) (22). IP3 is the second messenger that binds to IP3-sensitive Ca^{2+}-release channels (IP3 receptor, IP3R) and mobilizes Ca^{2+} from ER. DAG promotes cytosolic protein kinase C (PKC) translocation to plasmalemma and hence activates cellular kinase activity (23, 24).

On depolarization of the surface membrane of excitable cells such as neurons, the high $[Ca^{2+}]_o$ provides the chemical gradient needed for rapid Ca^{2+} influx through voltage-gated Ca^{2+} channels and ionotropic glutamate receptors of the NMDA type (22). Activation of intracellular ryanodine-sensitive Ca^{2+} release channels (ryanodime receptor, RyR) on ER membrane by influx Ca^{2+} further exaggerates the rapid rise of $[Ca^{2+}]_i$ through a phenomenon known as *Ca^{2+}-induced Ca^{2+} release* (CICR). Ca^{2+} released through CICR at sites spatially distinct from IP3R appears to work in concert with IP3-induced Ca^{2+} release from ER to provide both spatially and temporally diverse cellular Ca^{2+} signals (25).

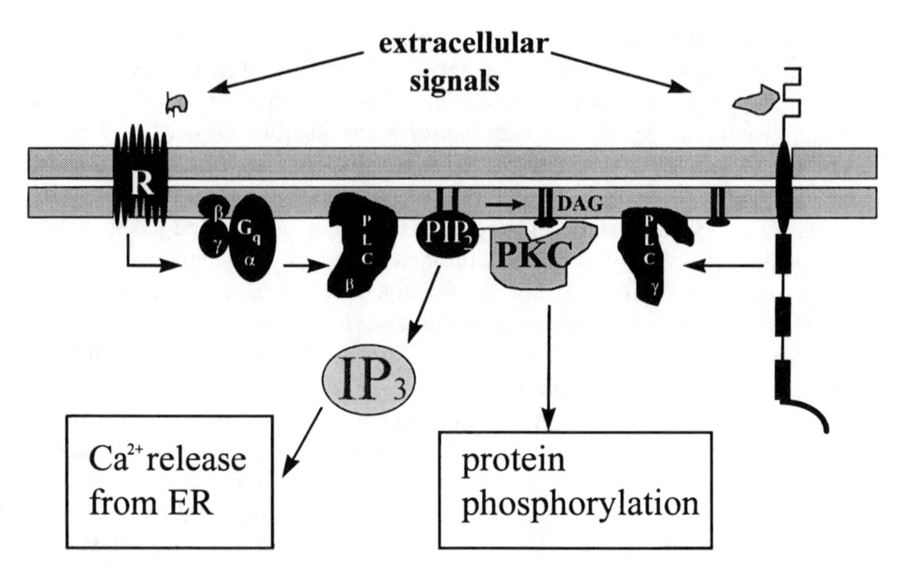

FIG. 1. Cellular signaling involving the phosphatidylinositide cascade. Ligand binding to either a receptor (R) coupled to a GTP-binding protein (G_q) or a receptor tyrosine kinase (TK) initiates hydrolysis of phospatidylinositol 4,5 bisphosphate (PIP_2) as a result of phospholipase C β or γ (PLC β or PLC γ) respectively.

Finally, cytosolic Ca^{2+} binds to calmodulin and hence activates kinase and phosphatase activities, which in turn alter a broad spectrum of cellular enzymatic activities essential for responses ranging from release of neurotransmitter from neurons (26), interleukin production in activated T cells, degranulation of mast cells (27–29), and contraction of muscle (30).

Consistent with its function as a second messenger, $[Ca^{2+}]_i$ level is tightly regulated by several Ca^{2+} regulatory proteins (21). First, with the expense of ATP, Ca^{2+}-dependent ATPase on plasmalemma extrudes Ca^{2+} out of the cells with stoichiometry of one Ca^{2+} for each ATP hydrolyzed. Second, a $Ca^{2+}(Mg^{2+})$-dependent ATPase on ER/SR membrane (sarco/endoplasmic reticulum calcium [SERCA] pumps) sequesters Ca^{2+} with higher efficiency than the plasmalemma. The stoichiometry of SERCA pumps is two Ca^{2+} translocated for each ATP hydrolyzed. Third, a Na^+/Ca^{2+} exchanger on plasmalemma extrudes Ca^{2+} through an electrogenic exchange system that transports three Na^+ inward and concomitantly extrudes one Ca^{2+}. This exchanger system possesses lower affinity for Ca^{2+} than Ca^{2+} ATPases and works in complement, especially in excitable cells, to maintain $[Ca^{2+}]_i$ within physiologic nanomolar concentration. Fourth, Ca^{2+} binding proteins in ER/SR lumen (calsequestrin, calreticulin, calbinding, etc.) bind to lumenal Ca^{2+} and enhance the capacity of the stores. Finally, the cytosolic free Ca^{2+} is further modulated through binding to cytosolic proteins, which have high capacity and low affinity for Ca^{2+} (e.g., calmodulin in brain, troponin C in skeletal muscle).

Calcium Transport in Mitochondria

In addition to ER/SR, mitochondria can accumulate Ca^{2+} through a uniporter system (18). Recent studies, however, have demonstrated that the affinity of mitochondrial uniporter for Ca^{2+} ($K_d \sim 10\ \mu M$) (21) is well above the $[Ca^{2+}]_i$ under normal physiologic conditions. Furthermore, the amount of Ca^{2+} typically accumulated in mitochondria is not sufficient to function as significant Ca^{2+} sink or store (31–33). The rate of Ca^{2+} accumulation by mitochondria is much slower than that of the ER/SR store (34). These results suggest that under physiologic conditions, mitochondria do not function as a major Ca^{2+}-regulatory organelle (35). In fact, Ca^{2+} in mitochondria has been suggested to be a metabolic mediator. The activities of several dehydrogenases involved in electron transport chain (ETC), ATP synthase, and critical elements of the tricarboxylic acid (TCA) cycle are regulated by Ca^{2+} (35). Hence, spatiotemporal changes in $[Ca^{2+}]_i$ (multiple Ca^{2+} pulses or single Ca^{2+} pulse) (36–38) exaggerate mitochondrial Ca^{2+} uptake (35). The increase in mitochondrial Ca^{2+} loading enhances the rate of oxidative phosphorylation, and thereby increases ATP production (21).

Ca^{2+} transport across inner mitochondrial membrane is very complex, and at least four different mechanisms have been identified (35) (Fig. 2): rapid influx through a Ca^{2+} uniporter; slow efflux through a $Ca^{2+}/2Na^+$ exchanger, slow efflux through a

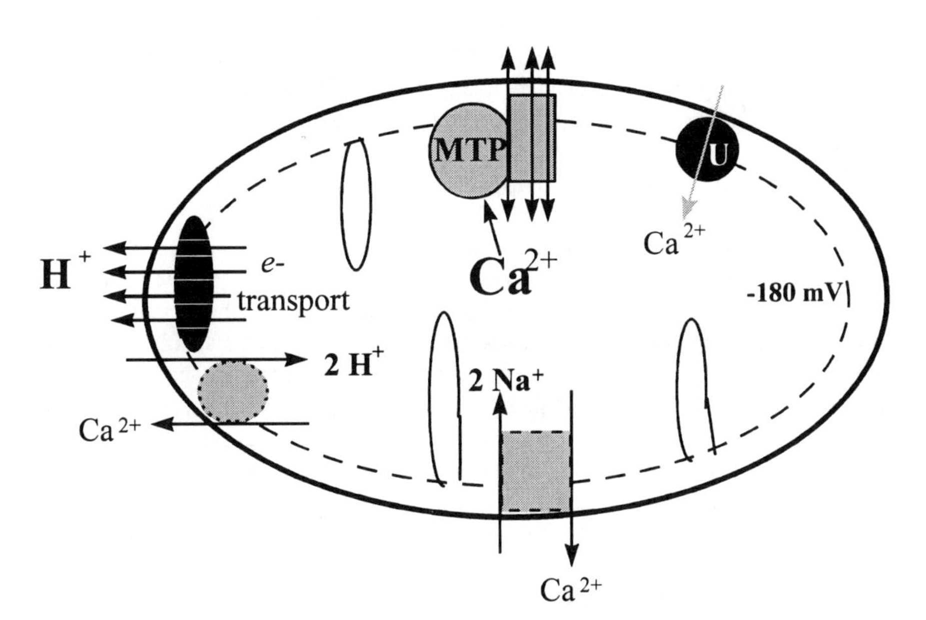

FIG. 2. Mechanisms regulating mitochondrial calcium transport. The mitochondrial transition pore (MTP) also can conduct several proteins including cytochrome oxidase. The calcium uniporter (U) is driven by the protein gradient, whereas the remaining pumps are driven by ATP from electron transport.

$Ca^{2+}/2H^+$ exchanger, and rapid efflux through the nonspecific permeability transition pore (PTP also termed mitochondrial transition pore; MTP) (39). Intramitochondrial Ca^{2+} is further buffered by calmitin (a 60-kDa mitochondrial matrix protein) and anions (e.g., cardiolipin). Cloning of cDNA-encoding calmitin suggests that calmitin is identical to calsequestrin in ER/SR, binding Ca^{2+} with high capacity and low affinity (40).

In mitochondria, an internally negative membrane potential is generated and maintained by active outward H^+ pumping within the electron transport chain (41). At negative membrane potential, the uniporter possesses an active conformation that permits Ca^{2+} influx (energetically downhill) (35). Ca^{2+} binding to external activation sites also activates the uniporter, whereas negative membrane potential provides the electrochemical driving force for inward diffusion of Ca^{2+} into mitochondria through the opened uniporter. Ca^{2+} influx through the uniporter is noncompetitively inhibited by ruthenium red (42).

Ca^{2+} efflux pathways in mitochondria are energy-dependent (43), because they are against the electrochemical gradient. The $Ca^{2+}/2H^+$ exchanger (dominant in liver, kidney) mediates slow efflux of Ca^{2+} (1.2 nmol/mg/min) compared with rapid Ca^{2+} influx through the uniporter (>1200 nmol/mg/min) (35, 44). By simultaneously exchanging two H^+ down its chemical gradient, the exchanger gains energy to extrude one Ca^{2+} against its electrochemical gradient (35, 45). The $Ca^{2+}/2Na^+$ exchanger (dominant in heart, skeletal muscle, brain, and other tissues) effluxes Ca^{2+} at a much faster rate (about 2 nmol/mg/min) than that of the $Ca^{2+}/2H^+$ exchanger. Under physiologic conditions, it cotransports two Na^+ (down its electrochemical gradient), providing enough energy for the exchanger to efflux one Ca^{2+} against its electrochemical gradient (46, 47). Hence, these mitochondrial Ca^{2+} efflux pathways are thought to be passive (35).

At rest, Ca^{2+} influx through the uniporter is well balanced by Ca^{2+} efflux through the exchanger systems. During multiple Ca^{2+} spiking or a persistent rise of cytosolic Ca^{2+} (micromolar), however, a large amount of Ca^{2+} can be taken up rapidly through the uniporter. Slow Ca^{2+} efflux through exchanger systems is inadequate to eliminate intramitochondrial Ca^{2+}. Excessive Ca^{2+} accumulation in mitochondria uncouples oxidative phosphorylation and leads to cell death (48). A rapid Ca^{2+} efflux pathway through MTP is activated under such a condition (39). Similar to voltage-gated channels of plasmalemma, MTP is modulated by membrane potential and matrix pH (39). The pore is closed at resting mitochondrial membrane potential (about -180 mV) (35). On mitochondrial membrane depolarization, oxidative stress, or binding of Ca^{2+} at internal sites, MTP becomes active resulting in an increase in open probability (49). Conversely, in the presence of adenine nucleotide, acidic pH, or binding of Ca^{2+} at external site, MTP closes (ceases to conduct Ca^{2+}) (39). Cyclosporin A also is known to selectively inhibit the pore indirectly through an interaction with a mitochondrial cyclophilin (50, 51). The mechanism appears to be independent of calcineurin inhibition (52). The increase in mitochondrial Ca^{2+} loading during cell stimulation (or oxidative stress) sensitizes MTP and causes Ca^{2+} efflux through the pore, which is followed by mitochondrial membrane depolarization. Accumulation of ADP in cytosol, resulting from Ca^{2+} extrusion through plasmalemma and sequestration into ER/SR, feeds back to inhibit the activity of MTP (39). As a result, MTP provides the rapid Ca^{2+} efflux to prevent intramitochondrial Ca^{2+} overload. Under pathophysiologic conditions, prolonged opening of MTP abolishes the membrane potential, arrests oxidative phosphorylation, and results in energy depletion. Swelling of mitochondrial matrix (53), loss of matrix proteins (54), irreversible cell injury, and necrosis are toxicologic sequelae.

Calcium Regulation in ER/SR

In general, Ca^{2+} is released from ER/SR through two pathways, IP3Rs and RyRs. Multiple isoforms of these receptors are differentially expressed in various cell types and tissues (55–57). The IP3R consists of four identical subunits, each with a molecular weight (Mr) of 313,000. Binding of IP3 to IP3R activates the channel. The channel activity is allosterically regulated by ATP and Ca^{2+}. Ca^{2+} at a concentration greater than 0.3 μM provides feedback inhibition of IP3R, whereas nanomolar Ca^{2+} is a coagonist. Caffeine and heparin are known pharmacologic inhibitors for IP3R, which block the channels by a mechanism different from Ca^{2+} (58). Recent studies have demonstrated that through an association with FKBP12 (FK506 binding protein, 12 kDa), calcineurin is linked to IP3R (and RyR) and modulates Ca^{2+} efflux through the channel (59).

Hormonal stimulation of nonexcitable cells results in rise of $[Ca^{2+}]_i$ through IP3 cascade. The rise of $[Ca^{2+}]_i$ can further trigger Ca^{2+} release through RyRs (CICR). The spatiotemporal changes in $[Ca^{2+}]_i$ provide meaningful signals for various cellular activities (22). Ca^{2+} measurements of fura 2–loaded bovine chromaffin cells suggest that Ca^{2+} release from ryanodine-sensitive stores (and IP3-sensitive stores) are quantal in nature (60). Submaximal stimulation produces responses by inducing rapid Ca^{2+} release from functionally discrete stores (all-or-none release), instead of partial Ca^{2+} release from all stores. Graded responses produced on increasing stimulus intensity are attributed to progressive recruitment of discrete stores. Differences in phosphorylation states of the Ca^{2+} channels, allosteric interaction between channel subunits, or interactions with accessory proteins have been suggested to contribute to the heterogeneity in sensitivity of discrete stores.

In resting cells, ER/SR channels release Ca^{2+} in the form of localized sparks involving only a single or several localized channel openings, known as *spontaneous miniature outward currents* (SMOCs) or *spontaneous transient outward currents* (STOCs) (22). Similarly, spontaneous opening of clusters or an individual voltage-gated Ca^{2+} channel on plasmalemma results in a rise of intracellular Ca^{2+} in a localized region, known as a *quantum emission domain*. Activation of local Ca^{2+} release channels by quantum emission domain result in a local Ca^{2+} transient originating at junctional sarcoplasmic reticulum (SR) in cardiac myocytes (22, 61). On membrane depolarization, plasmalemmal voltage-gated Ca^{2+} channels are recruited, resulting in a global rise of intracellular Ca^{2+}. Depending on the distribution of these channels, various forms of spatial Ca^{2+} waves are observed in different cell types. The excitability of neighboring ER Ca^{2+} release channels is enhanced by increasing stimulus intensity. Hence Ca^{2+} influx and Ca^{2+} release from Ca^{2+} release channels result in the global Ca^{2+} rise in cytosol, which is the summation of SMOC.

Studies of arterial smooth muscle reveal that spontaneous Ca^{2+} sparks produced by individual RyR units modulate muscle tension at rest by activating K^+ channels on plasmalemma, which hyperpolarize the muscle cells. These results illustrate the physiologic role of spontaneous Ca^{2+} sparks in maintaining muscle relaxation at rest. In addition, a local rise of Ca^{2+} also provides important information for cellular activities such as mitochondrial Ca^{2+} uptake and ATP generation. The local release of Ca^{2+} provides spatial information in the form of Ca^{2+} signals that initiate cellular responses in restricted regions. While the global Ca^{2+} wave produces a general rise in $[Ca^{2+}]_i$, the product of summation of several SMOCs, it is able to provide graded responses corresponding to different levels of stimulation. Because of differences in Ca^{2+} affinity of various effector systems, spatiotemporal Ca^{2+} signals can selectively activate target systems. For example, calmodulin (high affinity for Ca^{2+}) is activated by submicromolar cytosolic Ca^{2+}, whereas secretion (involving a mediator with low Ca^{2+} affinity) is triggered by local Ca^{2+} spikes (62).

SERCA pumps actively accumulate Ca^{2+} into ER/SR (21). In skeletal muscles, SERCA pumps are transmembrane proteins with Mr of approximately 110,000, constituting the major SR protein and localized on both the junctional and longitudinal SR. In cardiac muscle, phospholamban, a proteolipid consisting of five identical subunits each with Mr of 6000, is known to associate with SERCA pumps and modulate their activities. Phosphorylation of phospholamban with protein kinase (e.g., CaM II kinase) alters its conformation and induces dissociation from SERCA pumps and thereby relieves the inhibitory effect of phospholamban on the pumps (63).

Within the ER/SR lumen, Ca^{2+} is in millimolar concentration. Most of the Ca^{2+}, however, is bound to high-capacity and low-affinity binding proteins (calsequestrin and calreticulin) (21). Calsequestrin, an extrinsic lumenal hydrophobic protein with Mr of 45,000, binds Ca^{2+} with very high capacity (about 40 mole/mole) and low affinity (apparent K_d, 0.8 mM). It is differentially clustered at junctional SR in skeletal muscle and suggested to be involved in Ca^{2+} release upon activation of RyRs on SR. Calreticulin, a 60-kDa protein, is found colocalized with calsequestrin. Ca^{2+} binding to calreticulin causes protein aggregation and precipitation (64).

CALCIUM REGULATION IN MUSCLE: EXCITATION CONTRACTION COUPLING

The plasma membrane of excitable cells depolarizes on stimulation. Two types of excitable cells in mammals are neurons and muscle cells. Skeletal muscle transforms a chemical signal in the form of a neurotransmitter at the surface of the fiber into mechanical work (muscle contraction), in a process known as *excitation–contraction coupling* (65).

At the neuromuscular junction, an efferent motoneuron synapses with muscle fibers to form a motor endplate (65, 66). Membrane depolarization at the nerve terminal activates voltage-dependent Ca^{2+} channels on the presynaptic membrane. Rise of cytosolic Ca^{2+} triggers the release of acetylcholines (ACh) through exocytosis. Simultaneous release of hundreds or even thousands of quanta of ACh results in excitatory postsynaptic potentials. ACh binds to nicotinic ACh receptors and nonselective cation channels and activates inward current (primarily carried by Na^{+}), thereby depolarizing the muscle cells. If excitatory postsynaptic potentials sum to threshold, an action potential is propagated from the sarcolemma to the transverse tubule. Meanwhile, acetylcholinesterase (AChE) present in the synaptic cleft catalyzes rapid breakdown of ACh into choline and acetic acid (66). Choline is actively taken up by the nerve terminal to regenerate ACh. Rapid removal of ACh from the synaptic cleft enables the motor unit to be ready for another stimulus within a few milliseconds.

Within the transverse tubule of skeletal muscle, L-type voltage-gated Ca^{2+} channels (dihydropyridine receptor, DHPR) are clustered in tetrads and are in close proximity to the terminal cisternae of sarcoplasmic reticulum (junctional SR). DHPRs form physical couplings with RyR_1 (skeletal isoform of RyR), which constitute the junctional foot structures of triadic junctions (67). DHPRs function as voltage sensors on sarcolemma and convey a mechanical signal to RyR_1 on SR. It has been long believed that Ca^{2+} influx through DHPR activates RyR_1 by CICR. Recent studies show, however, that membrane depolarization induces conformational changes in DHPR. A direct interaction of the cytoplasmic loop between the second and third transmembrane domains (repeats II and III) within the α_1 subunit of DHPR and the N-terminal domain of RyR_1 is thought to transmit a signal across the 20-nm gap at the triadic junction (68, 69). Although RyR_1 on SR is activated through conformation coupling with DHPR on sarcolemma, CICR may play an important modulatory function in skeletal muscle.

The rapid and robust increase in cytosolic Ca^{2+} (micromolar) promotes binding of Ca^{2+} to troponin C, a calmodulin analog found on myosin (65, 66). Binding of Ca^{2+} to troponin C induces conformational changes in troponin I and tropomyosin on the actin filament. Exposure of myosin binding sites on actin F chains promotes complex formation between actin chains and myosin heads. Hydrolysis of ATP bound to myosin head groups induces conformational changes in myosin and leads to sliding of the myosin fiber along the actin filament, shortening of sarcomere, and finally muscle contraction. Muscle contraction is terminated by the release of ADP from myosin heads and binding of another ATP molecule to the newly exposed ATP site on myosin head. In the presence of low cytosolic Ca^{2+}, binding of ATP to myosin head promotes bridge-breaking between actin and myosin, thereby permitting the muscle to relax (65, 66).

Cytosolic Ca^{2+} is rapidly removed through active transport by SERCA pumps located on junctional and longitudinal SR (21). Calsequestrin inside the lumen of SR binds to Ca^{2+} (21) and further enhances Ca^{2+} loading within SR. Cytosolic Ca^{2+} can be brought back to basal nanomolar concentration within 30 milliseconds of muscle contraction (65). The rapid removal of cytosolic Ca^{2+} is essential during muscle relaxation (66).

In cardiac muscle, spontaneous action potentials originate from pacemaker cells (Purkinje cells) and propagate through gap junctions (containing intercalated discs) between cardiac cells (65, 66). Ca^{2+} influx through DHPR initiates cardiac muscle contraction (67). Ca^{2+} influx also triggers release of SR Ca^{2+} through RyR_2 (cardiac isoform of RyR) via the process of CICR. Similar to skeletal muscle, rise of cytosolic Ca^{2+} signals muscle contraction in heart.

In smooth muscle, depending on the tissue type, rise of cytosolic Ca^{2+} can be elicited by an action potential, neurotransmitters (e.g., ACh), or hormones (e.g., adrenaline) (65). For example, binding of ACh to muscarinic AChR on intestinal smooth muscle triggers cytosolic Ca^{2+} signaling through IP3 cascade. Increase in cytosolic Ca^{2+} enhances Ca^{2+} binding to calmodulin (65), which in turn binds to myosin light chain kinase and activates its kinase activity. Phosphorylation of one of the two myosin-II light chains promotes complex formation between myosin head and actin filament and muscle contracts.

CALCIUM REGULATION IN NEURONS: EXCITATION RESPONSE COUPLING

Ca^{2+} signaling in neurons is extremely complicated (Fig. 3). Spatiotemporal fluctuation of $[Ca^{2+}]_i$ mediates various neuronal activities (26). Although Ca^{2+} influx through plasmalemmal voltage-gated and ligand-gated Ca^{2+} channels is known to be essential in numerous physiologic processes, significant evidence for a major contribution of intracellular Ca^{2+} stores in neuronal signal transduction has emerged. Basal $[Ca]_i$ regulates expression and activity of tyrosine hydroxylase (70–72), a key enzyme involved in catecholamine synthesis. CICR mediates neuronal differentiation of cultured spinal nerves (73), long-term depression (LTD) induction in cerebellar Purkinje cells (74), and growth-cone migration during neurodevelopment (75).

The use of selective drugs and toxins has permitted identification of four classes of voltage-gated Ca^{2+} channel termed *T-*, *L-*, *N-* and *P-types* (76, 77). These Ca^{2+} channels are differentially expressed in various brain regions and neuron types. Based on the diversity observed in their pharmacology, they may participate in different aspects of neuronal function. Despite the fact that there is no direct evidence for differential localization of voltage-gated Ca^{2+} channel types in particular regions of a neuron, clusters of a particular channel type have been found localized to distinct neuronal structures. The close proximity between these channel clusters and specific Ca^{2+}-dependent neuronal processes may be critical in eliciting prompt and appropriate responses during stimulation.

T-type Ca^{2+} channels are characterized by their low activation threshold, fast inactivation, and small unitary conductance. The low activation threshold makes T-type channels especially sensitive to neuronal pacemaking activity. T-type channels are expressed in hippocampal, thalamic, and hypothalamic neurons. Because of the lack of a selective blocker for the T-type channel, however, little is known about its function in the central nervous systems. L-type Ca^{2+} channels, also known as DHPRs,

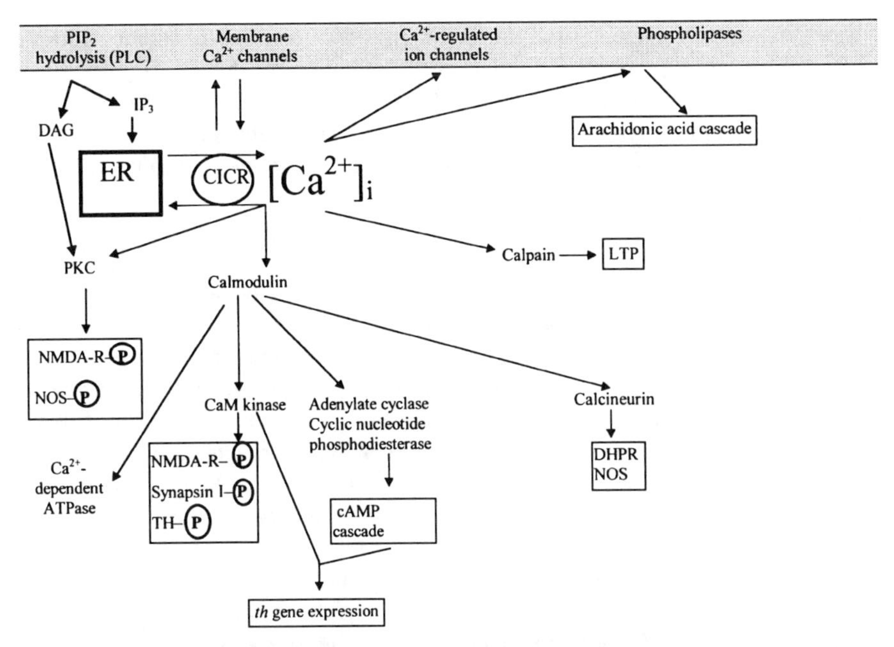

FIG. 3. Diagram showing the various biochemical targets of intracellular calcium concentration $[Ca^{2+}]_i$ in neurons. DAG, diacyl glycerol; IP3, inositol 3 phosphate; ER, endoplasmic reticulum; CICR, calcium induced calcium release; $[CA^{+2}]_i$, intracellular calcium concentration; LTP, long-term potentiation; PKC, protein kinase C; NMDA-R-P, phosphorylated N-methyl-D-aspartate receptor; NOS-P, phosphorylated nitric oxides synthase; TH-P, phosphorylated tyrosine hydroxylase; DHRP, diphydropyridine receptor.

conduct large unitary current with high activation threshold and almost no inactivation. They are widely expressed in excitable cells (including muscle). Their activity is selectively inhibited by dihydropyridine-type blockers (e.g., nifedipine) and enhanced by dihydropyridine-type agonists (e.g., Bay-K 8644). In a manner similar to L-type Ca^{2+} channels, N-type Ca^{2+} channels activate with high threshold. N-type Ca^{2+} channels are the primary current carrier in rat sympathetic neurons. The activity of N-type Ca^{2+} channels is selectively inhibited by ω-conotoxin. Finally, P-type Ca^{2+} channels are characterized by their unusually high activation thresholds. They are found predominately in dendrites of cerebellar Purkinje neurons. The activity of P-type channels is selectively blocked by ω-agatoxin IVA. The functions of various types of voltage-gated Ca^{2+} channels are further modified by neurotransmitters through mechanisms involving direct interaction between channel proteins and G proteins, G protein–mediated second-messenger systems (e.g., cAMP and DAG), or phosphorylation of channel proteins by protein kinases.

Voltage-clamp measurements made from mammalian central nervous systems have shown that at least two types of Ca^{2+} channels coexist in a given type of neuron. Their relative contributions to excitability vary, however. It is known that all neurons posses at least a certain level of L- and N-type Ca^{2+}-channel expression. Both

L- and N-type Ca^{2+} channels also coexist on growth cones of sympathetic neurons. Little contribution of L- and N-type channels is found in cerebellar Purkinje neurons, where P-type channels predominate. L- and N-type channels are involved in excitation response coupling, whereas P-type Ca^{2+} channels in cerebellar Purkinje neurons are necessary for generation of LTD. T-type Ca^{2+} channels respond to pacemaking activity of neurons.

In addition to voltage-gated Ca^{2+} channels, the NMDA-type receptor, an ionotropic glutamate receptor, also contributes significantly to Ca^{2+} influx during membrane depolarization (78, 79). Physiologic concentration of Mg^{2+} provides a voltage-dependent blockade of the receptor by binding to a site within the ionophore. Activation of NMDA receptor requires binding of glutamate and concomitant membrane depolarization. Binding of glutamate and glycine fully activates the receptor and increases the permeability of Ca^{2+}, Na^{+}, and K^{+}. The NMDA receptor is further modulated by kinase/phosphatase activity (80). The channel also is modulated allosterically by endogenous agents like spermine, redox status, and nitric oxide. Various subtypes of NMDA receptor are differentially expressed in mammalian brain. In general, NMDA receptors are found in cerebral cortex, hippocampal formation, and cerebellar granular cells. The receptor plays an important role in induction of LTP on tetanizing stimulation of hippocampus, and recent studies have further confirmed the role of NMDA receptors in LTD formation. During anoxia and truma, NMDA receptors are involved in glutamate-induced neurotoxicity. Massive Ca^{2+} influx through NMDA receptor has been suggested to be the underlying mechanism for cell death within susceptible regions in which NMDA receptors are highly expressed, although this interpretation has been challenged recently (81).

Activation of ionotropic receptors or voltage-gated Ca^{2+} channels induces Ca^{2+} influx (25). The rise of cytosolic Ca^{2+} acts as a second messenger for mobilization of intracellular Ca^{2+} from ER. Functional coupling of plasmalemmal L-type Ca^{2+} channel and RyR on ER, however, recently has been shown in cultured cerebellar granular cells (82). Activation of metabotropic glutamate receptors also can mobilize ER Ca^{2+} through the phosphatidyl inositide cascade. Depending on the filling capacity of the intracellular stores, Ca^{2+} influx through plasmalemma can either activate a Ca^{2+} uptake pathway through SERCA pumps (if stores are empty) or mobilize Ca^{2+} through opening of Ca^{2+}-releasing channels on ER (if stores are replete). The excitability of the Ca^{2+}-release channels on ER is found to be directly related to the filling state of the stores (25).

In the presence of Ca^{2+}, binding of IP3 to its receptor fully activates the receptor and mobilizes Ca^{2+} from ER. Similarly, Ca^{2+} also activates RyR through CICR. Both IP3-sensitive and Ry-sensitive Ca^{2+}-release channels are differentially expressed in neurons of mammalian central nervous systems (57). There is evidence, however, suggesting the presence of distinct stores within the same neutron. Through interaction between the IP3 and CICR pathways, Ca^{2+} mobilization from ER can be further modified. Ca^{2+} release through RyRs and Ca^{2+} influx through plasmalemma enhances the phosphatidyl inositide cascade by modulating the activity of phospholipase C and sensitizing the IP3R towards IP3. On the other hand, Ca^{2+} release from IP3-sensitive stores activates RyR by CICR. Mobilization of ER Ca^{2+} plays a significant role in neuronal functions such as neurogenesis (73, 75) and LTD in cerebellar Purkinje cells (74), as well as in neurotoxicity induced by glutamate (81). The spatial localization of the intracellular stores and Ca^{2+}-release channels and differential sensitivity of these channels in different regions of the brain (different isoform or filling states of the stores) contribute to the spatiotemporal nature of Ca^{2+} signaling in neurons.

Rise of cytosolic Ca^{2+} towards micromolar concentration triggers a series of neuronal activities (26). At the plasmalemma, cytosolic Ca^{2+} directly regulates the activity of K^+ channels, cation channels, and Cl^-channels. High $[Ca^{2+}]_i$ also activates PLC and phospholipase (PLA_2), thereby increasing IP3 turnover and promoting arachidonic acid cascade, respectively. In cytosol, a rise of $[Ca^{2+}]_i$ enhances PKC activity (synergized by DAG), levels of calpain (a Ca^{2+} dependent protease), and Ca^{2+} binding to calmodulin, which in turn trigger a series of Ca^{2+}-dependent processes.

Translocation of PKC to the inner leaflet of plasmalemma facilitates phosphorylation of membrane-bound proteins such as NMDA receptor (resulting in enhanced channel activity) and NO synthase (resulting in decreased NO production). Studies on the function of calpain suggest activation of calpain by Ca^{2+} plays a significant role in induction of LTP in hippocampus (83).

Calmodulin, a 17-kDa cytosolic protein containing two "EF-hand" domains, binds four Ca^{2+} ions allosterically with low micromolar affinity. The Ca^{2+}-bound calmodulin enhances activity of calmodulin-dependent protein kinases (CaM kinases), calcineurin, adenylate cyclase and cyclic nucleotide phosphodiesterase, and Ca^{2+}-dependent ATPase on plasmalemma (26).

The most predominant CaM kinase in neurons is type II CaM-dependent protein kinase (CaM II kinase). It is highly expressed throughout the neuronal cytosol, but especially high levels are found within postsynaptic regions. Phosphorylation of NMDA receptors by CaM II kinase is a critical step in generation of LTP. Phosphorylation of non-NMDA receptors by CaM II kinase (PKC also may be involved) also is essential in generation of LTD and LTP. Phosphorylation of synapsin I (a synaptic vesicle–associated protein that also binds to the cytoskeleton) by CaM II kinase reduces its affinity for vesicle association and hence facilitates the release of neurotransmitter at nerve terminals.

Calcineurin, a protein phosphatase expressed at a high level in mammalian brain, is activated by Ca^{2+}-bound calmodulin. Dephoshorylation of L-type Ca^{2+} channel within plasmalemma and NO synthase by calcineurin results in channel inactivation and enhancement in NO production. Activation of plasmalemmal Ca^{2+}-dependent ATPase by calmodulin functions as the feedback mechanism for elimination of Ca^{2+} from cytosol after neuronal excitation (84).

Calmodulin also modulates the cAMP-signaling cascade through its action on the enzymatic activities of adenylate cyclase and cyclic nucleotide phosphodiesterase, the key enzymes involved in cAMP metabolism. Together with CaM kinase, cAMP is involved in the regulation of tyrosine hydroxylase activity at both enzymatic and genetic levels. Phosphorylation of a cAMP response-element binding protein by CaM kinase enhances binding of this transcription factor to a cAMP response element upstream of the promoter region of the tyrosine hydroxylase gene. As a result, transcription of this gene is turned on, and hence synthesis of catecholamine is unregulated (70, 85).

The rise of cytosolic Ca^{2+} triggers exocytosis of synaptic vesicles, modulates synaptic plasticity, induces growth-cone migration during neurodevelopment, and regulates ion-channel activity through direct interaction with the mediator or kinase/phosphatase activity (75).

RYANODINE RECEPTORS

Ryanodine Binding to Conformationally Sensitive Sites on ER/SR Channels

The toxic plant alkaloid ryanodine first was purified and characterized from the powdered stemwood and roots of Ryania speciosa Vahl, by Rogers and coworkers in 1948 (86). As an insecticide, this alkaloid causes a characteristic flaccid paralysis of insect muscle (87). Isolation of 9,21-dehydroryanodine from ryania has facilitated the synthesis of radiolabeled ryanodine ([^{3}H]ryanodine) and permitted direct studies aimed at understanding the mechanism of this muscle poison (88, 89). With the availability of 9,21[^{3}H]ryanodine, studies of the molecular mechanism of ryania insecticide toxicity led to the identification of a ryanodine "receptor" (RyR), which in fact turned out to be synonymous with the Ca^{2+} release channel of SR. [^{3}H]Ryanodine binds to RyRs with high affinity and selectivity (67, 90). Receptor binding studies using [^{3}H]ryanodine have demonstrated the existence of both high-affinity (less than 10 nM) and low-affinity (up to 5 μM) binding sites on the RyR protein complex. As many as four allosterically interacting ryanodine-binding sites have been proposed on each receptor tetramer (91, 92). At nanomolar concentrations, ryanodine enhances channel activity by increasing the open probability (93). At micromolar concentrations, ryanodine promotes subconductance behavior of the channel. Finally, the channel is persistently blocked with approximately 200 μmol ryanodine. Another unique characteristic is that low-nanomolar ryanodine binds to RyR only if the channel is open, permitting [^{3}H]ryanodine to be used as a conformationally sensitive probe for studies of RyRs (94). Under physiologic conditions, binding of [^{3}H]ryanodine to RyR requires concomitant activation of the channel with Ca^{2+}. The dose-dependent and use-dependent characteristics of ryanodine binding to this Ca^{2+}-release channel make it a valuable tool for studies on Ca^{2+} mobilization from ER/SR in various tissues, including skeletal muscle, cardiac muscle, neurons, and immune cells.

Structure and Tissue Distribution

To date, three distinct isoforms of RyR have been identified, the skeletal (RyR_1), cardiac (RyR_2) and brain (RyR_3) isoforms. Each RyR consists of four identical subunits.

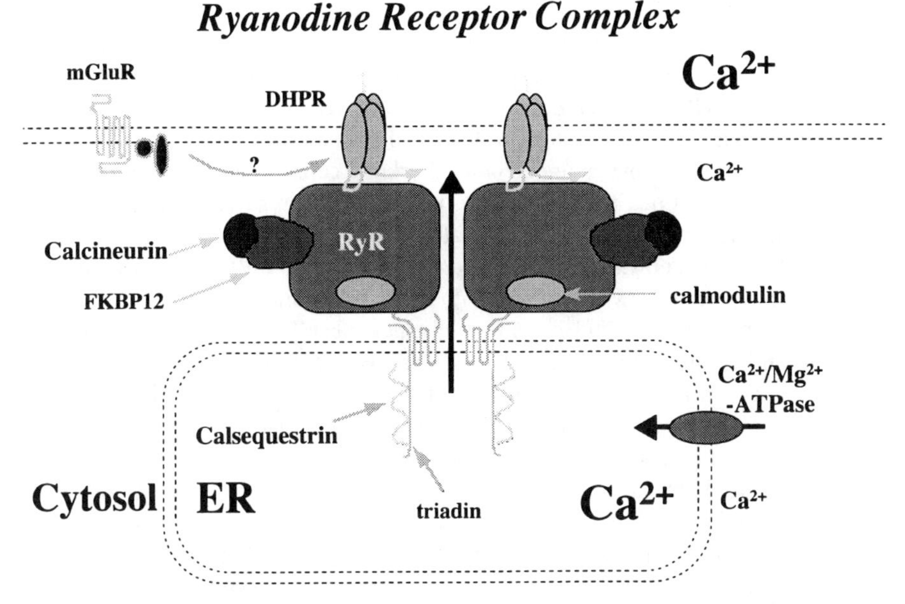

FIG. 4. Model depicting the organization of the ryanodine receptor type 1 (RyR) at the triad junction in mammalian skeletal muscle. For clarity, only two of the four subunits of the functional channel complex are shown. A similar organization is thought to exist within cerebellar Purkinje neurons in which metabotropic glutamate receptors (mGluR) are suspected of enhancing coupling between the L-type voltage-dependent Ca^{2+} channel (DHPR) and RyR.

Along with accessory proteins, the homotetrameric RyR protein forms the SR/ER Ca^{2+} release–channel complex (Fig. 4). In-situ hybridization studies have revealed that three genes encode the three RyR isoforms. The genes coding for RyR_1, RyR_2 and RyR_3 are located on human chromosome 19q13.1, 1q42.1–q43, and 15q14–q15 (87). Based on cDNA sequence, each RyR subunit is composed of 5032–5037, 4968–4976, and 4872 amino residues, with calculated molecular weights of 564 to 565, 565, and 552 kDa for RyR_1, RyR_2 and RyR_3, respectively. In general, a sequence homology of 66% to 70% is observed between two conspecific isoforms. Nonetheless, RyRs also are highly conserved in the same tissue among different species (>95% sequence homology of RyRs in mammalian skeletal muscle among human, pig, and rabbit) (87).

In rabbit skeletal muscle, the ryanodine binding site has been suggested to be located within the 76-kDa tryptic digested fragment from the C terminus of RyR_1. The hydrophobic segment in residues 6985–4362 forms the M1–M4 transmembrane domains, which enable the RyR_1 subunits to span the ER/SR membrane four times. Based on the criteria taken by MacLennan for structure elucidation of the SR Ca^{2+}-ATPase, however, Zorzato and coworkers (69) have proposed the presence of an additional eight transmembrane domains. The M1–M4 domains of RyR_1 have high sequence homology to the analogous domain of IP3 receptors, suggesting a possible role in forming the Ca^{2+} channel pore (95, 96).

RyR_1 and RyR_2 are found predominately in fast-twitch skeletal muscle and cardiac muscle, respectively. RyR_3, first identified in neurons, is found widely expressed in various tissues including central nervous system, smooth muscle and lung epithelial cells. Recently, RyR_3 has been found to be the major form expressed in parotid acini (98). In the central nervous system, a distinct distribution of the three isoforms is observed. RyR_1 predominates in cerebellar Purkinje cells, whereas RyR_2 is widely expressed throughout the brain, and RyR_3 is localized in restricted regions such as the hippocampus, basal ganglia, and thamalus (55, 99, 100).

Although little is known about the coexistence of multiple RyR isofoms within a cell, IP3Rs have been shown to coexist with RyRs in certain cell types (67). Cardiac myocytes coexpress RyR_2 and IP3R with distinct subcellular localization. RyR_2 is found within junctional and extrajunctional regions of SR, whereas IP3Rs are clustered on SR in close proximity to intercalated discs (101). The differential localization of these channels suggests distinct roles for these intracellular Ca^{2+} release channels in cardiac tissue. In this respect, Ca^{2+} is thought to be the second messenger responsible for CICR, whereas IP3 is involved in cell–cell signaling through gap junctions (62). Another typical example of coexpression of IP3R and RyR_1 is the cerebellar Purkinje cell. Both receptors are found throughout the Purkinje neuron, with the exception that only IP3Rs are found in the dendritic spines. Normally, low levels of Ca^{2+} in the Purkinje cytosol represses IP3R in the dendritic spines despite input from the parallel fiber. Stimulation from the climbing fiber activates RyR_1 and leads to Ca^{2+} mobilization from ER. The cytosolic Ca^{2+} elevation spreads from dendrite to dendritic spines, where it sensitizes IP3Rs to activation. Therefore, input from the climbing fiber at the dendrite sensitizes IP3R in dendritic spines and facilitates activation of IP3R via input from the parallel fiber (102). This synergistic mobilization of Ca^{2+} from distinct intracellular Ca^{2+} stores is suggested to mediate the induction of LTD in Purkinje cells, a process involved in motor learning (74, 103, 104). The evidence to date suggests no complete overlap between IP3- and ryanodine-sensitive stores but instead points to partial overlap of the two stores, especially in smooth muscle cells and PC12 cells (67).

Accessory Proteins

Direct coupling of α_1 subunit of DHPR and RyR_1 has been suggested to be involved in the initial steps of excitation–contraction coupling in skeletal muscle (68) and excitation response coupling in cerebellar granule cells (82). In addition, channel gating of RyRs is modulated by several accessory proteins (105–110).

Calmodulin interacts directly with RyRs in skeletal muscle, cardiac muscle, and brain (111) with a stoichiometry of two to three calmodulin per RyR subunit (114, 115). Through a mechanism independent of kinase activity (114), calmodulin (micromolar to submicromolar) inactivates RyRs by reducing channel open time (111).

Calsequestrin, the major Ca^{2+} binding protein within the SR lumen, links indirectly to the lumenal segment of RyR_1 (106, 116). The conformational change in RyR_1 also conveys information to the SR lumen through calsequestrin and may be essential in regulating the Ca^{2+}-release process. Recently, functional interactions between RyR and calsequestrin also have been suggested as playing an important role in regulating excitability of RyRs in response to different filling states of SR (105, 106).

Triadin (111, 117), a 95-kDa highly basic glycoprotein (composed of 25.4% basic residues) (118), was suggested to be associated with both RyRs and the α_1 subunit of DHPR. Cloning and amino acid analysis of triadin, however, disfavor a direct linkage between triadin and DHPR. As the N terminus of the cytosolic domain of triadin contains only 47 amino acids, it is too short to link directly to sarcolemmal DHPR (111). In cardiac SR, triadin associates with RyR_2 monomer in a 1:1 stoichiometry. Primary sequence of triadin indicates that the protein processes only one transmembrane domain. The extremely high density of basic amino residues in the lumenal terminus of triadin (an excess of 46 basic residues spread throughout the lumenal domain) (118) may be critical in interacting with acidic moiety of RyRs (119). In fact, direct evidence has been provided for the interactions between RyR_1 and triadin in skeletal muscle (116), as well as RyR_2 and triadin in cardiac muscle (117). Through disulfide linkages, a high–molecular weight complex consisting of RyR_1 and triadin is stabilized during channel activation. It has been shown that redox reaction between hyperreactive sulfhydryls on RyR_1 and triadin are involved in the allosteric modulations of the high–molecular weight complex by physiologic and pharmacologic effectors (116). The linkage between RyRs and triadin also is proposed to provide an anchorage site for binding of calsequestrin inside the SR lumen (106, 116).

Calcineurin (59), a calmodulin-dependent phosphatase, recently has been found to tightly associate with RyR_1 (and IP3R) and may be involved in regulating gating behavior of the channel. FKBP12 (FK506 binding protein of 12 kDa), the major T-cell immunophilin, is tightly associated with RyR_1 (120, 121). In cardiac muscle, FKBP12.6 associates with RyR_2. The immunosuppressive drug FK506 has revealed a family of immunophilins was identified and found to be differentially expressed in all cell types. These proteins, with their cis–trans peptidylprolyl isomerase activity, play a significant role in post-translational protein processing. Recent studies have reported that together with heat-shock protein 90, certain immunophilins function as molecular chaperones and may direct proper folding of proteins (122).

Two novel proteins, one 60 kDa (123) and one 90 kDa (124), also are found to be associated with the RyR complex; one possesses kinase activity and the other is the substrate of this kinase. Finally, another 150 to 160-kDa protein is found to be associated with RyR_1 (125). Phosphorylation of this protein by casein II kinase inhibits RyR_1 channel activity. These accessory proteins associate with RyRs to form the Ca^{2+} release channel complex at the triad junction and modulate channel function through allosteric protein–protein interactions.

Immunophilin: FKBP12

Immunophilins are a family of proteins that bind to immunosuppressant drugs with appreciable affinity. FK506 binding proteins (FKBP) contain at least five members with Mr 12,000 to 52,000. Although FKBP12 was first identified in immune cells

(120, 121), the protein is expressed in virtually all cell types, especially in neurons. FK506, a potent immunosuppressant, is used clinically in preventing graft rejection after liver transplant. Immunosuppression with FK506 is thought to be mediated by direct binding of the drug to FKBP12. Formation of an FK506/FKBP12 complex promotes a trimeric complex with calcineurin at the phosphatase active site of calcineurin. The cytoplasmic component of a nuclear factor of activated T cells (NF-ATc) is a substrate of calcineurin. Dephosphorylation of NF-ATc by calcineurin induces nuclear translocation of NF-ATc. In the nucleus, dimerization of NF-ATc and NF-ATc (the nuclear component of NF-AT) turns on Interleukin2 (IL2) transcription (120). IL2 is a cytokine that mediates mast-cell degranulation during immunological responses. Inhibition of calcineurin with FK506 prevents production of IL2 and elicits an immunosuppressive effect (126).

The high affinity (K_d, low nanomolar) and specificity of FK506 (and some analogs) toward FKBP12 makes the drug a valuable pharmacologic tool for studies on functions of FKBP12 (120). Recently, FKBP12 has been found to tightly associate with IP3R in cerebellum, and RyR_1 in skeletal muscle and cerebellum, and RyR_2 in cardiac muscle (FKBP12.6, a 12.6-kDa isoform of FKBP12), as well as type I transforming growth factor–receptor in yeast (120). Association of FKBP12 to these receptors has been suggested to regulate receptor-mediated signaling in these systems.

In skeletal muscle, FKBP12 associates with RyR_1 in a stoichiometry of 4:1. Binding of FKBP12 appears to stabilize the closed conformation of the RyR_1-channel complex and full conductance transitions. Introduction of FK506 promotes dissociation of FKBP12 from RyR_1 by competing with a common binding site essential for protein–protein interaction of the FKBP12/RyR_1 channel complex. The resulting FKBP12-deficient channel conducts current with multiple subconductance states. In the presence of channel activators like Ca^{2+} and caffeine, activity of the FKBP12-deficient RyR is further enhanced by increasing the mean open time and open probability of the channel. Association of FKBP12 to RyR_1 may promote cooperation among RyR_1 subunits. Dissociation of FKBP12 with FK506 increases maximal binding capacity of [^{3}H]ryanodine with lowered binding affinity, suggesting loss of negative allosteric interaction between high- and low-affinity [^{3}H]ryanodine binding sites (127). In cardiac muscle, association of FKBP12.6 modulates RyR_2 channels in a similar manner as FKBP12 modulation of RyR_1 in skeletal muscle (120). Furthermore, the association of FKBP12 to RyR_1 complex may be involved in promoting cooperation between neighboring channels. The "coupled gating" behavior of multiple channels has been reported in measurements with multiple channels reconstituted in membrane lipid bilayer using recombinant RyR_1 (coexpressed with FKBP12), as well as native SR. Introduction of FK506 dissociates FKBP12 from recombinant RyR_1 complex and eliminates the coupled gating behavior of multiple channels (128). The cooperation between neighboring RyR_1 channels may contribute significantly to the robust release of Ca^{2+} from SR during excitation–contraction coupling.

In the central nervous system, FKBP12 expression is elevated during neuronal regeneration, whereas introduction of FK506 into the culture medium of plasmocytoma12 cells and sensory ganglia promotes differentiation and process extension

(129, 130). FKBP12 is highly expressed and colocalized with calcineurin in neurons (131), suggesting a functional linkage between the two proteins. Calcineurin is involved in regulation of various neuronal functions including NOS activity. NO is an important mediator in the generation of LTP and LTD (132, 133). It also is known that NO mediates glutamate-induced neurotoxicity. Administration of FK506 as a neuroprotective agent has been suggested under ischemic conditions (132).

FKBP12/RyR MECHANISM FOR ORTHO-SUBSTITUTED PCB NEUROTOXICITY

Polychlorinated biphenyls (PCBs) are a family of widely distributed environmental contaminants. In general, PCBs can be classified according to their molecular structure as coplanar or noncoplanar. Because of steric hindrance between ortho-substituted chlorines on opposite phenyl rings of the biphenyl structure, PCB congeners possessing more than one ortho-substituted chlorine favor noncoplanar conformation. In contrast, PCB congeners with less than two ortho-chlorine substituents generally are considered coplanar (134, 135).

Coplanar PCBs share several structural similarities to structure as 2,3,7,8-tetrachlorodibenzo-p-dioxin (TCDD) and hence compete for a common site on the aryl hydrocarbon receptor (AhR). Coplanar 3,3′,4,4′,5-pentachlorobiphenyl (PCB 126) binds to AhR with affinity comparable to TCDD. Through an AhR-mediated pathway, PCB 126 elicits TCDD-like toxicity including wasting syndrome, immune suppression, and teratogenicity. In contrast, noncoplanar PCBs, because they lack coplanarity (necessary to bind AhR) fail to bind AhR with high affinity and thereby induce a different spectrum of toxicity (135).

Epidemiologic studies have revealed that in-utero and lactational exposure to PCBs is related to lower IQ scores in children. Statistically significant differences have been observed in children exposed to PCBs including impaired learning and memory, poor attention, as well as lowered reading comprehension (136–139). These results suggest that PCBs may possess subtle neurotoxic effects in humans, especially if exposure is during neurodevelopmental stages. Behavioral studies using rats and nonhuman primates have demonstrated that neonatal and in-utero exposure to PCB mixtures (Aroclors 1248 and 1016) is correlated with altered activity patterns, impaired learning and memory, and delayed reflex development (140–143). The PCB congeners responsible for the behavioral impairments observed in both animals and human, however, are still unknown, and the mechanisms underlying neurotoxicity are poorly understood.

Perinatal exposure of ortho-substituted PCB congeners to rats has been found to cause impairment in learning a delayed spatial alternation task (141). Exposure of nonhuman primates to lightly chlorinated ortho-substituted PCB congeners, but not coplanar PCB congeners, has been shown to lower dopamine concentrations in specific regions of the brain in vivo and in tissue culture (144). Studies of the structure–activity relationship of catecholamine depletion in PC12 cells have further

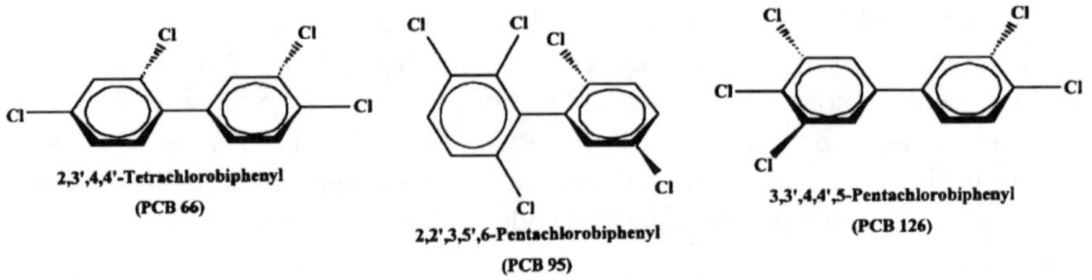

2,3',4,4'-Tetrachlorobiphenyl (PCB 66)

2,2',3,5',6-Pentachlorobiphenyl (PCB 95)

3,3',4,4',5-Pentachlorobiphenyl (PCB 126)

FIG. 5. Structures of noncoplanar PCB 95 and coplanar PCB 66 and PCB 126. PCB 95 has been shown to possess potent activity altering neuroplasticity in the hippocampal slice preparation (151) and performance of rats on an in-vivo hippocamapal task (152). These actions are correlated with a receptor-mediated mechanism involving the FKBP12/RyR complex (149, 150, 154).

demonstrated a stringent structural requirement for neurotoxicity, with the ortho-substituted PCB congeners being the most active. Of the congeners tested, the highly ortho-substituted PCB congeners potently decrease dopamine levels in PC12 cells in culture (144, 145). Thus, results from animal studies suggest that ortho-substituted PCB congeners may be responsible for the behavioral impairments observed in humans.

In order to understand the cause of the neurotoxicity of PCBs, Kodavanti and coworkers performed mechanistic studies using primary cultured cerebellar granular cells (146–148). They showed that noncoplanar PCB, but not coplanar PCB, decreases Ca^{2+} sequestration in synaptosomes and mitochondria preparations. Together with the results from measurements of inorganic phosphate liberation consequent to PCB treatment, they have proposed that a general inhibition of Ca^{2+}-dependent ATPase may mediate the neurotoxicity of PCB.

Recently Wong and others (149, 150) provided evidence for a stringent structure–activity relationship among PCBs possessing two or more chlorine substitutions in the ortho positions for activation of ryanodine-sensitive Ca^{2+} channels of mammalian striated muscle (skeletal and cardiac) and the central nervous system (Fig. 5), revealing an AhR-independent mechanism through which PCBs disrupt Ca^{2+} signaling. The most potent congener at the receptor yet identified, PCB 95 (2,2′,3,5′,6-pentachlorobiphenyl), was found to alter Ca^{2+} transport across neuronal microsomal membrane vesicles by an RyR-mediated pathway, instead of an IP3R-mediated pathway. These actions of PCB 95 at RyRs may underlie its ability to alter neuronal excitability in the rat hippocampal slices in vitro (151) and both locomotor activity and spatial learning in an in-vivo rat model (152). More generally, an RyR-mediated mechanism could account for the ability of noncoplanar PCBs to alter PKC translocation and phosphoinositide metabolism in primary cerebellar granular cell cultures (147, 153).

More detailed mechanistic studies have revealed that the actions of PCB 95 on microsomal Ca^{2+} transport and RyRs are mediated through the major T-cell immunophilin FKBP12, which is tightly associated with RyRs in muscle and brain (154). The toxicologic significance of an immunophilin-mediated mechanism by which

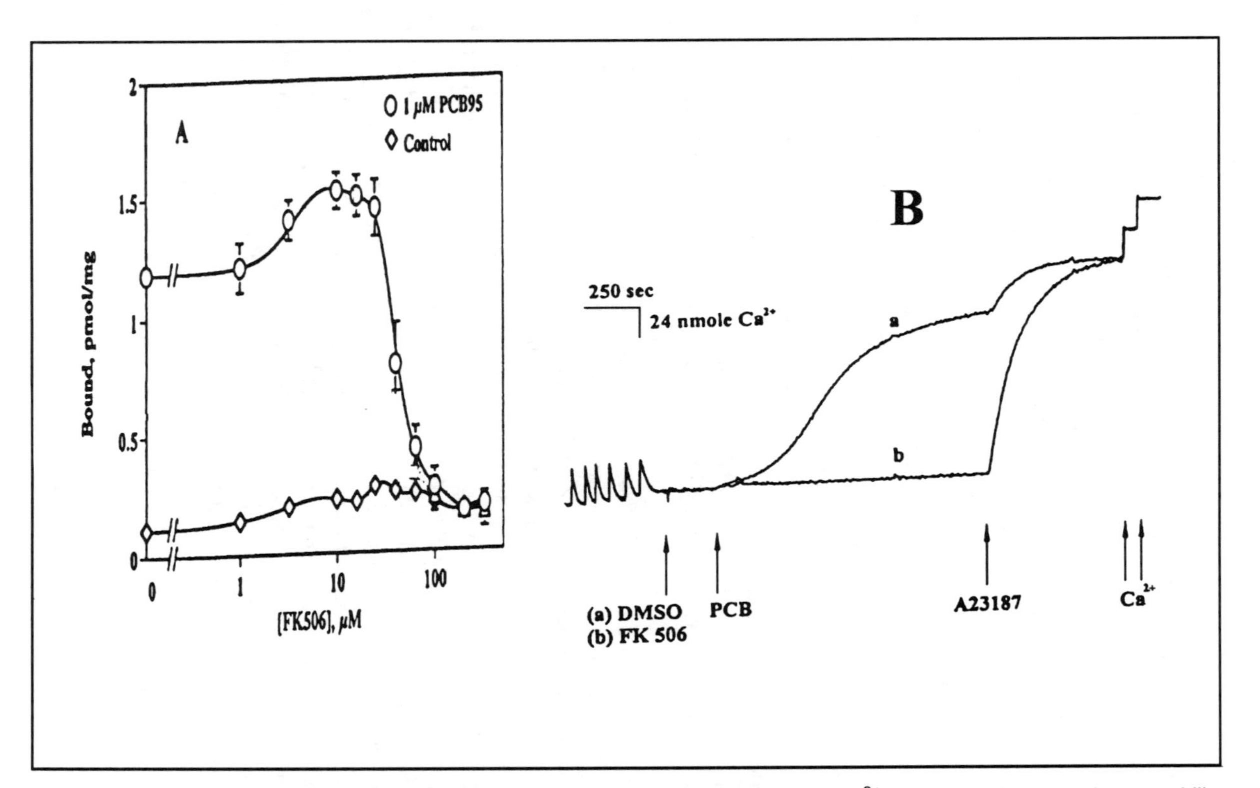

Fig. 6. PCB 95 disrupts the binding of [^{3}H]ryanodine to RyR (*A*) and microsomal Ca^{2+} transport (*B*) by an immunophilin FKBP12-dependent mechanisms. See text for details.

ortho-substituted PCBs alter microsomal Ca^{2+} signaling and Ca^{2+}-dependent cascades could provide the first molecular link between the actions of PCBs on the immune system and the nervous system. The evidence for an immunophilin-mediated mechanism comes from biochemical studies with [^{3}H]ryanodine and Ca^{2+} transport studies using isolated microsomes. The level of high-affinity binding of [^{3}H]ryanodine (1 nmol) to microsomal preparations is low if assayed in the presence of physiologic concentrations of monovalent cations. Incubation with 10M of the noncoplanar congener PCB 95 (2,2′,3,5′,6-pentachlorobiphenyl) enhanced the specific occupancy of [^{3}H]ryanodine to RyR dose-dependently (Fig. 6*A*). Although FK506 did not significantly alter the high affinity binding of [^{3}H]ryanodine to RyR, FK506 did eliminate PCB 95–enhanced [^{3}H]ryanodine occupancy with an IC_{50} of 40 μmol. These results demonstrated that at concentrations known to dissociate FKBP12 from RyR, FK506 eliminated PCB 95–enhanced binding of [^{3}H]ryanodine to RyR, suggesting an immunophilin-dependent mechanism. Ca^{2+} transport measurements revealed that addition of 1 μmol PCB 95 induced a net Ca^{2+} efflux from actively loaded SR vesicles (Fig. 6*B*, trace *a*). FK506 (50 μmol) introduced approximately 3 minutes prior to addition of 1 μmol PCB 95 completely eliminated the response to PCB 95 (Fig. 6*B*, trace *b*). FK506 selectively eliminated PCB 95–induced Ca^{2+} release from SR, because RyR_1s maintained responsiveness to caffeine and Ca^{2+}-induced Ca^{2+} release (154).

Marks (120) has shown that the high affinity interaction between the RyR oligomer and FKBP12 is essential for stabilizing the native full-conductance gating behavior of the SR Ca^{2+} release channel, because RyR expressed heterologously in the absence of FKBP12 exhibits several channel subconductances if reconstituted in bilayer lipid membranes. Further support of the functional importance of the immunophilin in stabilizing the RyR channel complex comes from pharmacologic studies with immunosuppressant FK506 and its analogs. Studies from several laboratories (120) have revealed that FK506 is sufficient to dissociate FKBP12 from RyR, although it is not clear whether complete dissociation of the immunophilin is achieved. In the present study, FK506 completely eliminated PCB 95–induced Ca^{2+} release and PCB 95–enhanced binding of [^{3}H]ryanodine to RyR in the same concentration range required to dissociate FKBP12 from RyR, indicating a strong correlation between the activity of PCB 95 towards RyR and the integrity of the FKBP12/RyR_1 complex.

By virtue of their unique and specific activity, certain ortho-substituted PCBs are potent and invaluable new probes to understand how FKBP12 regulates microsomal Ca^{2+} buffering under physiological and pathophysiological conditions. Further, the newly identify mechanism may underlie the seemingly diverse toxicity that has been attributed recently to noncoplanar PCBs.

ACKNOWLEDGMENTS

Supported by grants from the National Institutes of Health ES05002, ES05707, and ES07059.

REFERENCES

1. Rasmussen H, Barrett PQ. Calcium messenger system: an integrated view. *Physiol Rev* 1984;64:938–984.
2. Brown EM, Gamba G, Riccardi D, Lombardi M, Butters R, Kifor O, Sun A, Hediger MA, Lytton J, Hebert SC. Cloning and characterization of an extracellular Ca^{2+}-sensing receptor from bovine parathyroid. *Nature* 1993;366:575–580.
3. Ruat M, Molliver ME, Snowman AM, Snyder SH. Calcium sensing receptor: molecular cloning in rat and localization to nerve terminals. *Proc Natl Acad Sci USA* 1995;92:3161–3165.
4. Stewart AF, Broadus AE. Mineral metabolism. In: Endocrinology and Metabolism. Felig P, Baxter JD, Broadius AE, Frohman LA, eds. New York: McGraw-Hill, 1987;1317–1453.
5. Nakanishi S. Molecular diversity of glutamate receptors and implications for brain function. *Science* 1992;258:597–603.
6. Brown EM, Vassilev PM, Hebert SC. Calcium ions as extracellular messengers. *Cell* 1995; 83:679–682.
7. Brown EM, Pollak M, Seidman CE, Seidman G, Chou YH, Riccardi D, Hebert SC. Calcium-ion-sensing cell-surface receptors. *N Engl J Med* 1995;333:234–240.
8. Heath HD, Jackson CE, Otterud B, Leppert MF. Genetic linkage analysis in familial benign (hypocalciuric) hypercalcemia: evidence for locus heterogeneity. *Am J Hum Genet* 1993;53:193–200.
9. Lundgren S, Hjalm G, Hellman P, Ek B, Juhlin C, Rastad J, Klareskog L, Akerstrom G, Rask L. A protein involved in calcium sensing of the human parathyroid and placental cytotrophoblast cells belongs to the LDL-receptor protein superfamily. *Exp Cell Res* 1994;212:344–345.
10. Saito A, Pietromonaco S, Loo AK, Farquhar MG. Complete cloning and sequencing of rat gp330/"megalin," a distinctive member of the low density lipoprotein receptor gene family. *Proc Natl Acad Sci USA* 1994;91:9725–9729.
11. Menon GK, Grayson S, Elias PM. Ionic calcium reservoirs in mammalian epidermis: ultrastructural localization by ion-capture cytochemistry. *J Invest Dermatol* 1985;84:508–512.
12. Heinemann U, Lux HD, Gutnick MJ. Extracellular free calcium and potassium during paroxsmal activity in the cerebral cortex of the cat. *Exp Brain Res* 1977;27:237–243.
13. Stabel J, Arens J, Lambert JD, Heinemann U. Effects of lowering $[Na^+]_o$ and $[K^+]_o$ and of ouabain on quisqualate-induced ionic changes in area CA1 of rat hippocampal slices. *Neurosci Lett* 1990;110:60–65.
14. Chebabo SR, Hester MA, Jing J, Aitken PG, Somjen GG. Interstitial space, electrical resistance and ion concentrations during hypotonia of rat hippocampal slices. *J Physiol (Lond)* 1995;487:685–697.
15. Smith JB, Dwyer SD, Smith L. Decreasing extracellular Na^+ concentration triggers inositol polyphosphate production and Ca^{2+} mobilization. *J Biol Chem* 1989;264:831–837.
16. Williams RJ. Calcium ions: their ligands and their functions. *Biochem Soc Symp* 1974;39:133–138.
17. Fiskum G. Intracellular levels and distribution of Ca^{2+} in digitonin-permeabilized cells. *Cell Calcium* 1985;6:25–37.
18. Gunter TE, Zuscik MJ, Puzas JE, Gunter KK, Rosier RN. Cytosolic free calcium concentrations in avian growth plate chondrocytes. *Cell Calcium* 1990;11:445–457.
19. Blaustein MP, Di Pola R, Reeves JP. Sodium-calcium exchange: Blaustein MP, Di Pola R, Reeves JP, eds. *Proceedings of the Second International Conference.* New York: New York Academy of Science, 1991.
20. Sheu SS, Blaustein MP. Sodium/calcium exchange and control of cell calcium and contractility in cardiac and vascular smooth muscle. In: Fozzard HA, Haber E, Jennings RB, Katz AM, Morgan HE, eds. *The Heart and Cardiovascular System, Scientific Foundations.* New York: Raven 1992;903–943.
21. Carafoli E. Intracellular calcium homeostasis. *Annu Rev Biochem* 1987;56:395–433.
22. Bootman MD, Berridge MJ. The elemental principles of calcium signaling. *Cell* 1995;83:675–678.
23. Puceat M, Vassort G. Signalling by protein kinase C isoforms in the heart. *Mol Cell Biochem* 1996;157:65–72.
24. Alberts B, Bray D, Lewis J, Raff M, Roberts K, Watson JD. Cell signaling. In: Adams R, Walker A, eds. *Molecular Biology of the Cell.* New York: Garland Publishing 1989;704.
25. Simpson PB, Challiss RA, Nahorski SR. Neuronal Ca^{2+} stores: activation and function. *Trends Neurosci* 1995;18:299–306.
26. Kennedy MB. Regulation of neuronal function by calcium. *Trends Neurosci* 1989;12:417–420.

27. Bang H, Muller W, Hans M, Brune K, Swandulla D. Activation of Ca^{2+} signaling in neutrophils by the mast cell-released immunophilin FKBP12. *Proc Natl Acad Sci USA* 1995;92:3435–3438.
28. Fruman DA, Klee CB, Bierer BE, Burakoff SJ. Calcineurin phosphatase activity in T lymphocytes is inhibited by FK 506 and cyclosporin A. *Proc Natl Acad Sci USA* 1992;89:3686–3690.
29. Spencer DM, Wandless TJ, Schreiber SL, Crabtree GR. Controlling signal transduction with synthetic ligands. *Science* 1993;262:1019–1024.
30. Franzini-Armstrong C, Jorgensen AO. Structure and development of E-C coupling units in skeletal muscle. *Annu Rev Physiol* 1994;56:509–534.
31. Somlyo AP, Somlyo AV, Shuman H. Electron probe analysis of vascular smooth muscle: composition of mitochondria, nuclei, and cytoplasm. *J Cell Biol* 1979;81:316–335.
32. Somlyo AP. Excitation-contraction coupling and the ultrastructure of smooth muscle. *Circ Res* 1985;57:497–507.
33. Somlyo AP, Urbanics R, Vadasz G, Kovach AG, Somlyo AV. Mitochondrial calcium and cellular electrolytes in brain cortex frozen in situ: electron probe analysis. *Biochem Biophys Res Commun* 1985;132:1071–1078.
34. Vercesi A, Reynafarje B, Lehninger AL. Stoichiometry of H^+ ejection and Ca^{2+} uptake coupled to electron transport in rat heart mitochondria. *J Biol Chem* 1978;253:6379–6388.
35. Gunter TE, Gunter K, Sheu SS, Gavin CE. Mitochondrial calcium transport: physiological and pathological relevance. *Am J Physiol* 1994;267:C313–339.
36. Berridge MJ. Spatiotemporal aspects of calcium signalling. *Jpn J Pharmacol* 1992;58 (suppl 2):142P–149P.
37. Berridg MJ. Cytoplasmic calcium oscillations: a two pool model. *Cell Calcium* 1991;12:63–72.
38. Dupont G, Berridge MJ, Goldbeter A. Signal-induced Ca^{2+} oscillations: properties of a model based on Ca^{2+}-induced Ca^{2+} release. *Cell Calcium* 1991;12:73–85.
39. Bernardi P, Petronilli V. The permeability transition pore as a mitochondrial calcium release channel: a critical appraisal. *J Bioenerg Biomembrs* 1996;28:131–138.
40. Lestienne P, Bataille N, Lucas-Heron B. Role of the mitochondrial DNA and calmitine in myopathies. *Biochim Biophys Acta* 1995;1271:159–163.
41. Gunter TE, Pfeiffer DR. Mechanisms by which mitochondria transport calcium. *Am J Physiol* 1990;258:C755–786.
42. Reed KC, Bygrave FL. The inhibition of mitochondrial calcium transport by lanthanides and ruthenium red. *Biochem J* 1974;140:143–155.
43. Puskin JS, Gunter TE, Gunter KK, Russell PR. Evidence for more than one Ca^{2+} transport mechanism in mitochondria. *Biochemistry* 1976;15:3834–3842.
44. Wingrove DE, Gunter TE. Kinetics of mitochondrial calcium transport: I. Characteristics of the sodium-independent calcium efflux mechanism of liver mitochondria. *J Biol Chem* 1986;261:15159–15165.
45. Wingrove DE, Gunter TE. Kinetics of mitochondrial calcium transport: II. A kinetic description of the sodium-dependent calcium efflux mechanism of liver mitochondria and inhibition by ruthenium red and by tetraphenylphosphonium. *J Biol Chem* 1986;261:15166–15171.
46. Lukacs G, Fonyo A. Ba^{2+} ions inhibit the release of Ca^{2+} ions from rat liver mitochondria. *Biochim Biophys Acta* 1985;809:160–166.
47. Li W, Shariat-Madar Z, Powers M, Sun X, Lane RD, Garlid KD. Reconstitution, identification, purification, and immunological characterization of the 110-kDa Na^+/Ca^{2+} antiporter from beef heart mitochondria. *J Biol Chem* 1992;267:17983–17989.
48. Farber JL. The role of calcium in cell death. *Life Sci* 1981;29:1289–1295.
49. Petronilli V, Cola C, Massari S, Colonna R, Bernardi P. Physiological effectors modify voltage sensing by the cyclosporin A-sensitive permeability transition pore of mitochondria. *J Biol Chem* 1993;268:21939–21945.
50. Broekemeier KM, Dempsey ME, Pfeiffer DR. Cyclosporin A is a potent inhibitor of the inner membrane permeability transition in liver mitochondria. *J Biol Chem* 1989;264:7826–7830.
51. McGuinness O, Yafei N, Costi A, Crompton M. The presence of two classes of high-affinity cyclosporin A binding sites in mitochondria: evidence that the minor component is involved in the opening of an inner membrane Ca^{2+}-dependent pore. *Eur J Biochem* 1990;194:671–679.
52. Bernardi P, Broekemeier KM, Pfeiffer DR. Recent progress on regulation of the mitochondrial permeability transition pore: a cyclosporin-sensitive pore in the inner mitochondrial membrane. *J Bioenerg Biomembr* 1994;26:509–517.
53. Vercesi AE. Dissociation of $NAD(P)^+$-stimulated mitochondrial Ca^{2+} efflux from swelling and membrane damage. *Arch Biochem Biophys* 1984;232:86–91.

54. Igbavboa U, Zwizinski CW, Pfeiffer DR. Release of mitochondrial matrix proteins through a Ca^{2+}-requiring, cyclosporin-sensitive pathway. *Biochem Biophys Res Commun* 1989;161:619–625.
55. Giannini G, Conti A, Mammarella S, Scrobogna M, Sorrentino V. The ryanodine receptor/calcium channel genes are widely and differentially expressed in murine brain and peripheral tissues. *J Cell Biol* 1995;128:893–904.
56. Morgan JM, De Smedt H, Gillespie JI. Identification of three isoforms of the InsP3 receptor in human myometrial smooth muscle. *Pflugers Arch* 1996;431:697–705.
57. Sharp AH, McPherson PS, Dawson TM, Aoki C, Campbell KP, Snyder SH. Differential immunohistochemical localization of inositol 1,4,5-trisphosphate- and ryanodine-sensitive Ca^{2+} release channels in rat brain. *J Neurosci* 1993;13:3051–3063.
58. Ehrlich BE, Kaftan E, Bezprozvannaya S, Bezprozvanny I. The pharmacology of intracellular Ca^{2+}-release channels. *Trends Pharmacol Sci* 1994;15:145–149.
59. Cameron AM, Steiner P, Roskams AJ, Ali SM, Ronnett GV, Snyder SH. Calcineurin associated with the inositol 1,4,5-trisphosphate receptor-FKBP12 complex modulates Ca^{2+} flux. *Cell* 1995;83:463–472.
60. Cheek TR, Berridge MJ, Moreton RB, Stauderman KA, Murawsky MM, Bootman MD. Quantal Ca^{2+}mobilization by ryanodine receptors is due to all-or-none release from functionally discrete intracellular stores. *Biochem J* 1994;301:879–883.
61. Shacklock PS, Wier WG, Balke CW. Local Ca^{2+} transients (Ca^{2+} sparks) originate at transverse tubules in rat heart cells. *J Physiol (Lond)* 1995;487:601–608.
62. Allbritton NL, Meyer T. Localized calcium spikes and propagating calcium waves. *Cell Calcium* 1993;14:691–697.
63. Jackson WA, Colyer J. Translation of Ser16 and Thr17 phosphorylation of phospholamban into Ca^{2+}-pump stimulation. *Biochem J* 1996;316:201–207.
64. Heilmann C, Spamer C, Leberer E, Gerok W, Michalak M. Human liver calreticulin: characterization and Zn^{2+}-dependent interaction with phenyl-sepharose. *Biochem Biophys Res Commun* 1993;193:611–616.
65. Keynes RD, Aidley DJ. Nerve and muscle. Keynes RD, Aidley DJ, eds. New York: Cambridge University Press, 1991;1–178.
66. Despopoulos A, Silbernagl S. Nerve and muscle. In: Despopoulos A, Silbernagl S, eds. *Color Atlas of Physiology*. New York: Thieme Medical, 1991;22–59.
67. Berridge MJ, Cheek TR, Bennett DL, Bootman MD. Ryanodine receptors and intracellular calcium signaling. In: Sorrentino V, ed. *Ryanodine Receptors*. Boca Raton: Florida. CRC Press, 1996;120–153.
68. Lu X, Xu L, Meissner G. Activation of the skeletal muscle calcium release channel by a cytoplasmic loop of the dihydropyridine receptor. *J Biol Chem* 1994;269:6511–6516.
69. Zorzato F, Fujii J, Otsu K, Phillips M, Green NM, Lai FA, Meissner G, MacLennan DH. Molecular cloning of cDNA encoding human and rabbit forms of the Ca^{2+} release channel (ryanodine receptor) of skeletal muscle sarcoplasmic reticulum. *J Biol Chem* 1990;265:2244–2256.
70. Chae HD, Kim KT. Cytosolic calcium is essential in the basal expression of tyrosine hydroxylase gene. *Biochem Biophys Res Commun* 1995;206:659–666.
71. Houchi H, Hamano S, Masuda Y, Ishimura Y, Azuma M, Ohuchi T, Oka M. Stimulatory effect of pituitary adenylate cyclase-activating polypeptide on catecholamine synthesis in cultured bovine adrenal chromaffin cells: involvements of tyrosine hydroxylase phosphorylation caused by Ca^{2+} influx and cAMP. *Jpn J Pharmacol* 1994;66:323–330.
72. Bunn SJ, Sim AT, Herd LM, Austin LM, Dunkley PR. Tyrosine hydroxylase phosphorylation in bovine adrenal chromaffin cells: the role of intracellular Ca^{2+} in the histamine H1 receptor-stimulated phosphorylation of Ser8, Ser19, Ser31, and Ser40. *J Neurochem* 1995;64:1370–1378.
73. Holliday J, Adams RJ, Sejnowski TJ, Spitzer NC. Calcium-induced release of calcium regulates differentiation of cultured spinal neurons. *Neuron* 1991;7:787–796.
74. Kohda K, Noue T, Mikoshiba K. Ca^{2+} release from Ca^{2+} stores, particularly from ryanodine-sensitive Ca^{2+} stores, is required for the induction of LTD in cultured cerebellar Purkinje cells. *J Neurophysiol* 1995;74:2184–2188.
75. Gomez TM, Snow DM, Letourneau PC. Characterization of spontaneous calcium transients in nerve growth cones and their effect on growth cone migration. *Neuron* 1995;14:1233–1246.
76. Mori Y. Molecular biology of voltage-dependent calcium channels. In: Peracchia C, ed. *Handbook of Membrane Channels*. New York: Academic Press, 1994;163–176.
77. Tsien RW, Lipscombe D, Madison DV, Bley KR, Fox AP. Multiple types of neuronal calcium channels and their selective modulation. *Trends Neurosci* 1988;11:431–438.

78. Mori H, Mishina M. Structure and function of the NMDA receptor channel. *Neuropharmacology* 1995;34:1219–1237.
79. Barnar EA. Receptor classes and the transmitter-gated ion channels. *Trends Biochem Sci* 1992;17:368–374.
80. Lieberman DN, Mody I. Regulation of NMDA channel function by endogenous Ca^{2+}-dependent phosphatase. *Nature* 1994;369:235–239.
81. Mody I, MacDonald JF. NMDA receptor-dependent excitotoxicity: the role of intracellular Ca^{2+} release. *Trends Pharmacol Sci* 1995;16:356–359.
82. Chavis P, Fagni L, Lansman JB, Bockaert J. Functional coupling between ryanodine receptors and L-type calcium channels in neurons. *Nature* 1996;382:719–722.
83. Denny JB, Polan-Curtain J, Ghuman A, Wayner MJ, Armstrong DL. Calpain inhibitors block long-term potentiation. *Brain Res* 1990;534:317–320.
84. Carafoli E, Krebs J, Chiesi M. Calmodulin. In: Cohen P, Klee CB, eds. *Molecular Aspects of Cellular Regulation.* Elsevier, 1988;297–312.
85. Sim LJ, Selley DE, Tsai KP, Morris M. Calcium and cAMP mediated stimulation of Fos in cultured hypothalamic tyrosine hydroxylase-immunoreactive neurons. *Brain Res* 1994;653:155–160.
86. Rogers EF, Koniuszy FP, Shavel JJ, Folkers K. Plant insecticides: I. Ryanodine, a new alkaloid from Ryania speciosa Vahl. *J Am Chem Soc* 1984;70:3086–3088.
87. Ogawa Y. Role of ryanodine receptors. *Crit Rev Biochem Mol Biol* 1994;29:229–274.
88. Waterhouse AL, Holden I, Casida JE. Didehydroryanodine: a new principal toxic constituent of the botanical insecticide ryania. *J Chem Soc Chem Commun* 1984;19:1265–1266.
89. Waterhouse AL, Pessah IN, Francini AO, Casida JE. Structural aspects of ryanodine action and selectivity. *J Med Chem* 1987;30:710–716.
90. Pessah IN, Waterhouse AL, Casida JE. The calcium-ryanodine receptor complex of skeletal and cardiac muscle. *Biochem Biophys Res Commun* 1985;128:449–456.
91. Pessah IN, Zimanyi I. Characterization of multiple [^{3}H]ryanodine binding sites on the Ca^{2+} release channel of sarcoplasmic reticulum from skeletal and cardiac muscle: evidence for a sequential mechanism in ryanodine action. *Mol Pharmacol* 1991;39:679–689.
92. Zimanyi I, Buck E, Abramson JJ, Mack MM, Pessah IN. Ryanodine induces persistent inactivation of the Ca^{2+} release channel from skeletal muscle sarcoplasmic reticulum. *Mol Pharmacol* 1992;42:1049–1057.
93. Buck E, Zimanyi I, Abramson JJ, Pessah IN. Ryanodine stabilizes multiple conformational states of the skeletal muscle calcium release channel. *J Biol Chem* 1992;267:23560–23567.
94. Pessah IN, Stambuk RA, Casida JE. Ca^{2+}-activated ryanodine binding: mechanisms of sensitivity and intensity modulation by Mg^{2+}, caffeine, and adenine nucleotides. *Mol Pharmacol* 1987;31:232–238.
95. Furuichi T, Yoshikawa S, Miyawaki A, Wada K, Maeda N, Mikoshiba K. Primary structure and functional expression of the inositol 1,4,5-trisphosphate-binding protein P400. *Nature* 1989;342:32–38.
96. Ferris CD, Snyder SH. Inositol 1,4,5-trisphosphate-activated calcium channels. *Annu Rev Physiol* 1992;54:469–488.
97. Wagenknecht T, Grassucci R, Frank J, Saito A, Inui M, Fleischer S. Three-dimensional architecture of the calcium channel/foot structure of sarcoplasmic reticulum. *Nature* 1989;338:167–170.
98. DiJulio DH, Watson EL, Pessah IN, Jacobson KL, Ott SM, Buck ED, Singh JC. Ryanodine receptor type III (Ry3R) identification in mouse parotid acini: I. Properties and modulation of [^{3}H]ryanodine-binding site. *J Biol Chem* 1996;272:15687–15696.
99. Ledbetter MW, Preiner JK, Louis CF, Mickelson JR. Tissue distribution of ryanodine receptor isoforms and alleles determined by reverse transcription polymerase chain reaction. *J Biol Chem* 1994;269:31544–31551.
100. Furuichi T, Furutama D, Hakamata Y, Nakai J, Takeshima H, Mikoshiba K. Multiple types of ryanodine receptor/Ca^{2+} release channels are differentially expressed in rabbit brain. *J Neurosci* 1994;14:4794–4805.
101. Kijima Y, Saito A, Jetton TL, Magnuson MA, Fleischer S. Different intracellular localization of inositol 1,4,5-trisphosphate and ryanodine receptors in cardiomyocytes. *J Biol Chem* 1993;268:3499–3506.
102. Berridge MJ. Cell signalling: a tale of two messengers. *Nature* 1993;365:388–389.
103. Ito M. Long-term depression. *Annu Rev Neurosci* 1989;12:85–102.
104. Kano M. Calcium-induced long-lasting potentiation of GABAergic currents in cerebellar Purkinje cells. *Jpn J Physiol* 1994;44(suppl 2):S131–136.
105. Hidalgo C, Donoso P. Luminal calcium regulation of calcium release from sarcoplasmic reticulum. *Biosci Rep* 1995;15:387–397.

106. Guo W, Campbell KP. Association of triadin with the ryanodine receptor and calsequestrin in the lumen of the sarcoplasmic reticulum. *J Biol Chem* 1995;270:9027–9030.
107. Takeshima H, Nishimura S, Matsumoto T, Ishida H, Kangawa K, Minamino N, Matsuo H, Ueda M, Hanaoka M, Hirose T, Numa S. Primary structure and expression from complementary DNA of skeletal muscle ryanodine receptor. *Nature* 1989;339:439–445.
108. Sorrentino V, Volpe P. Ryanodine receptors: how many, where and why? *Trends Pharmacol Sci* 1993; 14:98–103.
109. Herrmann-Frank A, Richter M, Lehmann-Horn F. 4-Chloro-m-cresol: a specific tool to distinguish between malignant hyperthermia-susceptible and normal muscle. *Biochem Pharmacol* 1996;52:149–155.
110. Mack MM, Zimanyi I, Pessah IN. Discrimination of multiple binding sites for antagonists of the calcium release channel complex of skeletal and cardiac sarcoplasmic reticulum. *J Pharmacol Exp Ther* 1992;262:1028–1037.
111. Coronado R, Morrissette J, Sukhareva M, Vaughan DM. Structure and function of ryanodine receptors. *Am J Physiol* 1994;266:C1485–1504.
112. Hohenegger M, Suko J. Phosphorylation of the purified cardiac ryanodine receptor by exogenous and endogenous protein kinases. *Biochem J* 1993;296:303–308.
113. Witcher DR, Kovacs RJ, Schulman H, Cefali DC, Jones LR. Unique phosphorylation site on the cardiac ryanodine receptor regulates calcium channel activity. *J Biol Chem* 1991;266:11144–11152.
114. Menegazzi P, Larini F, Treves S, Guerrini R, Quadroni M, Zorzato F. Identification and characterization of three calmodulin binding sites of the skeletal muscle ryanodine receptor. *Biochemistry* 1994;33:9078–9084.
115. Guerrini R, Menegazzi P, Anacardio R, Marastoni M, Tomatis R, Zorzato F, Treves S. Calmodulin binding sites of the skeletal, cardiac, and brain ryanodine receptor Ca^{2+} channels: modulation by the catalytic subunit of cAMP-dependent protein kinase? *Biochemistry* 1995;34:5120–5129.
116. Liu G, Pessah IN. Molecular interaction between ryanodine receptor and glycoprotein triadin involves redox cycling of functionally important hyperreactive sulfhydryls. *J Biol Chem* 1994;269:33028–33034.
117. Guo W, Jorgensen AO, Jones LR, Campbell KP. Biochemical characterization and molecular cloning of cardiac triadin. *J Biol Chem* 1996;271:458–465.
118. Knudson CM, Stang KK, Moomaw CR, Slaughter CA, Campbell KP. Primary structure and topological analysis of a skeletal muscle-specific junctional sarcoplasmic reticulum glycoprotein (triadin). *J Biol Chem* 1993;268:12646–12654.
119. Liu G, Abramson JJ, Zable AC, Pessah IN. Direct evidence for the existence and functional role of hyperreactive sulfhydryls on the ryanodine receptor-triadin complex selectively labeled by the coumarin maleimide 7-diethylamino-3-(4′-maleimidylphenyl)-4-methylcoumarin. *Mol Pharmacol* 1994;45:189–200.
120. Marks AR. Cellular functions of immunophilins. *Physiol Rev* 1996;76:631–649.
121. Snyder SH, Sabatini DM. Immunophilins and the nervous system. *Nat Med* 1995;1:32–37.
122. Bose S, Weikl T, Bugl H, Buchner J. Chaperone function of Hsp90-associated proteins. *Science* 1996; 274:1715–1717.
123. Leddy JJ, Murphy BJ, Qu Y, Doucet JP, Pratt C, Tuana BS. A 60 kDa polypeptide of skeletal-muscle sarcoplasmic reticulum is a calmodulin-dependent protein kinase that associates with and phosphorylates several membrane proteins. *Biochem J* 1993;295:849–856.
124. Guo W, Jorgensen AO, Campbell KP. Characterization and ultrastructural localization of a novel 90-kDa protein unique to skeletal muscle junctional sarcoplasmic reticulum. *J Biol Chem* 1994;269:28359–28365.
125. Orr I, Shoshan-Barmatz V. Modulation of the skeletal muscle ryanodine receptor by endogenous phosphorylation of 160/150-kDa proteins of the sarcoplasmic reticulum. *Biochim Biophys Acta* 1996;1283:80–88.
126. Liu J. FK506 and cyclosporin, molecular probes for studying intracellular signal transduction [published erratum appears in *Immunol Today* 1993;14:399]. *Immunol Today* 1993;14:290–295.
127. Mack MM, Molinski TF, Buck ED, Pessah IN. Novel modulators of skeletal muscle FKBP12/calcium channel complex from Ianthella basta: role of FKBP12 in channel gating. *J Biol Chem* 1994;269:23236–23249.
128. Ondrias K, Marks AR. A role for FKBP in coupled gating of the recombanant ryanodine receptor/calcium release channel [abstract]. *Circulation* 1995;92:I–235.

129. Lyons WE, George EB, Dawson TM, Steiner JP, Snyder SH. Immunosuppressant FK506 promotes neurite outgrowth in cultures of PC12 cells and sensory ganglia. *Proc Natl Acad Sci USA* 1994;91:3191–3195.
130. Lyons WE, Steiner JP, Snyder SH, Dawson TM. Neuronal regeneration enhances the expression of the immunophilin FKBP-12. *J Neurosci* 1995;15:2985–2994.
131. Steiner JP, Dawson TM, Fotuhi M, Glatt CE, Snowman AM, Cohen N, Snyder SH. High brain densities of the immunophilin FKBP colocalized with calcineurin. *Nature* 1992;358:584–587.
132. Dawson TM, Steiner JP, Dawson VL, Dinerman JL, Uhl G, Snyder SH. Immunosuppressant FK506 enhances phosphorylation of nitric oxide synthase and protects against glutamate neurotoxicity. *Proc Natl Acad Sci USA* 1993;90:9808–9812.
133. Izumi Y, Clifford DB, Zorumski CF. Inhibition of long-term potentiation by NMDA-mediated nitric oxide release. *Science* 1992;257:1273–1276.
134. Safe S. Toxicology, structure-function relationship, and human and environmental health impacts of polychlorinated biphenyls: progress and problems. *Environ Health Perspect* 1993;100:259–268.
135. Safe S. Polychlorinated biphenyls (PCBs), dibenzo-p-dioxins (PCDDs), dibenzofurans (PCDFs), and related compounds: environmental and mechanistic considerations which support the development of toxic equivalency factors (TEFs). *Crit Rev Toxicol* 1990;21:51–88.
136. Jacobson JL, Jacobson SW. Intellectual impairment in children exposed to polychlorinated biphenyls in utero. *N Engl J Med* 1996;335:783–789.
137. Koopman-Esseboom C, Weisglas-Kuperus N, de Ridder MA, Van der Paauw CG, Tuinstra LG, Sauer PJ. Effects of polychlorinated biphenyl/dioxin exposure and feeding type on infants' mental and psychomotor development. *Pediatrics* 1996;97:700–706.
138. Chen YC, Guo YL, Hsu CC, Rogan WJ. Cognitive development of Yu-Cheng ("oil disease") children prenatally exposed to heat-degraded PCBs. *JAMA* 1992;268:3213–3218.
139. Chen YC, Guo YL, Hsu CC. Cognitive development of children prenatally exposed to polychlorinated biphenyls (Yu-Cheng children) and their siblings. *J Formos Med Assoc* 1992;91:704–707.
140. Schantz SL, Levin ED, Bowman RE, Heironimus MP, Laughlin NK. Effects of perinatal PCB exposure on discrimination-reversal learning in monkeys. *Neurotoxicol Teratol* 1989;11:243–250.
141. Schantz SL, Moshtaghian J, Ness DK. Spatial learning deficits in adult rats exposed to *ortho*-substituted PCB congeners during gestation and lactation. *Fundam Appl Toxicol* 1995;26:117–126.
142. Schantz SL, Seo BW, Moshtaghian J, Peterson RE, Moore RW. Effects of gestational and lactational exposure to TCDD or coplanar PCBs on spatial learning. *Neurotoxicol Teratol* 18:305–313.
143. Schantz SL. Neurotoxic food contaminants: Polychlorinated biphenyls (PCBs) and related compounds. In: Niesink R, Jasper R, eds. *Behavioral Toxicology and Addiction: Food Drugs and Environment.* Open University of Netherlands, in press.
144. Seegal RF, Bush B, Shain W. Lightly chlorinated *ortho*-substituted PCB congeners decrease dopamine in nonhuman primate brain and in tissue culture. *Toxicol Appl Pharmacol* 1990;106:136–144.
145. Shain W, Bush B, Seegal R. Neurotoxicity of polychlorinated biphenyls: structure-activity relationship of individual congeners. *Toxicol Appl Pharmacol* 1991;111:33–42.
146. Kodavanti PR, Shin DS, Tilson HA, Harry GJ. Comparative effects of two polychlorinated biphenyl congeners on calcium homeostasis in rat cerebellar granule cells. *Toxicol Appl Pharmacol* 1993;123:97–106.
147. Kodavanti PR, Ward TR, McKinney JD, Tilson HA. Increased [^{3}H]phorbol ester binding in rat cerebellar granule cells by polychlorinated biphenyl mixtures and congeners: structure-activity relationships. *Toxicol Appl Pharmacol* 1995;130:140–148.
148. Kodavanti PR, Ward TR, McKinney JD, Waller CL, Tilsonb HA. Increased [^{3}H]phorbol ester binding in rat cerebellar granule cells and inhibition of $45Ca^{2+}$ sequestration in rat cerebellum by polychlorinated diphenyl ether congeners and analogs: structure-activity relationships. *Toxicol Appl Pharmacol* 1996;138:251–261.
149. Wong PW, Pessah IN. *Ortho*-substituted polychlorinated biphenyls alter calcium regulation by a ryanodine receptor mediated mechanism: structural specificity toward skeletal and cardiac type microsomal calcium release channels. *Mol. Pharmacol* 1996;49:740–751.
150. Wong PW, Brackney WR, Pessah IN. *Ortho*-substituted polychlorinated biphenyls (PCBs) alter microsomal calcium transport by direct interaction with ryanodine receptors of mammalian brain. *J Biol Chem* 272:15145–15153.

151. Wong PW, Joy RM, Albertson TE, Schantz SL, Pessah IN. *Ortho*-substituted 2,2′3,5′,6-pentachlorobiphenyl (PCB 95) alters rat hippocampal ryanodine receptors and neuroplasticity *in vitro*. Neurotoxicology, 18:443–456.
152. Schantz SL, Seo BW, Wong PW, Pessah IN. Long-term effects of developmental exposure to 2,2′3,5′,6-pentachlorobiphenyl (PCB 95) on locomotor activity, spatial learning and memory and brain ryanodine receptors. *Neurotoxicology* 18:457–467.
153. Shafer TJ, Mundy WR, Tilson HA, Kodavanti PR. Disruption of inositol phosphate accumulation in cerebellar granule cells by polychlorinated biphenyls: a consequence of altered Ca^{2+} homeostasis. *Toxicol Appl Pharmacol* 1996;141:448–455.
154. Wong PW, Pessah IN. Non-coplanar PCB 95 alters microsomal Ca^{2+} transport by an immunophilin FKBP12-dependent mechanism. *Mol Pharmacol* 1997;51:693–702.

PART IV

Analysis of Xenobiotic-Inducible Responses

Toxicant–Receptor Interactions
Edited by Michael S. Denison and William G. Helferich

10

Approaches for the Identification of Xenobiotic-Inducible Genes

John P. Vanden Heuvel

Department of Veterinary Science, Pennsylvania State University, University Park, Pennsylvania, USA

Differential gene expression is a key component of many complex phenomena including cellular development, differentiation, maintenance, and injury or death. In fact, the subset of genes that is being expressed determines to a large extent the phenotype of a cell. Also, a loss of control of differential gene expression underlies many disease states, not the least of which is cancer. The identification of genes that are being expressed in one cell type versus another (e.g., control versus treated, tumor versus normal) can help in explaining the function of those genes as well as lend insight into the system being examined. For this reason, the identification of differentially expressed genes has been pursued for diverse stimuli such as responses to biologic programs (developmental and circadian cues), physical agents (e.g., UV irradiation, x rays), and chemical agents (hormones and xenobiotics). In fact, over 300 examples of these pursuits can be found in the literature (1) and have resulted in a much greater understanding of cellular biology and our responses to physical and chemical insult.

This chapter discusses approaches that are used to identify the genes being regulated by key receptor systems. The general concepts of model development and molecular approaches for identifying genes responsive to receptor-mediated xenobiotics and hormones are discussed. Subsequently, specific examples of how these techniques have been used in identifying genes responsive to phorbol esters, estrogen, dioxin, and peroxisome proliferators are discussed. Finally, the difficult process of linking altered gene expression with a penultimate biologic effect is addressed.

CLONING STRATEGIES AND DEVELOPMENT OF A MODEL SYSTEM

Basic Considerations

Before embarking on a project to clone differentially expressed genes, it is important that a basic understanding of the complexity of gene expression be realized. Three factors that contribute to this complexity and difficulty in cloning differentially expressed genes are that genes are not present at the same abundance, that the intensity of response varies greatly from gene to gene, and that there are many indirect ways to alter gene expression. We briefly discuss these parameters as they pertain to cloning genes induced by receptor-mediated xenobiotics, and how these factors impinge on developing an optimal model system.

mRNA Abundance

The mammalian genome of 3×10^9 base pairs (bp) has enough DNA to code for approximately 300,000 genes, assuming a length of 10,000 bp per gene (2). Obviously, not every gene is expressed per cell (also, not every segment of DNA is associated with a gene product). In fact, in a mammalian cell, hybridization experiments have shown that approximately 1% to 2% of the total sequences of nonrepetitive DNA are represented in mRNA (2). Thus, if 70% of the total genome is nonrepetitive, 10,000 to 15,000 genes are expressed at a given time.

The average number of molecules of each mRNA per cell is called its *representation* or *abundance*. Of the 10,000 to 15,000 genes being expressed, the mass of RNA being produced per gene is highly variable. In fact, a few sequences usually provide a large proportion of the total mass of mRNA. Hybridization and kinetic experiments

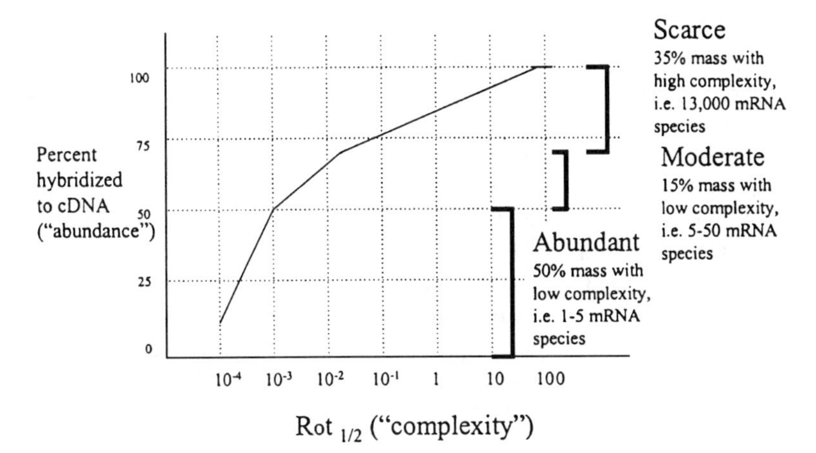

FIG. 1. Kinetics of hybridization of cDNA and its mRNA template. Complementary DNA is prepared from polyadenylated mRNA in the presence of a radioactive nucleotide and is hybridized with total mRNA from the same tissue. Aliquots are removed at different times and digested with single strand–specific nuclease. The percentage hybridized is plotted against the product of the concentration of RNA in moles of nucleotide per liter × time (seconds) called R_0t. Adapted from references 2–4. From graphs similar to this, the existence of three distinct classes of mRNA was proposed, the abundant, moderate, and scarce components.

between excess mRNA and cDNA in solution identify several components of mRNA complexity, as shown in Fig. 1. (For a detailed description of these RNA hybridization experiments see reference [2].) The majority of the mass of RNA (50%) is accounted for by a component with few mRNA species (low Rot1/2). In fact, approximately 65% of the total mRNA may be accounted for in as few as 10 mRNA species. The remaining 35% of the total RNA represents the remaining 10,000 to 15,000 genes being expressed in that tissue. Of course, the genes present in each category may be present in very different amounts and represent a continuum of expression levels. For means of this discussion, we divide the three major components (3) into abundant, moderate, and scarce, representing approximately 100,000 copies, 5000 copies and, less than 10 copies per cell respectively.

There are several reasons for discussing the components of mRNA. First, in doing a differential screen (e.g., subtractive hybridization or differential display) what are actually being compared are two populations of mRNA, and the genes that overlap or form the intersection between groups are examined. Comparing two extremely divergent populations, such as liver and oviduct, as much as 75% of the sequences are the same (2), equating to 10,000 genes that are identical and approximately 3000 genes that are specific to the oviduct . This suggests that there may be a common set of genes, representing required functions, that are expressed in all cell types. These often are referred to as *housekeeping* or *constitutive genes*. Second, there are overlaps between all components of mRNA, regardless of the number of copies per cell. That is, differentially expressed genes may be abundant, moderate, or scarce. In fact, the scarce mRNA may overlap extensively from cell to cell, on the order of 90% for the liver and oviduct comparison (4). It is worthy to note, however, that a small number of differentially expressed genes are required to denote a specialized function to that cell, and the level of expression does not always correlate with importance of the gene product.

As to be discussed subsequently, the key to developing an effective model for the study of differential gene expression may be to keep the differences in the abundant genes to a minimum. This is due to the fact that a small difference in expression of a housekeeping gene, say two fold, will result in a huge difference in the number of copies of that message from cell type to cell type (i.e., an increase of 10,000 copies per cell). Also, it is important to have a screening method that can detect differences in the scarce component. If the two populations to be compared have little difference in the abundant genes and screening technique has been optimized to detect differences in the scarce population, the odds of cloning genes that are truly required for a specialized cellular function have increased dramatically.

Intensity of Response

A basic pharmacologic principle is that drugs and chemicals have different affinities for a receptor, and the drug–receptor complexes have different efficacies for producing a biologic response, or altering gene expression. A corollary of this principle states that not all genes being affected by the same drug–receptor complex will have identical dose–response curves. That is, comparing two responsive genes, the affinity of the drug–receptor complex for the DNA response elements found in the two genes and the efficacy of the drug–receptor–DNA complex at effecting transcription can be quite different. In fact, similar DNA response elements may cause a repression or an induction of gene expression, depending on the context of the surrounding gene. Therefore, if comparing two populations of mRNAs (i.e., control versus treated), there may be differences of orders of magnitude in the levels of induction and repression, regardless of the fact that all the genes are affected by the same drug–receptor complex.

Needless to say, the extent of change is important in the detection of these differences, but not the biological importance of that deviation. For technical reasons, it often is difficult to detect small changes in gene expression (less than two-fold). A two-fold change in a gene product, however, may have dramatic effects on the affected cell, especially if it encodes a protein with a very specialized or nonredundant function. Also, the detection of a difference between two cell populations is easier if the majority of the differences are in scarce mRNAs. Once again, this is due to technical aspects of analyzing gene expression whereby the change from 500 to 1000 copies per cell is a dramatic effect compared with a change from 1×10^5 to 2×10^5, an effect that may be virtually unnoticed.

Specificity of Response

The last factor to be discussed regarding the complexity of mRNA species is that regulation of gene expression is multifaceted. The analysis of differential gene expression is most often performed by comparing steady-state levels of mRNA. That is, the amount of mRNA that accumulates in the cell is a function of the rate of formation (transcription) and removal (processing, stability, degradation). If differences in protein products are being compared, one adds translation efficiency, processing, and degradation to the scenario. With all the possible causes for altered gene expression, the specificity of response must be questioned. Is the difference in mRNA or protein observed an important effect on expression or is it secondary to a parameter in that model system for which one has not controlled or accounted?

In the best-case situation, the key mechanism of gene regulation that results in the endpoint of interest should be known. One should have at least a criterion in mind for the type of response that is truly important. With most receptor systems, early transcriptional regulation may predominate as this key event. The other modes must be acknowledged, however, at least when clones are being characterized. By assuming that the key event is mRNA accumulation, the true initiating response, such as protein phosphorylation or processing, may be overlooked. Also, the extent and diversity of secondary events, those that require the initial changes in gene expression, may far exceed the primary events. The amplification of an initial signal (i.e., initial response "gene A" causes regulation of secondary response "gene B") can confuse the interpretation of altered mRNA accumulation. Once again, one must have a clear understanding of whether a primary or secondary event is the key response and design the model accordingly.

Summary: Key Components of Model Design

The comparison of two cell types, control versus treated tissue, represents a complex analysis of two different populations of mRNA species. The key to any cellular system to study receptor-mediated affects on gene expression in a mechanistically useful manner is to utilize the simplest, best-defined model possible. Figure 2 shows a basic model system that starts with drug–receptor binding and ultimately leads to a biologic response. If a segment of the response can be isolated, such as primary responses, the comparisons between populations will be more facile. A comparison of normal versus tumor cells requires the analysis of thousands of different mRNA species, most of which have little or no connection to the development of the disease (cancer). Comparing two identical populations of cells, however, one that has been treated with a hormone or xenobiotic and the other with the appropriate vehicle for a short period of time, may result in a very small subset of genes that represent some important initial response. The guidelines suggested here should aid in the design of a good system to examine genes being affected by a particular treatment as well as in interpreting data in the literature that were obtained by the methods described.

Utilize a model system that has a reproducible, biologically pertinent response. After genes are cloned, an attempt must be made to equate this change in mRNA levels with some biologic response. For example, if one is interested in cloning genes that are causally linked to altered patterns of growth and differentiation, one should choose

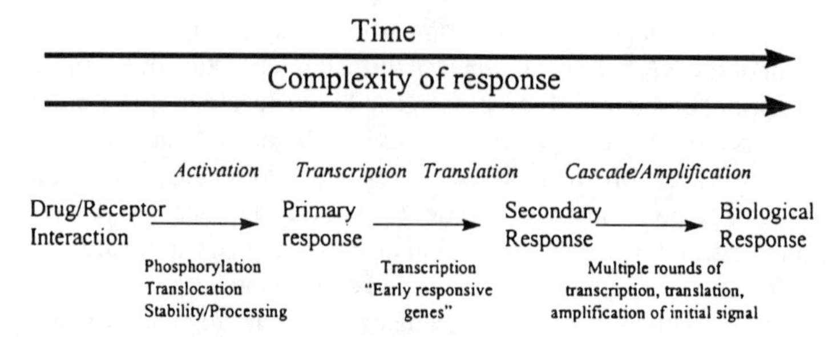

FIG. 2. Development of the appropriate model system. The development of a model system to study differentially expressed genes is critical for the studies to be successful. The narrower the time frame and level of complexity of response, the higher likelihood of success. Also, one must understand the basic mechanism of gene regulation and remove as many variables as possible.

cells or tissues that have the capacity to respond to treatment with the prototypical changes associated with this process, such as increased DNA synthesis. The novel genes one has cloned are only as pertinent as the model system will allow.

Compare two populations of mRNA in which the abundant component is unaffected. This can be accomplished by comparing similar types or populations of cells (not tumor versus normal) and by looking at earlier events. As mentioned previously, a small difference in expression of an abundant gene, say a two-fold induction in a 10,000-copy-per-cell gene, will result in a huge difference in mass of the message.The increase in the abundant gene may dilute the possibility of detecting more pertinent changes.

Identify the subset of genes one desires to pursue. Is the predominant interest in identifying genes that are primary or secondary events? Are genes that are regulated by transcription more important than those affected by mRNA stabilization? Many subsets of genes can be selected for in a model system. For example, if one is interested in primary events, early time points after treatment are warranted. If genes affected by mRNA stability are the investigator's forte, he or she should treat the two cell populations with a transcription inhibitor such as actinomycin D or α-amanitin prior to the receptor agonist.

Utilize a method that can detect differences in the scarce component of mRNA. Quite often the scarce mRNA component represents the types of genes that infer tissue-specific or chemical-specific responses. In the sections that follow, we will discuss the differences, strengths, and weaknesses of the various technical procedures to clone differential expressed genes. The need to clone genes of a particular expression level may determine which method is ultimately chosen.

Identify a positive control gene for verification of response. Prior to initiating a screen, one must make sure that the cells have responded as expected. A gene that is known to be regulated by the treatment in question can be checked by Northern blots, RNase protection, or reverse transcriptase–polymerase chain reaction (RT-PCR). Also, this gene should be one that is subsequently cloned from the differential screening.

EXPERIMENTAL APPROACHES

This section is not intended to discuss the details of different screening methods but to serve as a guide to the basic theory behind the more popular methods. For specific procedures the reader is directed toward laboratory manuals such as *Current Protocols in Molecular Biology* (5) and *Molecular Cloning* (6). Interestingly, a side-by-side comparison of three differential screening methods has been performed (1). The same treatment condition, interferon-4 induction of gene expression in HeLa cells, was examined by subtractive hybridization, electron subtraction, and differential-display PCR. The three methods all were capable of finding known interferon responsive genes. The amount of false positives and the number of novel response genes found, however, were quite different among the methods utilized. These authors suggest that differential display PCR may be the best overall approach. Each project and model system are different, however, and personal preference and experience may dictate the choice of method to clone novel responsive genes. We discuss four of the most common procedures to clone the differences between two cell populations, three of which are outlined in Fig. 3. A side-by-side comparison of the different methods is given in Table 1.

Differential Hybridization

For many techniques used to examine gene expression, one must be familiar with the construction of cDNA libraries. These libraries are essential for differential and

TABLE 1. *Comparison of differential hybridization, subtractive hybridization, differential display PCR and electronic subtraction*

	Differential hybridization	Subtractive hybridization	Differential display	Electronic subtraction
RNA required	1–5 μg poly(A)	1–5 μg poly(A)	5 μg total	1–5 μg poly(A)
Time required until clones indentified	Months	Months	Weeks	Weeks
Expertise required	Much	Much	Little	Little
Prevalence of mRNA surveyed	Abundant	Abundant and rare	Abundant and rare	Abundant
Types of differences identified	>Two-fold	All-or-none	>Two-fold	>Two-fold
Condiserations	Technically difficult	One-way comparisons, technically difficult	Reported to have high rate of false positives	High cost of operation

Adapted from Wan et al. (1).

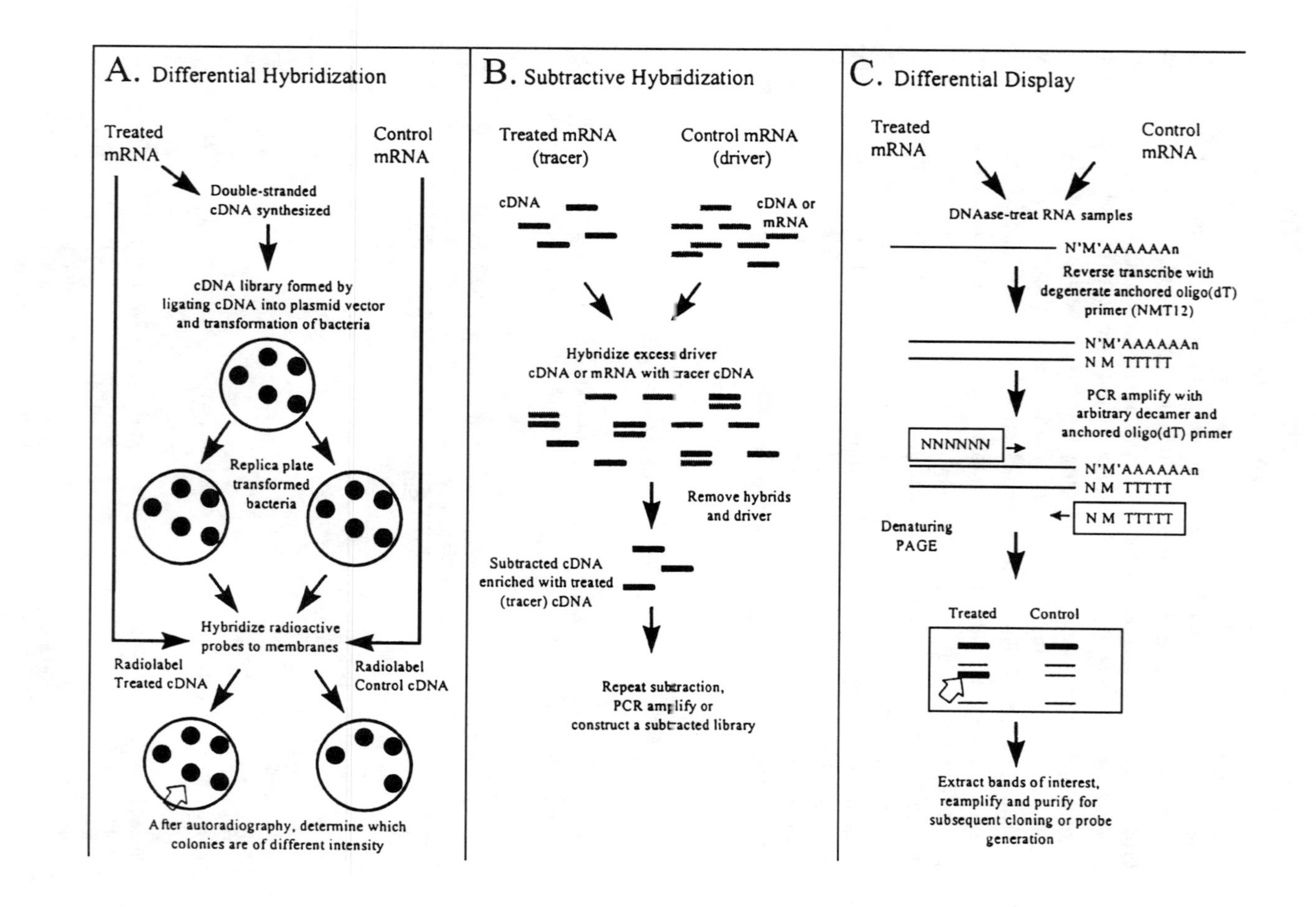

A. Differential Hybridization
Treated mRNA
Control mRNA
Double-stranded cDNA synthesized
cDNA library formed by ligating cDNA into plasmid vector and transformation of bacteria
Replica plate transformed bacteria
Hybridize radioactive probes to membranes
Radiolabel Treated cDNA
Radiolabel Control cDNA
After autoradiography, determine which colonies are of different intensity
B. Subtractive Hybridization
Treated mRNA (tracer)
Control mRNA (driver)
cDNA
cDNA or mRNA
Hybridize excess driver cDNA or mRNA with tracer cDNA
Remove hybrids and driver
Subtracted cDNA enriched with treated (tracer) cDNA
Repeat subtraction, PCR amplify or construct a subtracted library
C. Differential Display
Treated mRNA
Control mRNA
DNAase-treat RNA samples
N'M'AAAAAAn
Reverse transcribe with degenerate anchored oligo(dT) primer (NMT12)
N'M'AAAAAAn
N M TTTTT
PCR amplify with arbitrary decamer and anchored oligo(dT) primer
NNNNNN
N'M'AAAAAAn
N M TTTTT
N M TTTTT
Denaturing PAGE
Treated
Control
Extract bands of interest, reamplify and purify for subsequent cloning or probe generation

subtractive hybridization and electronic subtraction and also aid in the identification of full-length mRNAs in differential display PCR. The premise behind library construction is to convert mRNA from a given cell type into double-stranded cDNA using reverse transcriptase. Linkers or adapters are added to the cDNA and subsequently ligated into a bacteriophage vector. Following packaging of the virus and infection of bacteria, the library is amplified, titered, and characterized.

The experimental procedures for differential hybridization are shown in of Fig. 3*A*. A cDNA library of the "treated" cells is established using the basic procedures described above. Transformed bacteria at low density are replica-plated onto nylon membranes, fixed by heat or UV irradiation, and probed with ^{32}P-cDNA made to either the "treated" or "control" mRNA. The autoradiograms of these membranes are examined and compared. Alternatively, a color-coding method may be used to aid in the comparisons (7). The positions of the differences between the two hybridization conditions are marked and the clones removed from the original plate. Probes are generated from these clones for slot-blot, Northern blot, or RNase protection assays.

Differential hybridization was the first method routinely used to clone differentially expressed genes. This procedure has an advantage over subtractive hybridization in being able to examine either repressed or induced messages. The procedures, however, are technically challenging and may be biased toward more abundant sequences or for mRNAs that have a large difference in expression.

Subtractive Hybridization

Subtractive hybridization has proven to be one of the most useful methods to clone differentially expressed genes. Although many adaptations have been described, the basic procedures remain the same (see Fig. 3*B*). Double-stranded cDNA from treated ("tracer") cells is produced and hybridized to a molar excess of mRNA from control ("driver") cells. The conditions and efficiency of hybridization are critical and should be optimized for the desired abundance level of differentially expressed cDNAs (8). The mRNAs that are common to the two populations (the double-stranded hybrids) are separated from the differentially expressed sequences, usually using a streptavidin–biotin interaction (i.e., the driver mRNA is biotinylated). The resulting single-stranded, "subtracted" cDNA is either reselected, amplified, or used to construct a subtracted cDNA library (9).

Subtractive hybridization may represent the best chance to clone differentially expressed genes in the scarce component. This is due to the fact that the subtraction is an enrichment of the differences between two populations of mRNAs. The differences between the two populations, however, must be dramatic in order for the differentially expressed gene to be selected. Also, the comparison is "one-way" in that studying genes that are induced or repressed requires two separate experiments.

←

Fig. 3. Basic procedures for examining differentially expressed mRNAs. See text for details. Not shown is electronic subtraction, which involves the random selection and sequencing of a cDNA library derived from control and treated mRNA.

Differential-Display PCR

Differential-display PCR is the newest addition to the battery of common techniques used to study differential gene expression. Since its initial discription (10), there have been numerous modifications and improvements (11). The basic principle (as shown in Fig. 3*C*) is to reverse-transcribe mRNA from control and treated cells and to systematically amplify the 3′ terminals. This is accomplished by using a set of four anchored oligo (dT) primers (T12MN primers, where M is G, A, or C and N is any base) and an arbitrary decamer. The region between these primers is amplified by PCR in the presence of a radioactive nucleotide. Subsequently, the PCR products from the control and treated cDNA are resolved on a denaturing SDS–polyacrylamide gel and the differences in "fingerprints" determined. The differentially expressed products are eluted from the gel, reamplified, and used to make probes or are cloned for sequencing.

Differential-display PCR is a very versatile, technically facile procedure that can examine either repression or induction in the same set of samples. In addition, multiple comparisons can be made (e.g., a time course or triplicate samples) on the same gel. Also, because PCR is a sensitive procedure, it may be possible to examine the scarce component of mRNA. Differential-display PCR, however, has been associated with a high rate of false positives, and its ability to identify low expression genes has been questioned (12). Also, it identifies short fragments of cDNA at the 3′ end of the message, making identification of gene products and verification more difficult. This particular concern may be circumvented using an adaptation that utilizes two arbitrary decamers instead of one anchored primer and a decamer and does not require radioactivity, resulting in a larger fragment of cDNA from internal sequences at higher concentrations (13).

Electronic Subtraction and Serial Analysis of Gene Expression

This is the ultimate brute-force method of examining the differences between two populations of mRNA. Randomly selected cDNAs from libraries prepared from the two treatment conditions are sequenced. Usually 1000 to 3000 cDNAs are sequenced from each library (1). A particular mRNA is described as *differentially expressed* if its frequency in this random sampling is different between the two conditions. This type of approach has led to the establishment of the database for "expressed sequence tags" (dbEST), which contains thousands of short sequences from several species (14–17).

Recently, an adaptation of electonic subtraction was described that allows for a rapid analysis of thousands of transcripts. The serial analysis of gene expression (18) is based on the premise that a sequence as short as nine nucleotides is enough to uniquely identify the transcript (i.e., 4^9 or more than 260,000 transcripts can be distinguished). Using a combination of restriction enzymes and the ligation of PCR primers, a concatenation of these 9-bp segments and intervening 4-bp "punctuations" is produced. In this manner a continuous string of data with multiple sequences per clone (e.g., 40 tags/clone) can be obtained.

Electronic subtraction has several advantages over other methods, not the least of which is its ease of use. Automation of plasmid purification and sequencing makes it the least technically challenging method to examine differential gene expression. In addition, the data obtained is digital and reusable (1). The major disadvantage of electronic subtraction is the fact that it preferentially identifies abundant mRNA species. When sequencing 1000 templates for each treatment group, there is a very low likelihood of observing differences in the scarce component of mRNA. According to one calculation (1), identifying an mRNA of scarce abundance (e.g., 1 out of 20,000 mRNAs) requires sequencing of 126,000 cDNAs from each condition, and identifying all mRNAs (e.g., 1 out of 70,000 mRNAs) results in sequencing more than 400,000 templates for each population. Even utilizing SAGE technology this represents a monumental undertaking.

Miscellaneous Methods

In addition to the methods described previously, there are several other procedures that may be used, although they are less popular and often more challenging to perform. For example, two-dimensional polyacrylamide gel electrophoresis results in a "fingerprint" of proteins that is distinctive to a treatment condition or cell type (5, 6). Obtaining sequence information about the differentially expressed protein, however, is very difficult. The same can be said for Western blot analysis, although in this case there is lower resolution, and isolating the correct protein may be difficult. There are some exciting new technologies that may prove very useful in studying differentially regulated genes. Array-based methods have been developed that involve the spotting of multiple clones onto various supports (nitrocellulose, microscope slides) followed by hybridization with labeled cDNA and detection of expression (19). The negative aspect of this approach is the actual preparation and cataloguing of the probes to place on the support and the fact that only known sequences are examined. These negative aspects of array-based hybridization have been addressed by covalently attaching oligonucleotides (20-mers) chemically synthesized in a combined fashion directly to a solid support (20). Using this approach, the expression of previously uncharacterized mRNAs and thousands of known products may be examined in a rapid manner.

APPLICATIONS AND EXAMPLES

Dioxin and Aryl Hydrocarbon Receptor Ligands

The polychlorinated dibenzo-p-dioxins (PCDDs) and polychlorinated dibenzofurans (PCDFs) are two series of ubiquitous environmental pollutants. Of the 75 possible PCDDs and 135 PCDFs, by far the most extensively studied isomer is 2,3,7,8-tetrachlorodibenzo-p-dioxin (TCDD), commonly referred to as simply *dioxin*. In laboratory animals, dioxin is one of the most toxic chemicals ever described, with effects seen on carcinogenesis, immune function, reproduction, organogenesis, lipid metabolism, and behavior (21). The effects seen with PCDDs and PCDFs on a gross as well as molecular level are exquisitely tissue-, species-, and sex-dependent.

As discussed in chapter 1, dioxins and related chemicals bind to, activate, and regulate gene expression through the aryl hydrocarbon receptor (AhR). On activation, the AhR binds to specific DNA regions and regulates the transcription of a battery of genes. Most of these transcribed genes are involved in the metabolism of xenobiotics and include cytochromes P450 1A1 (CYP1A1) and 1A2 (CYP1A2), and UDP glucuronyltransferase (UGT) (22). In addition, there are many secondary effects on gene expression, primarily because of dioxin's effects on hormones and hormone systems such as the estrogen (see chapter 3) and epidermal growth factor (23) receptors.

Based on this brief overview, it is obvious that the choice of model system for studying differential gene expression by dioxins is critical. Because of the tissue- and species-specific responses and toxicity, various biologically relevant endpoints in many different cell types may be examined. In addition, the complexity of response, with numerous primary and secondary responses including transcriptional, post-transcription, and post-translational mechanisms of regulation, make the time after treatment a critical parameter to be considered.

The first, and most extensive, differential cloning project with dioxin was performed by Sutter et al. in 1991 (24). The model system in this case was very well defined, utilizing a responsive human keratinocyte cell line (in humans, the skin is very sensitive to dioxin exposure) and a short time of treatment. In addition, the cells were treated with a combination of dioxin and cycloheximide (CHX), a protein synthesis inhibitor. Therefore, the clones derived from this project were primary responses. A differential hybridization approach was used whereby the keratinocyte cell line SCC-12F subclone c12c2 was treated for 6 hours with TCDD (10 nmol) and CHX (10 mg/mL) and a library constructed. A total of 110,000 cDNA clones were screened and 16 TCDD-responsive clones, representing five different genes, were idenified. The genes that were cloned included the known TCDD-responsive gene CYP1A1 and two genes with known effects on cell growth and differentiation, plasminogen activator inhibitor 2 and interleukin 1β. In addition, one previously unknown clone was later found to be a novel cytochrome P450, CYP1B1 (25).

Although human keratinocytes represent a biologically relevant target site for the study of overt toxicity and altered differentiation, the liver is most appropriate tissue for the study of TCDD-induced carcinogenesis. Two studies using differential-display PCR have resulting in the cloning of novel dioxin-responsive genes (26, 27). In the human liver-derived cell line Hep G2, TCDD downregulated fibrinogen γ chain and plastin (26) in a time-dependent manner. Meanwhile, two mRNA sequences with little identity to known sequences were induced. The involvement of the AhR or the mechanism of gene regulation was not discussed. In rat liver treated in vivo with TCDD, roughly 30 differentially expressed reverse-transcriptase-PCR products were observed after amplification (27). One of these products was cloned (25-Dx) and its expression shown to be increased in a dose-dependent fashion in response to both acute and chronic TCDD exposure. The sequence of 25-Dx cDNA suggests it belongs to a new member of the cytokine or growth factor–receptor superfamily.

Estrogen/Estrogen-Receptor Ligands

Estrogen plays an important role in the growth, differentiation and function of target cells that express the estrogen receptor (ER). The ER is a member of the steroid-hormone receptor superfamily and acts as an estrogen-responsive transcription factor that recognizes and binds to estrogen-responsive elements found on target genes. Estrogen is predominantly associated with the growth and development of the uterus and mammary gland, but ERs have been identified in the central nervous, skeletal, and cardiovascular systems (28). Estrogen responsiveness is an important factor in determining effective treatment of breast cancer. Therefore, understanding which genes are regulated by estrogen and the mechanism of action of ER agonists and antagonists has important clinical applications.

Several attempts have been made to discover estrogen-responsive genes in human breast-cancer cells (29–31) with the predominant goal being to find good markers of hormone responsiveness for clinical evaluation. As a result of this bias, few of the mRNA species identified as differentially expressed actually have been sequenced or cloned. A fortunate aspect of using human breast-cancer cell lines is the availability of a cell line for which the ER status is well characterized. Therefore, responsive clones can be examined in ER-positive cell lines (T47D, MCF-7, ZR 75 1) and compared with ER-negative cell lines (BT20, HBL-100) to determine the role of the cognate receptor in altered gene expression.

Construction of a library from cDNA of estrogen stimulated MCF-7 cells and subsequent differential hybridization resulted in the isolation of four different mRNA sequences (29). Two of these isolates were induced (pNR-3 and pNR-4) and two repressed (pNR-1 and pNR-2). Similarly, by differential library screening of T47D cells treated for 24 hours with estradiol, several responsive clones were identified (pSYD2, 3, 5, and 6 were induced) (31). Comparing the responsiveness to estrogen in ER-positive and -negative cell lines clearly indicated that pNR-1 and pNR-2 as well as pSyd2 may be good clinical markers of hormone responsiveness in breast cancer. That is, these particular mRNAs were induced only in the cells that contained a functional ER.

A novel assay for the examination of estrogen-responsive genes called *genomic binding-site cloning* has been applied to this problem (32, 33). Although these procedures are not typically used to study xenobiotic-inducible genes, a brief discussion of the methodology and its strengths is warranted. First, total genomic DNA is digested by restriction endonucleases and mixed with the DNA-binding domain of the transcription factor, in this case the C domain of the ER. The DNA fragments that are associated with the protein are separated based on binding to a nitrocellulose filter. After several rounds of selection, these protein-associated pieces of DNA are cloned and sequenced. Using this technique a novel estrogen-responsive clone was identified that encodes a Zn finger motif and thus was named *estrogen-responsive finger protein* (EFP, 33). The 5′ regulatory region of EFP contains the perfect palindromic estrogen-responsive element, and its mRNA and protein is induced by estrogen in human breast-cancer cell lines.

Phorbol Esters and Protein Kinase C Ligands

Phorbol esters represent one of the most intensely studied groups of chemical tumor promoters. Concomitant with this tumor-promoting activity is the ability to facilitate the conversion of cells from a quiescent (G_0) state to G_1 and subsequent entry into the cell cycle. This mitogenic effect can be seen in a variety of cells, but T cells may be best characterized (34). One phorbol ester, tetradecanoyl phorbol acetate (TPA), is a potent tumor promoter and mitogen and often is used to typify this group of chemicals. TPA binds to certain protein kinase C subtypes leading to activation and subsequent effects on gene expression. Therefore, protein kinase C can be considered the cognate receptor for TPA and other phorbol esters. Early-response genes such as c-jun and c-fos, which together comprise the AP1 transcription factor, are induced directly in response to TPA stimulation (34).

Several attempts have been made to clone genes that are responsive to phorbol esters (35–37). In most cases, TPA is used to typify a mitogenic response, and genes isolated are considered to be involved in the generalized responses that start cellular proliferation. A good example of this approach is described in one of the early papers on differential hybridization by Lim et al. (36). The Swiss 3T3 mouse embryo cell line was utilized, as it responds to TPA with increased DNA synthesis and cell division. Genes that were induced as primary responses were isolated, utilizing CHX to inhibit protein synthesis. A library of recombinant phage, derived from density-arrested 3T3 cells exposed for 3 hours with TPA (50 ng/mL) and CHX (10 mg/mL), was screened with radioactive cDNAs made from cells treated with CHX or CHX and TPA. Of 50,000 plaques, seven families of clones were isolated that were induced by TPA. The known TPA-responsive gene c-fos was one of the clones identified, thus verifying the model and the methods utilized. The other unknown clones were induced by epidermal growth factor, fibroblast growth factor, and serum indicating that the genes were involved in the mitogenic response. Several of the six TPA-stimulated sequences (TIS) subsequently have been characterized. TIS8 and the early growth–responsive gene appear to be cDNAs from the same gene (38). TIS10 encodes a novel prostaglandin synthase mRNA (39). The remaining TPA-inducible clones that have been examined have little similarity to known proteins including TIS7 (40), TIS11 (41), and TIS21 (42).

Peroxisome Proliferators and Peroxisome Proliferator–Activated Receptor Ligands

Peroxisome proliferators (PPs) represent a diverse group of chemicals with a high likelihood of clinical, occupational, and environmental exposure to humans. A human health concern exists because of the fact that in laboratory rodents there is an association between peroxisome proliferation and liver cancer (43). Chemicals classified as PPs include fibrate hypolipidemic drugs, phthalate ester plasticizers, herbicides such as 2,4-D, and endogenous, long-chain fatty acids. Novel members of the steroid-hormone receptor superfamily have been cloned in several species, including humans, that help explain the molecular events involved in PP-dependent gene regulation. These receptors are activated by xenobiotic and naturally occurring peroxisome proliferators, hence the name *PP-activated receptors* (PPAR, 44). These are ligand-activated transcription factors that control gene expression by interaction with specific response elements located upstream of responsive genes (45). Genes containing response-element motifs include acyl-Coenzyme A oxidase (45) and other genes involved in fatty-acid metabolism. In addition, several messenger RNAs have been shown to be affected by PP administration, although whether this is due to transcriptional events is not known. These genes include the oncogenes c-myc, c-Ha-ras, and c-fos (46, 47).

Three separate attempts have been made to identify novel PP-responsive genes, each with a separate goal. In the first report, FitzPatrick et al. (48) used subtractive hybridization to identify cDNAs in the livers of rats treated with PPs clofibrate and diethylhexylphthalate. After four rounds of subtractive hybridization, a total of 14 known PP-responsive genes and four previously unknown genes were cloned. The known PP-responsive genes included those involved in lipid metabolism such as acyl-CoA oxidase, cytochrome P450 IVA, fatty-acid binding protein, and peroxisomal bifunctional protein. In addition, a novel enoyl-CoA hydratase and clones with homology to gastrin-binding protein and extensin precursor also were identified. The model and procedures utilized were optimal for identifying PP-responsive genes that are involved in peroxisome proliferation, because animals were treated in vivo for 7 consecutive days, at which time the number of peroxisomes is significantly increased.

Attempts also have been made to find PP-responsive genes that are causally linked to carcinogenesis. Corton et al. (49) used differential-display PCR to search for genes that are modulated by the peroxisome proliferator Wy14,643. In these studies, hepatic mRNA was isolated from rats at several times post-treatment. One advantage of differential display is the fact that multiple comparisons can be made and one can get an idea of a time-course for regulation of a particular product, well before the gene is actually cloned and identified. In this particular study, genes were cloned based on when they were induced, and those that were regulated early were preferentially selected. The sequence of one upregulated cDNA appears to be an α-hydroxysteroid dehydrogenase (HSD), an enzyme involved in sterol metabolism. Induction of this enzyme may explain the effects of certain PPs on serum estradiol levels and suppression of ovulation (49), and its link to carcinogenesis is being examined.

The previous studies performed with PPs focused on later events and may represent secondary responses. In recent studies, we utilized differential-display PCR to isolate and clone novel PP-responsive genes that may better explain the effects of these chemicals on DNA synthesis (Vanden Heuvel, unpublished data). A primary rat hepatocyte model system, previously shown to respond to PP-induced tritiated thymidine uptake into DNA, was utilized, and differentially expressed PCR products seen shortly after treatment by the peroxisome proliferator Wy14,643 were cloned by differential display. A gene nearly identical to the mouse Zn finger protein mZFP37 and with high homology to the Kruppel family of human transcription factors was shown to be responsive to PPs and subsequently named *rZFP37*. Because rZFP37 is transcriptionally regulated by Wy14,643, induced by serum and CHX, and maximally induced within 30 minutes, it supports the premise that this gene belongs to the immediate early-response gene family. The fact that this protein is a member of the immediate early-response gene family and putative transcription factor suggests that it may be important in responses such as increased cell proliferation.

ESTABLISHING A LINK BETWEEN EXPRESSION AND EFFECT

The ultimate goal of any analysis of differential gene expression is to tie a particular gene to the endpoint being examined. In fact, cloning a responsive gene is the easy part; being able to assign a causal relationship between the expression and effect may take years. Of course, there are some instances in which this relationship is easily fathomable. Take for example the cloning of an enoyl-CoA hydratase cDNA from PP-treated hepatocytes (48) or Egr-1 (TIS8) from TPA-treated fibroblasts (36). In these cases it is easy to state the effect of the induction of the gene, such as altered lipid metabolism with induction of enoyl-CoA hydratase or cell proliferation with altered Egr-1 production. In a preponderance of studies, however, there is an unclear connection between gene and response [e.g., interleukin 1β regulation by dioxin (24)] or for cloning a gene with no known homologues [e.g., TIS11 regulation by TPA (36) or pSyd expression by estrogen (31)]. In these cases, the most definitive means to assign a function to a gene is to artificially augment its expression by transiently transfecting an expression vector containing the cloned gene or decreasing its expression with antisense oligonucleotides, or to abrogate its expression in transgenic mice. The specific approaches used are obviously quite difficult and are gene-specific as well. These are the sorts of approaches that are currently being used to address many of the xenobiotic receptor–mediated responses described previously.

CONCLUSIONS AND FUTURE PERSPECTIVES

The identification of genes regulated by xenobiotics is an important avenue of study that has progressed rapidly in recent years. Through the examination of differentially expressed genes one can identify potential mechanisms of the biologic or toxicologic effects produced by the foreign chemical. One also can learn about basic cell biology and understand how a gene is involved in complex phenomena such as proliferation, differentiation, or carcinogenesis. The scientist has many tools at his or her disposal to pursue important regulatory genes including differential and subtractive hybridization, differential-display PCR, and electronic subtraction. Emerging techniques such as array-based hybridization may be the future of study on xenobiotic-inducible genes and will allow for a nearly instantaneous and universal access to genetic information. Although all of the methods described herein are powerful procedures, the data obtained are only as good as the model system can permit. The "garbage-in, garbage-out" theory of scientific study must be closely adhered to, both in the implementation of studies and in the interpretation of the current literature. A simple, reproducible, biologically pertinent model system always will result in more useful and valid information. The prospects of improved methodology and approaches to studying differential gene regulation are approaching rapidly. With these advances come the possibility of understanding critical issues such as developmental-, tissue-, species-, and chemical-specific responses to xenobiotics that have plagued and perplexed scientists for decades.

ACKNOWLEDGMENTS

J. P. Vanden Heuvel was supported by grants from the Public Health Service, National Institute of Health (DK 49009-01A1 and ES 07799-01A1). The author would also like to thank Dr. Tom Sutter (School of Hygiene and Public Health, The Johns Hopkins University) for providing his expertise and guidance.

REFERENCES

1. Wan JS, Sharp SJ, Poirier GM-C, Wagaman PC, Chambers J, Pyati J, Hom Y-L, Galindo JE, Huvar A, Peterson PA, Jackson MR, Erlander MG. Cloning differentially expressed mRNAs. *Nature Biotech* 1996;14:1685–1691.
2. Lewin B. *Genes IV*. Oxford University Press, Cambridge, MA, 1990.
3. Bishop JO, Morton JG, Rosbash M, Richardson M. Three abundance classes in HeLa cell messenger RNA. *Nature* 1974;270:199–204.
4. Axel R, Feigelson P, Schutz G. Analysis of the complexity and diversity of mRNA from chicken liver and oviduct. *Cell* 1976;7:247–254.
5. Ausubel FM, Brent R, Kingston RE, Moore DD, Seidman JG, Smith JA, Struhl K, eds. *Current Protocols in Molecular Biology*. New York: John Wiley and Sons, 1994.
6. Sambrook J, Fritsch EF, Maniatis T, eds. *Molecular Cloning: a Laboratory Manual*. Cold Spring Harbor, NY: Cold Spring Harbor Press, 1989.
7. Dunigan DD, Smart TE, Zaitlin MA. Differential color coding aid for colony plaque screening analysis. *Biotechniques* 1987;5:32–37.
8. Gastel JA, Suttter TR. A control system for cDNA enrichment reactions. *Biotechniques* 1996;20:870–875.
9. Wang Z, Brown DD. A gene expression screen. *Proc Natl Acad Sci USA* 1991;88:11505–11509.
10. Liang P, Pardee, AB. Differential display of eukaryotic messenger RNA by means of the polymerase chain reaction. *Science* 1992;257:967–971.
11. Liang P, Averboukh L, Pardee AB. Distribution and cloning eukaryotic mRNAs by means of differential display: refinements and optimization. *Nucleic Acids Res* 1993;21:3269–3275.
12. Debouck C. Differential display or differential dismay? *Curr Opin Biotech* 1995;6:597–599.
13. Sokolov BP, Prockop DJ. A rapid and simple PCR-based method for isolation of cDNAs from differentially expressed genes. *Nucleic Acids Res* 1994;22:4009–4015.
14. Boguski MS, Lowe TMJ, Tolstoshev CM. dbEST-Database for "expressed sequence tags". *Nat Genet* 1993;4:332–333.
15. Adams MD, Soares MB, Kerlavage AR, Fields C, Venter JC. Rapid cDNA sequencing (expressed sequence tags) from a directionally cloned human infant brain cDNA library. *Nat Genet* 1993;4:373–380.
16. Itoh K, Matsubara K, Okubo K. Identification of an active gene using large-scale cDNA sequencing. *Genet* 1994;40:295–296.
17. Okubo K, Hori N, Matoba R, Niiyama T, Fukushima A, Kojima Y, Matsubara K. Large scale cDNA sequencing for analysis of quantitative and qualitative aspects of gene expression. *Nat Genet* 1992;2:173–179.

18. Velculescu VE, Zhang L, Vogelstein B, Kinzler KW. Serial analysis of gene expression. *Science* 1995;270:484–487.
19. Caetano-Anolles G. Scanning of nucleic acids by *in vitro* amplification: new developments and applications. *Nat Biotech* 1996;14:1668–1674.
20. Lockhart DJ, Dong H, Byrne MC, Follettie MT, Gallo M, Chee MS, Mittmann M, Wang C, Kobayashi M, Horton H, Brown EL. Expression monitoring by hybridization to high density oligonucleotide arrays. *Nat Biotech* 1996;14:1675–1680.
21. Vanden Heuvel JP, Lucier G. Environmental toxicology of polychlorinated dibenzo-p-dioxins and polychlorinated dibenzofurans. *Environ Health Perspect* 1993;100:189–200.
22. Vanden Heuvel JP, Clark GC, Tritscher AM, Greenlee WF, Lucier GW, Bell, DA. Dioxin-responsive genes: examination of dose-response relationships using reverse-transcriptase polymerase chain reaction. *Cancer Res* 1994;54:62–68.
23. Cook JC, Gaido KW, Greenlee WF. Ah receptor: relevance of mechanistic studies to human risk assessment. *Environ Health Perspect* 1987;76:71–77.
24. Sutter TR, Guzman K, Dold KM, Greenlee WF. Targets for dioxin: genes for plasminogen activator inhibitor-2 and interleukin-1. *Science* 1991;254:415–418.
25. Sutter TR, Tang YM, Hayes CL, Wo YY, Jabs EW, Li X, Yin, H, Cody CW, Greenlee WF. Complete cDNA sequence of a human dioxin-inducible mRNA identifies a new subfamily of cytochrome P450 that maps to chromosome 2. *J Biol Chem* 1994;269:13092–13099.
26. Wang X, Harris PK, Ulrich RG, Voorman RL. Identification of dioxin-responsive genes in Hep G2 cells using differential mRNA display RT-PCR. *Biochem Biophys Res Commun* 1996;220:784–788.
27. Selmin O, Lucier GW, Clark GC, Tritscher AM, Vanden Heuvel JP, Gastel JA, Walker NJ, Sutter TR, Bell DA. Isolation and characterization of a novel gene induced by 2,3,7,8-tetrachlorodibenzo-p-dioxin in rat liver. *Carcinogenesis* 1996;17:2609–2615.
28. Orimo A, Inoue S, Ikeda K, Noji S, Muramatsu M. Molecular cloning, structure and expression of mouse estrogen-responsive finger protein Efp: co-localization with estrogen receptor mRNA in target organs. *J Biol Chem* 1995;270:24406–24413.
29. May FEB, Westely BR. Cloning of estrogen-regulated messenger RNA sequences from human breast cancer cells. *Cancer Res* 1986;46:6034–6040.
30. May FEB, Westely BR. Identification and characterization of estrogen-regulated RNAs in human breast cancer cells. *J Biol Chem* 1988;263:12901–12908.
31. Manning DL, Archibald LH, Ow KT. Cloning of estrogen-responsive messenger RNAs in the T-47D human breast cancer cell line. *Cancer Res* 1990;50:4098–4104.
32. Inoue S, Kondo S, Hashimoto M, Kondo T, Muramatsu M. Isolation of estrogen receptor-binding sites in human genomic DNA. *Nucleic Acids Res* 1991;19:4091–4096.
33. Inoue S, Orimo A, Hosoi T, Kondo S, Toyoshima H, Kondo T, Ikegami A, Ouchi Y, Orimo H, Muramatsu M. Genomic binding-site cloning reveals an estrogen-responsive gene that encodes a RING finger protein. *Proc Natl Acad Sci USA* 1993;90:11117–11121.
34. Scott JL, Dunn SM, Zeng T, Baker E, Sutherland GR, Burns GF. Phorbol ester-induced immediate early response gene by human T cells is inhibited by co-treatment with calcium ionophore. *J Cell Biochem* 1994;54:135–144.
35. Angel P, Poting A, Mallick U, Rahmsdorf HJ, Schorpp M, Herrlich P. Induction of metallothionein and other mRNA species by carcinogens and tumor promoters in primary human skin fibroblasts. *Mol Cell Biol* 1986;6:1760–1766.
36. Lim RW, Varnum BC, Herschman HR. Cloning of tetradecanoyl phorbol ester-induced "primary response" sequences and their expression in density-arrested Swiss 3T3 cells and a TPA non-proliferative variant. *Oncogene* 1987;1:263–270.
37. Shimizu N, Ohta M, Fujiwara C, Sagara J, Mochizuki N, Oda T, Utiyama H. Expression of a novel immediate early gene during 12-O-tetradecanoyl-phorbol-13-acetate-induced macrophagic differentiation of HL-60 cells. *J Biol Chem* 1991;266:12157–12161.
38. Varnum BC, Lim RW, Sukhatme VP, Herschman HR. Nucleotide sequence of a cDNA encoding TIS11, a message induced in Swiss 3T3 cells by the tumor promoter tetradecanoyl phorbol acetate. *Oncogene* 1989;4:119–120.
39. Kujubu DA, Fletcher BS, Varnum BC, Lim RW, Herschman HR. TIS10, a phorbol ester tumor promoter-inducible mRNA from Swiss 3T3 cells, encodes a novel prostaglandin synthase/cyclooxygenase homologus. *J Biol Chem* 1991;266:12866–12872.
40. Varnum BC, Lim RW, Herschman HR. Characterization of TIS7, a gene induced in Swiss 3T3 cells by the tumor promoter tetradecanoyl phorbol acetate. *Oncogene* 1989;4l:1263–1265.

41. Varnum BC, Ma Q, Chi T, Fletcher B, Herschman HR. The TIS11 primary response gene is a member if a gene family that encodes proteins with a highly conserved sequence containing an unusual Cys-His repeat. *Mol Cell Biol* 1991;11:1754–1758.
42. Fletcher BS, Lim BR, Varnum BC, Kujubu DA, Koslo RA, Herschman HR. Structure and expression of TIS21 a primary response gene induced by growth factors and tumor promoters. *J Biol Chem* 1991;266:14511–14518.
43. Moody DE, Reddy JK, Lake BG, Popp JA, Reese DH. Peroxisome proliferation and nongenotoxic carcinogenesis: commentary on a symposium. *Fundam Appl Toxicol* 1991;16:233–248.
44. Issemann I, Green S. Activation of a member of the steroid hormone receptor superfamily by peroxisome proliferators. *Nature* 1990;347:645–650.
45. Tugwood JD, Issemann I, Anderson RG, Bundell KR, McPheat WL, Green S. The mouse peroxisome proliferator activated receptor recognizes a response element in the 5′ flanking sequence of the rat acyl CoA oxidase gene. *EMBO J* 1992;11:433–439.
46. Cherkaoui MM, Lone YC, Corral-Debrinski M, Latruffe N. Differential proto-oncogene mRNA induction from rats treated with peroxisome proliferators. *Biochem Biophys Res Commun* 1990;173:855–861.
47. Ledwith BJ, Manam S, Troilo P, Joslyn DJ, Galloway SM, Nichols WW. Activation of immediate-early gene expression by peroxisome proliferators *in vitro*. *Mol Carcinogen* 1993;8:20–27.
48. FitzPatrick DR, Germain-Lee Emily, Valle D. Isolation and characterization of rat and human cDNAs encoding a novel putative peroxisomal enoyl-CoA hydratase. *Genomics* 1995;27:457–466.
49. Corton JC, Moreno ES, Merritt A, Marsman DS, Sausen PJ, Cattley RC, Gustafsson JA. Rat 17-β-hydroxysteroid dehydrogenase type IV is a novel peroxisome proliferator-inducible gene. *Mol Pharmacol* 1996:50;1157–1166.

Index